Springer Proceedings in Business and Economics

More information about this series at http://www.springer.com/series/11960

Michał Suchanek
Editor

Challenges of Urban Mobility, Transport Companies and Systems

2018 TranSopot Conference

Editor
Michał Suchanek
Faculty of Economics
University of Gdańsk
Sopot, Poland

ISSN 2198-7246 ISSN 2198-7254 (electronic)
Springer Proceedings in Business and Economics
ISBN 978-3-030-17742-3 ISBN 978-3-030-17743-0 (eBook)
https://doi.org/10.1007/978-3-030-17743-0

This Springer imprint is published by the registered company Springer Nature Switzerland AG
The registered company address is: Gewerbestrasse 11, 6330 Cham, Switzerland

Preface

The TranSopot 2017 conference is a continuation of a long series of conferences devoted to the topic of the transport sector development. The goal of the conference is to exchange views on the current trends in the transport growth and to spread the results of conducted research. The main purpose of the conference is to integrate researchers and practitioners in the field of transport, shipping, and logistics. With that point in mind, "TranSopot 2018 Conference: Transport Development Challenges in the 21st century" has been held from 28 to 30 May, 2018, at the Faculty of Economics, University of Gdańsk. The proceedings of the conference are presented in this book. Transport is a very specific area of social and economic life. It creates countless opportunities and allows to fulfill the need for mobility while also generating significant costs, either directly for the companies or indirectly for the societies in the form of externalities. This is why planning and organizing transport is a task which requires a multi-level approach with a focus on the operational, ecological, and financial aspects.

There is a need to analyze the transport solutions both at a micro-level, that of a single city or a single company, and at a macro-level of a whole transportation system. The transport decisions made by an individual in regard to the transport mode and route add up to the structure and efficiency of the whole system.

We are now at a point in which many transport systems cannot grow extensively anymore, due to lack of space or the amount of additional costs. This is why there is a deep need for new solutions, ones which are innovative and sustainable, while also increasing the efficiency of transport operations. These solutions are analyzed and presented in this book, which contains research performed at a scale of individual cities or companies but also in regard to the whole transport systems. The researchers, who are often also practitioners in the field of transport, provide not only the theoretical background for the transport analysis but also the empirical data and practical experience.

Sopot, Poland

Michał Suchanek

Contents

Part III Green and Innovative Transport Systems

Part I
Creation, Organisation and Evaluation of Urban Mobility

Determinants of Public Transport Integration in Cities and in the Region at the Example of Pomorskie Voivodeship

Krzysztof Grzelec and Olgierd Wyszomirski

Abstract Unintegrated public transport in cities and regions fails to meet the needs and preferences of the inhabitants. Disintegration can be observed mainly within the operational and economic areas. There are various types and forms of public transport integration in the cities and regions. Specific integration-related solutions can be observed in the cities and regions of France, the USA and Great Britain. The region where further public transport integration is necessary is the Pomorskie Voivodeship in Poland. The region is home to polycentric metropolis of Gdańsk with Gdańsk and Gdynia as its major cities. The earlier achievements related to public transport integration in this area cannot be regarded as satisfactory. They resulted from specific circumstances, the most important of which were political ones. Fares for services in terms of their amount, structure and collection play an important role in the process of public transport integration in the cities and the region. Modal split between public transport and individual trips by private cars can be recognized as a measure of success related to integration with impact on the attractiveness of public transport offer.

Keywords Public transport · City · Region · Transport integration · Determinants of integration

1 Introduction

Passengers of public transport expect high quality of services. It can be achieved by integrating particular modes of transport in cities and in the region. There are

K. Grzelec (✉)
Gdańsk University of Technology, Gdańsk, Poland
e-mail: krzgrzel@pg.edu.pl; krzgrzel@pg.gda.pl

O. Wyszomirski
University of Gdańsk, Sopot, Poland
e-mail: o.wyszomirski@wp.pl

M. Suchanek (ed.), *Challenges of Urban Mobility, Transport Companies and Systems*, Springer Proceedings in Business and Economics,
https://doi.org/10.1007/978-3-030-17743-0_1

numerous possible solutions related to public transport integration, covering the technical, operating, organizational and economic areas.

The region where public transport integration is necessary is the Pomorskie Voivodeship in Poland. The centre of this region is home to polycentric metropolis of Gdańsk inhabited by more than 1 million people. Railway transport managed by Voivodeship regional government plays a significant role in rendering transport services in the region. The transport is integrated only to a small extent with the local transport managed by municipalities.

This article aims to present the determinants of public transport integration in cities and in the region at the example of Pomorskie Voivodeship. The identification of determinants is important for the theoretical and practical reasons. The knowledge of determinants can facilitate the public transport integration process in cities and in the region, in particular in the metropolitan area.

2 The Essence and Scope of Public Transport Integration

Unintegrated system of public transport fails to meet the needs and preferences of the inhabitants, which, in turn, leads to a decrease in the number of passengers and the modal share. The lack of integrated system of public transport adversely affects [1]:

- the comfort of travel—one trip by various modes of public transport requires more than one ticket;
- clarity of information on services—customers must learn various tariffs, timetables presented in various forms, different regulations applicable to passengers as well as different rights to concession tickets:
- the comfort of travel: timetables and connections between operators are not coordinated, which extends waiting time between various modes of transport;
- costs—the cost of travel is the sum of transportation costs making up the journey.

Disintegration of public transport is mainly visible in the operational area and charging system.

The operational disintegration results from the lack of coordination in timetables on routes managed by various operators. Consequently, we can observe travel conditions arduous for passengers travelling on routes managed by more than one transport company. The tariff disintegration results from differently developed charging policy related to public transport services. Apart from different charges for public transport services in urban areas, agglomerations or metropolis, there are different, in terms of validity, types of season tickets and lack of season tickets for all modes of such transport.

The integration of urban transport, with other activities related to the transport policy, can lead to streamlining the modal split in urban transport. The integration activities and some activities related to the transport policy are significantly favoured by the inhabitants.

In a broader context, we can distinguish the following types and forms of transport integration in urban areas and in the region [2]:

- various modes of public transport;
- public and individual transport;
- transport policy with other policies related to spatial planning and city management;
- spatial integration, based on effective strategies for land use (e.g. multimodal terminals and transfer platforms, shared lanes for modes of public transport);
- infrastructural integration, based on the development of various technical solutions in the transport infrastructure (e.g. passages linking public transport stops, flyovers, underground passages, shared stops for public transport);
- organizational integration (e.g. coordinated timetables, metropolitan tickets for various modes of transport);
- economic integration, focused on the implementation of various instruments supporting the sustainable development and efficiency of public transport system;
- information integration (e.g. information systems for passengers, websites, electronic travel planners).

The integration of public transport may include [3]:

- instruments of transport policy;
- instruments related to developing the transport offer, including prices and information on services;
- planning the use of modes of transport and spatial planning;
- other policies, e.g. health and education policy.

Transport integration means coordination and optimization of timetables and routes of various operators, and the construction and use of transport interchanges including also individual transport: private cars and bicycles and trips on foot [4]. The organizational integration is based on agreements and contracts between stakeholders which guarantee effective operation of the system [5].

The integrated offer can be used to extend the public transport offer for both low-density markets and new segments of customers [6].

Integrated ticket can be defined as a ticket which makes it possible for passengers to travel by one or more modes of transport managed by one or more operators [7]. It is an important, though not the only crucial element of integrated transport broader concept.

Under formal and legal conditions in Poland, within the subjective scope, the integration of public transport may take different forms: ticket or tariff and ticket agreements concluded between the public transport operators, established municipal or metropolitan associations with specialized body—transport authority. The selected form of integration usually determines its level and effectiveness. In Poland, it is determined, in general, by political decisions, pushing the operational and economic results and the results of sustainable mobility policy into the background.

3 Public Transport Integration—Selected Examples

3.1 Nantes (France) [8]

Nantes is an example of broader approach to the issues of transport integration (not only public transport). The integration of particular transport sub-systems in the city involves:

- developing long-term vision and strategy of the city;
- implementing actions to integrate transport;
- consensus of opinion among neighbouring municipalities through political support and common implementation of integrated policy and transport plans;
- integration of various modes of transport, including sharing of cars and bicycles and walking;
- regular investments in high-quality infrastructure and vehicles constituting attractive and alternative offer for private car users;
- high quality of public transport services at reasonable price.

Long-term and unfailing implementation of policy on integrated approach to public transport planning within the last thirty years provided very good results. Many people consider the city the desirable place to live and work. The city received national and international awards, such as City of the Year granted by the European Commission in 2009, and was named the European Green Capital in 2013.

3.2 Cities in the USA [9]

The North American cities made impressive progress in integrating bicycle and public transport. Since 2000, the percentage of buses with bicycle racks has increased almost threefold. The integration of railway and bicycle transport has been developing as well. Most of the light rail, underground and suburban rail allows the transportation of bicycles, excluding peak hours, and more often provides special fastening for bicycles, such as: hooks, racks and stands in designated zones in the carriages.

In the North American cities, as in the European cities with significant share of bicycle trips, actions were centred on providing high-quality bicycle parking at the railway stations. The main transport interchanges cover multiservice stations which fully meet the needs of the cyclists. In order to encourage passengers to change their transport habits, the federal, state and local authorities have been more widely implementing laws encouraging the use of bicycle, although actions leading to bicycle transport integration with public transport differ in particular cities.

3.3 The Greater London Urban Area, Great Britain [10]

In accordance with the Greater London Authority Act, the body responsible for the management of transport is Transport for London (TfL). The organization subordinate to the mayor of London is divided into three main departments and also comprises a number of daughter-companies.

The first department in TfL is responsible for transport—bus transport, bicycle infrastructure, water transport, streets, bus station and taxis for passenger transport. All the functions related to transport management are integrated, and priority is given to environmentally friendly TfL modes of transport.

The second department is responsible for rail and underground transport covering:

- the Overground, i.e. similar to Warsaw-based Rapid Urban Rail (SKM)—municipal rail using the state infrastructure;
- Dockland Light Railway (DLR)—mode of transport as intermediate transport solution between the tram and underground;
- trams.

Under mutual agreements, TfL tickets are valid on railway routes within London, managed by the state authorities.

The third department of TfL is Croslink responsible for building and operation of the new railway tunnel off the city, constituting joint undertaking of TfL and the British Department for Transport.

The characteristic feature of London transport system involves significant emphasis on the privatization of transport services—bus and tram services, DLR and Overground, with state ownership of rail transport infrastructure and some depots used by private operators. However, attempts to introduce public and private partnership in the London Underground failed to succeed.

4 Objectives and Actions Related to Public Transport Integration in Pomorskie Voivodeship

Upon identifying the objectives of passenger transport integration in Pomorskie Voivodeship, we need to take into account two groups of factors specified under:

- research results related to the transport needs, demand and inhabitants' transport preferences which determine their transport behaviour;
- objectives of sustainable mobility policy.

The expectations of passengers and inhabitants should constitute grounds for integrated transport offer. Subsidizing the public transport by public authorities responsible for its management may sometimes lead to conviction that this is mainly some kind of social support. Therefore, there is a tendency to abandon the implementation of modern solutions related to management in favour of administration and "manual

Table 1 Ranking of transport demand in selected municipalities which constitute MZKZG

Centre of metropolis		City and municipalities surrounding the metropolis			
Gdańsk 2016	Gdynia 2015	Pruszcz Gd. municipality 2010	Pruszcz Gd.	Luzino municipality	Kolbudy municipality
Direct access	Direct access	Low cost	Punctuality	Direct access	Frequency
Punctuality	Punctuality	Punctuality	Direct access	Frequency	Reliability
Frequency	Frequency	Frequency	Frequency	Low cost	Direct access
Availability	Availability	Direct access	Availability	Availability	Punctuality
Travel time	Low cost	Availability	Low cost	Punctuality	Speed
Low cost	Reliability	Speed	Speed	Speed	Low cost

Source [11–16]

control" with the use of laws and regulations, meeting the needs of political and social nature. A common mistake involves failure to recognize the process-based nature of decision-making in the public transport and tendency to concentrate on the implementation of investment (which can bring spectacular achievements on the political level) and on the introduction of modern equipment (e.g. electronic fare collection systems). As a result, the importance of transport needs and customer expectations are downgraded in favour of the approach "we are doing what we can". Consequently, the obtained results make it possible to achieve the operational and tactical goals but not the strategic ones. We need to indicate that every investment, even the one meeting the expectations of passengers and inhabitants only to a minimal extent, will have usually small group of users. It results from the mass nature of the public transport services. In time, it becomes problematic to retain the said group as customers and increase the number of users or trips they take, not least because their expectations are rising. The phenomenon can be observed in Gdynia, where marketing research has been conducted every two–three years since 1994, and in Gdańsk, where in 2009 and 2016, Comprehensive Traffic Research was conducted.

The main objective of public transport integration in the context of assumptions and objectives of sustainable mobility policy involves raising competitiveness of its offer relative to trips made by private cars. The features which determine the attractiveness of public transport services are indicated under research results. The importance of particular attributes of public transport differs depending on the location of particular area relative to the centre of metropolis. Table 1 presents the ranking of most important transport demands in selected municipalities which constitute the Metropolitan Public Transport Association of Gdansk Bay (MZKZG).

Integrated public transport should, first of all, ensure competitive travel time relative to trips by private cars. Further, a particular area is located from the centre of metropolis, and the more important is the favourable cost of travel.

The enhanced competitiveness of public transport related to actions aiming at transport integration between 2018 and 2026 shall be achieved by:

- providing attractive, i.e. accepted by passengers conditions for the use of rapid public transport—(Urban Railway (SKM) and Metropolitan Railway (PKM) and trams), including the construction of interchange hubs;
- building bus lanes in the metropolis creating a network of arteries for road transport vehicles;
- priority for public transport vehicles implemented at all intersections.

In recent years, in Pomorskie Voivodeship a significant investment has been completed, i.e. the Pomeranian Metropolitan Railway which could compete with its offer with trips made by private cars. However, the objective of investment has not been fully achieved due to secondary importance attributed to the organization and management, including in particular the development of transport offer adjusted to the needs and preferences of inhabitants. Till today (March 2018), namely over two years after the launch, the management of transport offer has been constantly criticized by passengers and specialists. Relative success regarding the demand for such route, particularly from cities located at significant distance from the metropolis, results to a greater extent from limited potential of railway transport in the region, and not from adjusting the offer to the needs and expectations of inhabitants. We can also observe the phenomenon of cannibalization of demand—ca. 20% of PKM passengers amounts to people who previously travelled that distance by regional bus [17]. No improvement in PKM offer will make passengers look for alternative mode of transport to meet their needs, including in particular private cars.

The plan for integrated public transport in Pomorskie Voivodeship includes the standards of services characteristic of public transport offer. The most important standards include those related to the time of travel: access to metropolitan area (travel time to this area) and frequency of service [18].

Research results on transport preferences of the inhabitants of metropolitan area and planning assumptions related to the form of transport offer in Pomorskie Voivodeship clearly indicate that competitiveness of public transport depends on meeting the travel time demands. At implementing the solutions aiming at integrating the public transport, we need to design the network so that the public transport travel time is comparable to the travel time by private car, and in some circumstances shorter (in particular in the area of congestion). Therefore, in the metropolis we shall develop and renovate the rail and tram connections on separate tracks and lines, and build more bus lanes. Today, their total length in the two main cities in the metropolis fails to exceed 4 km.

The tariff and ticket integration is also important, although its impact on the implementation of sustainable development policy is smaller. Public transport is a mass transport service. The transport needs are secondary needs making it possible to meet the primary needs, such as: work, education, health protection and safety. Since the transport needs are mass and secondary in their nature, the public transport offer should involve user-friendly purchase of ticket and certainty in selecting the transport offer optimal for passengers. In such context, the transport integration should ensure using various modes of transport with one ticket which can be bought in various places. Therefore, the proper form of ticket distribution is intensive distribution.

The development of ICT and electronics create favourable conditions for tariff and ticket integration. However, in many European cities and agglomerations it was implemented earlier through proper actions related to the organization and management, and with the use of paper ticket [19].

The electronic ticket extended the types of tickets, purchase system and most importantly their availability (e.g. via Internet websites and applications dedicated to electronic payments). Common opinions of passengers (although often neglected in the ranking of transport demands) attributing significant importance to integrated tickets result mainly from the need to maximize the travel comfort through simplifying the process of selecting the correct ticket and its purchase. However, it is not a sufficient condition affecting the competitiveness of public transport compared to private car transport. The possibility to buy one ticket valid in different modes of public transport, provided by the public transport operator, must be accompanied by proper pricing policy which cannot be treated only as instrument of social policy but first of all as an element of service management procedure.

Still at the beginning of the 1990s, the cost-oriented approach dominated in Poland, where the costs of public transport and budget subsidies were treated as fixed parameters and the costs and income were balanced by the increase in ticket prices. When passenger transport was reorganized in Poland and private car traffic became significant competition, more and more operators at preparing their tariff projects began to focus on the demand side, taking into account ticket prices acceptable by passengers. Meanwhile, the objectives of sustainable mobility policy determine the necessity to include the competition side in calculating prices for public transport, where the competition does not come from other public transport operators, but from private cars. Taking into account the competition-based approach, it is advisable to refer the ticket prices to the costs of travel by car on the same route. Some experts advocate including in the ticket price economic analyses not only the cost for fuel but also the total running costs of cars. Such approach is correct from the methodological point of view; however, it fails to reflect the majority of actual elements regarding the inhabitants' choices which include the costs of fuel and more often the costs of parking fees. Other car running costs: insurance, costs of technical inspection, repairs, change of tyres are most frequently neglected by the inhabitants at analysing the attractiveness of private car compared to public transport.

Despite the long-term role of price in selecting particular mode of transport, it is some sort of catalyst for changes in the demand. The demand for urban transport services is not elastic; it is assumed that the rate of price elasticity of demand amounts to $e = -0.3$. The value of price elasticity of demand may differ depending on the analysed time period (long and short period of time). Elasticity rates take other values for peak hour and non-peak hour transport [20]; greater price elasticity will occur in areas with more private cars per household. The demand becomes also much more elastic at larger increases in prices. The elasticity of demand is positively correlated with the level of transport offer. The results of research conducted between 1997 and 2002 in Pomorskie Voivodeship in municipalities managed by the Urban Public Transport Authority in Gdynia (ZKM) indicted that in a difficult financial situation regarding municipal budgets, increasing the ticket prices and reducing the transport offer at the same time made the demand more elastic and the rate of price elasticity of demand might significantly exceed the value of -1 [21]. The phenomenon was initially related to the decrease in the transport mobility of passengers using single journey tickets. Over time, when the public transport offer is not improved, passengers may tend to give up on travelling by public transport under season tickets and choose travelling by own car.

Another feature of price related to the elasticity of demand is the fact that the increase and decrease in price by the same value frequently result in other changes in the demand. While the increase in price by 10% may lead to the decrease in demand by 3% (with model rate of $e = -0.3$), the decrease in price usually results in the increase in demand lower than 3%. Such situation has been observed in municipalities managed by ZKM in Gdynia. This phenomenon can be explained by the fact that the decrease in ticket prices with unchanged other parameters of the transport offer, mainly related to the travel time and comfort, cannot attract new passengers due to fairly minor importance of price among all transport demands. This statement is confirmed by the above-mentioned results of marketing research from Gdynia conducted in 2015 and KBR from Gdańsk conducted in 2016 which indicate full coherence as for the main reasons for using cars by the inhabitants. They indicated the following order: more comfort, shorter travel time by car, transport of belongings or shopping and no need to wait. Therefore, for a group of car users, the public transport ticket price fails to constitute an important determinant for choosing a particular mode of transport.

Upon analysing the importance of price in the public transport integration process, we need to indicate that the integrated ticket price cannot equal the sum of tickets from particular transport operators whose services are covered by integration. The price of integrated ticket must be acceptable to passengers and encourage car users to switch to public transport. It involves higher budget subsidies towards integrated ticket. We shall advocate the system where the central government in Poland like other countries in the European Union participate in financing the integrated public transport. It is also advisable to unify at the same time the rights to price concessions in urban and regional transport (simplification of tariff and ticket systems) and provide operators with refunds on integrated ticket concessions. At present, under legal provisions in Poland, integrated ticket must be divided into two components: railway and municipal ticket. Otherwise, the railway operator is not entitled to concession refund.

In conclusion, the ticket price fails to constitute the main factor affecting the choice of mode of transport. The reduction in price without significant improvement in transport offer regarding shorter travel time and enhanced competitiveness of public transport compared to individual one will not bring expected results concerning the implementation of fundamental goal of sustainable mobility policy—increase in the public transport modal share—whereas the increase in prices may have influence on the changes in transport behaviour as a catalyst, encouraging passengers to switch to private transport.

Taking into account the goals of sustainable mobility policy in Pomorskie Voivodeship included in the strategic documents, we can assume that the most important element is the modal split between the public and individual transport in metropo-

Table 2 Modal split in Gdańsk, Gdynia and Pomorskie Voivodeship

Type of travel	Travel share [%]		
	Gdańsk 2016	Gdynia 2016	Pomorskie Voivodeship
By private car	41.2	51.5	72.8
By public transport	32.1	35.6	23.0
On foot	20.8	10.9	–
By bicycle	5.9	1.6	0.2
Other	–	0.4	–
Sum	100.0	100.0	100.0

Source [11, 12, 18]

lis and cities outside the metropolis with the proportion of 50:50 and outside these areas in the proportion of 25:75, respectively. Taking into consideration the structure of travel in the largest cities in the metropolis and Voivodeship recorded in recent years (Table 2), we can assume that first the aim should be to weaker tendency to increase the modal share of cars and only then the reverse of this tendency.

The share of public transport in travels within cities in Pomorskie Voivodeship agglomeration has been decreasing. Between 2009 and 2016, in Gdańsk the share decreased by 7% points, whereas in Gdynia, between 2008 and 2015, by 12% points. Meanwhile, the data of the European Union for Public Transport indicate that modal share of public transport in urban travels in the developed countries between 1995 and 2012 rose from 34.1 to 39.7%, i.e. by 5.6% points [22]. This implies incorrect specification of priorities in the cities within Voivodeship by concentrating on the implementation of new investments and extending the scope of rights to concessions and free tickets without measurable improvement in the quality of services experienced by passengers and affecting changes in their transport behaviour. It results from downgrading the quality of services (transport offer) to secondary instrument and failure to provide proper financial resources to restore the adequate level of quality.

5 Conclusion

1. The objectives of public transport integration require standardization of actions shaping the widely understood transport offer. It is indispensable to develop proper organization and management structure equipped with modern instruments for managing services, prices, distribution and promotion.
2. The public transport integration process is not often perceived by public authorities as an instrument of sustainable mobility but as a tool of political game.

Nearly 30 years of ineffective integration of transport in Pomorskie Voivodeship confirms this view.

3. The increase in the number of passengers in public transport, regarding in particular the newly established connections, is recognized as the success of sustainable mobility policy, whereas in most cases it results from the new service life cycle and does not have to lead to permanent changes in modal split.
4. Public authorities responsible for operation and integration of public transport in Poland seem to downgrade the importance of fundamental demands related to transport in favour of those of secondary importance. Most often, they take administrative and political decisions which give them false sense of well-completed mission within the sustainable mobility policy.
5. Technological innovation creates favourable conditions for the process of integration. However, the belief that modern technologies themselves will solve problems and overcome barriers to integration proves ignorance of the role of public transport organization and management process.
6. The previous transport policy in some cities in Poland, which involved extending the scope of rights to free ticket, shall be recognized as irresponsible, leading in a longer period of time to the shortage of resources for adjusting the transport offer to the real expectations of passengers.

References

1. Public Transport Integration. Guidelines in market organisation. Strategies for public transport in cities. http://documents.rec.org/publications/SPUTNIC2MO_ptintegration_AUG2009_ENG.pdf
2. Solecka, K., Żak, J.: Integration of the urban public transportation system with the application of traffic simulation. 17th Meeting of the EURO Working Group on Transportation, EWGT2014, 2–4 July 2014, Sevilla, Spain. Trans. Res. Procedia **3**, 260 (2014)
3. May, A.D., Kelly, C., Shepherd, S.: The principles of integration in urban transport strategies. Transp. Policy **13**(4), 2 (2006)
4. Poliaková, B.: Key success factors of integrated transport systems. In: The 13th International Conference "Reliability and Statistics in Transportation and Communication (2013). http://www.tsi.lv/sites/default/files/editor/science/Publikacii/RelStat_13/session_3_ed_poliakova_ok.pdf
5. Saliara, K.: Public transport integration: the case study of Thessaloniki, Greece. Transp. Res. Procedia **4**, 537 (2014)
6. Häll, C.H: A framework for evaluation and design of an integrated public transport system. Linköping Studies in Science and Technology Licentiate Thesis No. 1257, Norrköping, p. 1 (2006)
7. Dotter, F.: CIVITAS Insight integrated ticketing and fare policy for public transport (2015). http://civitas.eu/sites/default/files/civitas_insight_12_integrated_ticketing_and_fare_policy_for_public_transport.pdf
8. Allen, H.: Integrated public transport. Nantes France (2013). https://unhabitat.org/wp-content/uploads/2013/06/GRHS.2013.Case_.Study_.Nantes.France.pdf
9. Purcher, J.: Integrating bicycling and public transport in North America. J. Publ. Transp. **12**(3) (2009)
10. Błaszczak, A.: Biuletyn Komunikacji Miejskiej. IGKM, Warszawa, nr 10, pp. 8–16 (2012)

11. Gdańskie badania ruchu wraz z opracowaniem modelu symulacyjnego Gdańska. Raport 2 (2016). http://www.brg.gda.pl/attachments/article/282/Raport-II.pdf
12. Preferencje i zachowania komunikacyjne mieszkańców Gdyni w 2015 r. Gdynia. Raport z badań (2017)
13. Preferencje i zachowania komunikacyjne mieszkańców Gminy Pruszcz Gdański w 2010 r. Gdańsk. Raport z badań (2011)
14. Preferencje i zachowania komunikacyjne mieszkańców Pruszcza Gdańskiego w 2010 r. Gdańsk., Raport z badań (2011)
15. Preferencje i zachowania komunikacyjne mieszkańców Gminy Luzino w 2012 r. Gdańsk. Raport z badań (2013)
16. Preferencje i zachowania komunikacyjne mieszkańców Gminy Kolbudy w 2014 r. Gdańsk. Raport z badań (2015)
17. Preferencje i zachowania komunikacyjne mieszkańców Gminy Żukowo w 2017 r. Gdańsk. Raport z badań (2018)
18. Plan zintegrowanego transportu publicznego dla województwa pomorskiego. Załącznik do Uchwały nr 788/XXXVII/14 Sejmiku Województwa Pomorskiego z 24 lutego 2014 r., s. 77–88 (2014)
19. Wyszomirski, O.: Funkcjonowanie transportu miejskiego w Leeds i Sheffield. Cz. I i II. Transport Miejski nr 5 i 6 (2001)
20. Litman, T.: Transit price elasticities and cross-elasticities. J. Publ. Transp. **7**(2), 52 (2004)
21. Grzelec, K.: Funkcjonowanie transportu miejskiego w warunkach konkurencji regulowanej, Fundacja Uniwersytetu Gdańskiego. Gdańsk, p. 305 (2011)
22. http://www.uitp.org/sites/default/files/cck-focus-papers-iles/MCD_2015_synthesis_web_0.pdf. Accessed 10.02.2018

Developing a Metropolitan Transport System—Exemplified by the Gdansk–Gdynia–Sopot Metropolitan Area

Kazimierz Jamroz, Krzysztof Grzelec, Lech Michalski, Romanika Okraszewska and Krystian Birr

Abstract As regional centres with metropolitan functions, metropolises grow and develop depending on how well they can meet the area's transport needs internally and externally, a result of their socio-economic relations. The transport system of a metropolitan area must ensure that people and goods can move efficiently, effectively and environmentally friendly. Over the last ten years, the transport system of the Gdańsk–Gdynia–Sopot Metropolitan Area has seen a clear increase in investment. The aim has been to increase capacity and to create a more balanced modal split between the different sub-systems. This sets the context for the Transport and Mobility Strategy for the Gdańsk–Gdynia–Sopot Metropolitan Area until 2030. Using primary research on the transport system, it was diagnosed and assessed. Next, suggestions and conditions were presented on how the transport system should develop and how it fits in with the policy of sustainable transport and mobility. The objective of this article is to present the main principles and objectives of strategic transport planning within the metropolitan area, the methods for developing parts of the strategy, main results and conclusions that feed into the discussion on how the Gdańsk–Gdynia–Sopot Metropolitan Area's transport system should develop.

Keywords Metropolitan transport system metropolitan · Strategy planning · Transport planning

K. Jamroz (✉) · K. Grzelec · L. Michalski · R. Okraszewska · K. Birr
Department of Highway and Transportation Engineering, Gdansk University of Technology, Gdańsk, Poland
e-mail: kazimierz.jamroz@pg.edu.pl

K. Grzelec
e-mail: krzysztof.grzelec@pg.edu.pl

L. Michalski
e-mail: lech.michalski@pg.edu.pl

R. Okraszewska
e-mail: romanika.okraszewska@pg.edu.pl

K. Birr
e-mail: krystian.birr@pg.edu.pl

M. Suchanek (ed.), *Challenges of Urban Mobility, Transport Companies and Systems*, Springer Proceedings in Business and Economics,
https://doi.org/10.1007/978-3-030-17743-0_2

1 Introduction

1.1 Metropolitan Areas

The urbanisation of the world is one of the most impressive facts of modern times. Urbanisation involves a complex process with multiple aspects: economic, demographic, social, legal, environmental and obviously spatial [1]. Since the mid-twentieth century, we have seen cities becoming strongly polarised and differentiated with urban units featuring different levels of complexity and spatial scale. What ensues from the changes are new types of spatial structures with development focussing on selected parts of space giving them an advantage over other units nationally, continentally and worldwide. What we are seeing are economic, social and cultural processes concentrating in major urban centres against a backdrop of economic ties between metropolitan areas and the regions around them becoming weaker and even ceasing to exist. Instead relations are built with the rest of the world. With jobs and higher level services such as education, finance, insurance, business services and culture concentrating in metropolitan centres, a dependency is formed between these centres and the areas around them. The centres become the destination for inward trips.

As well as changing economic and cultural ties, the process of metropolisation contributes to a strong differentiation of the space within the metropolitan centre's catchment area. When land use in urban and suburban areas changes, new and large settlement complexes are formed which are polycentric in character with blurred lines between urban and suburban zones. The spatial, social and cultural processes create new challenges for transport services in metropolises and metropolitan areas.

1.2 Transport Challenges Involved in Metropolisation Processes

As nodes within the urban settlement network, metropolitan centres are faced with the challenges of ensuring adequate access to them [2]. Metropolitan centres are surrounded by metropolitan areas which come with their own social, economic and spatial characteristics. This combined with the growth of metropolitan centres themselves requires a high level of social mobility and internal links. Strategic transport plans and mobility management pose a challenge to today's metropolises and metropolitan areas. Because metropolises often cross their administrative boundaries and metropolitan areas need functional and spatial links whose planning must be part of an integrated effort, local authorities must formalise the principles of cooperation and ensure adequate coordination of efforts.

Planning the transport development in metropolitan areas requires the preparation of a strategic document using the correct methodology and appropriate tools. Having a transport model may benefit the process of planning. The labour-intensive and costly stage of the planning process is a diagnosis of the existing condition demanding collecting data, especially regarding behaviour and transport preferences. The plan should identify various scenarios for the development of transport. The transport model, if used, helps to forecast how the transport system will behave for the different scenarios. The implementation of the scenario of sustainable transport development is difficult to implement due to political, social and economic reasons.

2 Characteristics of the Gdańsk–Gdynia–Sopot Metropolitan Area

The Gdańsk–Gdynia–Sopot Metropolitan Area (MA) is made up of eight counties and three cities and has a population of 1.5 million. Metropolitan transport functions are primarily delivered by the TEN-T trans-European transport network and its infrastructure (sea ports, airport, rail, express roads and motorway), other national roads, regional roads and rail lines. The MA's internal transport services between counties and municipalities are delivered primarily by the remaining transport infrastructure.

When assessing the MA's spatial structure for its transport, account should be taken of its suburbanisation, location within the sea ports' urban structures and their hinterland with industry and services, concentration of higher-order education services and jobs within the MA's core. Suburbanisation processes **increase trip times**. With gaps in the infrastructure, insufficient integration between the MA's transport sub-systems, slow pace of transport hub development, incomplete integration of ticketing and tariff systems, cycling not fully recognised as a mode for commuting to transport hubs, poor availability of public transport in areas outside the core of the MA and insufficient quality of transport services between the Tri-City and the MA's other municipalities, **cars have a growing share** in trips within the MA's core and to and from the core.

To meet the challenges of building a cohesive and integrated strategy for urban transport, a document was developed called the Transport and Mobility Strategy for the Gdańsk–Gdynia–Sopot Metropolitan Area until 2030.

3 Methodology for Developing the Gdańsk–Gdynia–Sopot MA Strategy

The Transport and Mobility Strategy for the Gdańsk–Gdynia–Sopot Metropolitan Area until 2030 [3] is the key strategy paper which sets out a plan for the development of the Gdańsk–Gdynia–Sopot Metropolitan Area's transport system. As well as a comprehensive analysis of the area's transport and its development, the document includes a diagnosis of the transport system based on the results of primary research on traffic and trips, a programme of the metropolitan area's transport development in the 2014–2020 financial perspective, analyses of how the metropolitan area's transport system can grow, prognostic transport model of trips for the metropolitan area and an environmental assessment.

Conducted specifically for the purposes of the strategy, the analytical material for the diagnosis is based on passenger transport studies and metropolitan area's residents' transport behaviour studies. The analysis also includes the results of a comprehensive traffic study conducted in 2009 by the city of Gdańsk.

The study of transport behaviour was carried out using direct interviews (F2F) and the PAPI technique at respondents' homes. The study's population was made up of people aged 13 and above residing in the counties of Lęborski, Pucki, Wejherowski, Kartuski, Gdański, Nowodworski, Tczewski and Malborski and in Gdańsk, Gdynia and Sopot. The sample included a total of eight thousand respondents. The survey's study instrument was the so called trip logbook. The study and analysis of the results provided data and the residents' mobility, reasons to travel, spatial and temporal distribution of trips, modes used and the modal split.

When diagnosing transport systems, the reliability of primary research is usually problematic. Even if full compliance is ensured, there is no guarantee that the result will be fully reliable and representative because of Poland's current formal and legal requirements. There are constraints on how raw and processed data can be used with cities using different scopes of research and different data aggregation procedures. For each scope of analysis, however, representativeness is always a problem. If it is to be achieved, research must cover an entire population or a properly selected sample [4, 5].

Representative samples are difficult to obtain in Poland. The reason is the personal data protection law which hampers and in some cases prevents (except for scientific studies) researchers from obtaining a list of the population for sampling purposes. Official lists only inform about registered residence rather than where people actually live. As a result, researchers will occasionally come across residents who do not meet the characteristics such as gender and age, because the flat was rented out by the registered owner. To address this problem, researchers must use reserve lists of people living nearby and meeting the characteristics. Because samples based on residents' list did not quite work well, the researchers decided to change how they selected households and use the random route where samples are selected randomly and based on quotas. Households are selected randomly and by quotas in proportion to gender and age of respondents in the survey areas.

The scope of the Traffic Complex Survey includes traffic volume and vehicle speed, modal and freight split, directions of passenger trips and how they are made, purpose of trips, reasons why specific modes are used and others. The main objective of the TCS is to diagnose the city's transport situation and build a traffic model to support traffic forecasts for the transport policy's scenarios. In addition, the traffic model is a tool to support decisions on how the area should develop in terms of its transport and land use. As a result, the consequences of a development policy can be better understood, a key aspect of strategic documents. Next, the model can and should be used for traffic analyses which are indispensable for transport projects' feasibility studies. The analyses should be made with the same official model provided by the city or metropolis and that way the quality and reliability of the results are ensured. They remain consistent with analyses made for the purpose of other transport projects.

The G-G-S MA's transport system diagnosis looked at the following problem areas: trip and transport behaviour determinants, identification of transport infrastructure and assessing its condition, consistency and access, assessing transport safety, especially road transport safety, the impact of transport on the environment and the quality of life.

The diagnosis for the G-G-S MA transport strategy draws on other strategy papers that address the operation and development of local, regional and national transport. These include primarily:

- Strategy of Integrated Territorial Investments for the Gdańsk–Gdynia–Sopot Metropolitan Area until 2020 [6];
- Pomorskie's plan for the sustainable development of public transport [7];
- Pomorskie Regional Development Strategy 2020 [8].

and transport plans of the municipalities and counties that form the G-G-S MA, municipality land use plans and more than twenty other strategies that have a local, regional or national coverage.

As regards the methodology, the G-G-S MA was prepared in conformity with the documents above adding detail to the projects set out in those documents.

As required by EU documents [9] which have been transposed into Polish law [10], the document was assessed for its strategic impact on the environment (EIA). It identified the local nature of the consequences. Many of the settlements were set to benefit from improved conditions of living and quality of life. While some of the changes and transformations may have adverse effects on the environment in the particular spots, compensatory measures will help to alleviate them [11].

4 Programme for the Development of the MA's Transport System

4.1 Goals of the MA's Transport System Development

The basic goal of the MA's transport system development is to support the effective delivery of social, economic and environmental goals of the MA set out in strategic regional and municipal documents that are linked with the MA. If the MA is to become a competitive place, it needs a high-quality transport system. Having such a system is a key to fostering public engagement and innovation, spatial order and sustainable urban mobility.

To improve the quality of the MA's transport system until 2020, the focus should be on transport objectives (general and modal, investment and organisational) whose delivery will support long-term development:

- Goal 1: Improve the quality of external transport links to MA;
- Goal 2: Improve the quality of internal transport links in MA;
- Goal 3: Improve the competitiveness of public transport and active mobility in MA.

If the external transport links (road, rail, air and water) are to be improved, more domestic and international centres should have direct connectivity to the MA and the MA should ensure better access (including the ports of Gdańsk and Gdynia) from domestic, international and Pomorskie's centres outside the MA. This objective ties in with transport infrastructure which determines access to the MA from other parts of Pomorskie, other regions, metropolitan areas both national and international, network infrastructure (roads, rail, water) and specific sites (sea ports) of European, national and regional significance.

Improving the quality of internal road and rail connectivity will involve efforts designed to reduce access time to the MA's centre from the other municipalities, especially using public transport, improve road safety and reduce the harmful consequences of transport on the MA's environment and provide access to new development sites that have metropolitan functions. This objective relates to transport infrastructure which is decisive for the Tri-City's (Gdańsk, Gdynia and Sopot) connectivity in relation to other MA settlement units and to network infrastructure (roads, rail and water) and point infrastructure (local sea ports, airports and airfields) that is of urban and metropolitan character.

Improving the operation of public transport and enhancing active mobility will involve efforts designed to reduce commute time by public transport for work, education and services, increase access to public transport, enhance the role of cycling in urban transport and its share in trips to hubs, increase internal integration of public transport and external integration with private transport (cars, cycling) and maintain adequate modal split as appropriate for sustainable development goals.

Goal three will be achieved through investment and organisational development of urban and metropolitan transport (buses, trolleybuses, trams, trains), the development of the cycle network (roads, lanes, trails, etc.) and space which promotes active mobility (pedestrian zones, traffic calming zones, limited access to heavy traffic).

4.2 MA's Transport System Model

Transport planning and taking relevant reasonable strategic decisions require the most modern tools such as a macroscopic trip model for the MA. It provides information about the area's current and forecasted transport. To check the effectiveness of the plans, a simulation model was used. It helped to analyse a variety of steps and scenarios of how the MA's transport and spatial planning will develop. The results were the basis for identifying sets of measures which, if delivered, would help to achieve the goals.

Table 1 Transport development scenarios within the G–G–S MA [1]

Factors impacting the scenarios		Social and economic situation measured with GDP	
		Bad	Good
Transport policy effectiveness	Low	Stagnation scenario	Pro-motorisation scenarios
	High	Restrictions scenario	Balanced scenario

The MA's transport model of trips is also useful for studying and analysing measures designed to:

- optimise the development of the transport network and arrange road projects by priority and maximise the benefits of delivering the projects;
- prepare a strategy for managing car journeys and public transport journeys;
- build traffic management scenarios in case of incidents, road works, large scale events and plan routes (lines) or change routes (lines) of public transport vehicles and analyse timetable changes;
- develop forecasts and operational analyses for cars and public transport;
- provide data for forecasting the effects of transport on the environment: noise analysis, emissions;
- provide data for microsimulation analysis for selected parts of the metropolis to conduct a detailed analysis of traffic conditions.

4.3 MA Transport System Development Scenarios

To identify the possible scenarios of MA transport development and compare them, a criterion was adopted, i.e. the modal split (of trips). It is largely affected by two factors: the socio-economic situation of the people and the effectiveness with which the transport policy achieves the goals of sustainable development. Based on this, four scenarios were identified (Table 1): stagnation, pro-motorisation, restrictive and sustainable.

When forecasting metropolitan trips in the MA transport scenarios in question, the following parameters of transport operation were considered:

- number of inter-municipality or inter-district trips [passengers];
- kilometres travelled [passenger kilometres];
- global travel time [passenger hours];
- share of inter-municipality or inter-district trips made by public transport [%].

Analyses have shown that in the strategic period until 2030, GDP is likely to grow and consequently the socio-economic situation is likely to be good. Availability of funding, including European Union funding, is decisive for whether infrastructure projects can be delivered or not. Considering the modern day expectations of transport systems, the **sustainable scenario** should be seen as the preferred scenario and the most difficult one to deliver. The other scenarios should be seen as undesirable, yet likely, and showing the consequences of abandoning the sustainable scenario. It is expected that if adopted, the sustainable scenario for the MA's transport will increase overall trips by 17.4%, i.e. almost up to 4.0 million trips daily, of which nearly 1.0 million will be metropolitan trips in 2030.

5 Discussion and Conclusion

Three years into the process, the strategy's scenario for sustainable transport development in the G-G-S MA until 2030 can now be evaluated. The results should feed into a discussion on how to implement the recommendations effectively:

1. As anticipated, the sustainable development scenario turned out to be the most difficult facing major political, formal, legal, social and economic barriers.
2. Political barriers have the biggest impact on public transport integration processes within the metropolitan area. It is safe to say that political reasons played a major role in reducing the integration of MA's transport sub-systems to an integration of tariffs and ticketing systems or ticketing systems alone, as could be seen in the last two years. Because the municipalities responsible for transport followed their particular interests only, a joint metropolitan public transport organiser was never appointed [12]. Political considerations are also to blame for failure to establish the Regional Transport Authority.
3. Formal and legal barriers also have an impact on public transport. A full integration of public transport tariffs and ticketing systems is not possible, due to the variety of refund schemes [13].
4. The parking policy with its extended pay and display zones and traffic calming is a welcome move. Political and social reasons (the residents objecting to proposals to designate lanes on streets with two lanes of traffic [14]) explain why so few bus lanes for public transport have been designated. (In the major cities of the MA, the length of bus lanes does not exceed 4 km.) It should be noted that giving preference to public transport vehicles, increasing the number of services and their frequency are the most effective instruments of achieving the goals of sustainable mobility.
5. The strategy provided an impetus to new activities which the metropolitan areas had not done before. Under the URBACT Freight Tails project, Gdynia, the MA's second largest city, is studying freight structure; this will initiate a change in the city's urban freight policy in the city centre.
6. Public transport services are faced with an economic barrier, i.e. not enough funding to make public transport more attractive by first of all increasing the frequency of trains (SKM and PKM). Funding is also scarce as regards public transport integration processes, especially integrating the various tariffs and tick-

eting systems. Local authorities, however, are moving funds around by giving more people the right to use public transport for free. As we know from European experience, this is a very ineffective instrument in reaching the goals of sustainable mobility policy [15].

7. The members of the MA are gradually implementing the strategy of sustainable mobility by running their own projects (developing infrastructure for cycling and walking) and taking up integrated tasks (metropolitan cycle). The local authorities work to maximise the use of modes of transport that are alternative to the car. Projects are being implemented and developed to improve access to ports. These involve new and upgraded road infrastructure (the Martwa Wisła River road tunnel, Gdynia's proposed Droga Czerwona Route and others) and rail infrastructure (ports of Gdańsk and Gdynia nodes, rail line 201 and others).

References

1. Worth, L.: Urbanism as a way of life. Am. J. Sociol. **44**, 97–104 (1938)
2. Michalski, L., Birr K.: Problems of planning the development of the transport system in metropolitan areas—an example of the Gdańsk-Gdynia-Sopot metropolitan area. Metropolitan **2**(8) (2017, in Polish)
3. Michalski, L., Jamroz, K., Grzelec, K., Grulkowski, S., Kaszubowski, D., Okraszewska, R., Birr, K., Kustra, W.: Strategy of transport and mobility for Gdańsk-Gdynia-Sopot metropolitan area until 2030. Gdańsk (2015, in Polish)
4. Kaczmarczyk, S.: Marketing Research. Methods and Techniques. PWE, Warsaw, p. 59 (2009, in Polish)
5. Churchill, G.A.: Marketing Research. Methodological Basis. PWN, Warsaw, p. 538 (2002, in Polish)
6. Okraszewska, R, Romanowska, A., Wołek, M., Oskarbski, J., Birr, K., Jamroz, K.: Integration of a multilevel transport system model into sustainable urban mobility planning. Sustainability **10**(2) (2018)
7. PBS, Gdańsk Development Office: Comprehensive Traffic Research Gdańsk. Sopot-Warsaw (2009, in Polish)
8. Gdańsk-Gdynia-Sopot Metropolitan Area: Strategy of Integrated Territorial Investments of the Gdańsk-Gdynia-Sopot Metropolitan Area until 2030, Gdańsk (2014–2016, in Polish)
9. Plan for the sustainable development of public transport for the Pomeranian Voivodship. Annex No. 1 to Resolution No. 788/XXXVII/14 of the Pomeranian Regional Assembly of 24 Feb 2014
10. Act of 3 October 2008 on access to information about the environment and its protection, public participation in environmental protection and environmental impact assessments. J. Laws **199**, item 1235 with later changes (2013)
11. Directive 2000/76/EC of the European Parliament and of the Council of 4 December 2000 on the incineration of waste. Off. J. L **332**, 28/12/2000
12. Grechuta, B.: Strategy of transport and mobility for Gdańsk-Gdynia-Sopot metropolitan area until 2030. Annex 6—Strategic Environmental Impact Assessment, Gdańsk (2015, in Polish)
13. Sierpiński, G.: Theoretical model and activities to change the modal split of traffic. In: Mikulski, J. (ed.) Transport Systems Telematics 2012. Telematics in the Transport Environment, CCIS, vol. 329, pp. 45–51. Springer, Heidelberg (2012)
14. Grzelec, K., Kołodziejski, H., Wyszomirski, O.: Determinants of tariff changes in urban public transport in the area of functioning of the Metropolitan Transport Union of the Gulf of Gdańsk. Sci. J. Univ. Gdańsk. Transp. Econ. Logist. **61**, 82 (2016)

15. Kołodziejski, Red, H.: The concept of the operation of high-speed city train in the Tri-City and the Pomeranian metropolitan railway in servicing the metropolitan area and integration of public transport in the metropolitan area and region, including tariff and ticketing integration in the OMT area and settlement rules between its participants. Foundation for the Development of Civil Engineering, p. 108, Gdańsk (2015, in Polish)

Selection of Public Transport Operator in Public Procurement System in Poland

Grzegorz Krawczyk

Abstract Regulated competition is one of the models for organising public transport market in Poland. This model assumes that operators are selected in a competition procedure, with the application of public procurement procedure. Tender procedure in the operator selection process is applied in the markets with high volume of operation work, i.e. in big cities and inter-municipal associations. Therefore, the public procurement system is an effectively used tool for the opening and liberalisation of public transport markets. The scope and efficiency of liberalisation of urban transport depend to a large extent on the transport organiser. It is the transport organiser that has the possibility of opening the market by applying tender procedures in the operator selection process. The application of tender procedures provides the possibility of influencing market concentration and, consequently, the intensity of competition between operators. The purpose of this article is to present the public transport operator selection process with the application of public procurement system in Poland. The article synthetically presents the results of tender proceedings carried out by the public transport organiser in the Silesian Metropolis area, which is the largest transport organiser that fully applies the competition rules in the operator selection process. The market behaviours of operators were identified, and the intensity of competition was determined on the basis of analyses of tender documentation.

Keywords Competition · Public transport market · Municipal transport

1 Introduction

Regulated competition is one of the models for organising public transport market in Poland. This model assumes that operators are selected in a competition procedure, with the application of public procurement procedure. The reduction in competitive solutions in the public transport market results from the existence of natural monopoly

G. Krawczyk (✉)
University of Economicsin Katowice, Katowice, Poland
e-mail: grzegorz.krawczyk@ue.katowice.pl

M. Suchanek (ed.), *Challenges of Urban Mobility, Transport Companies and Systems*, Springer Proceedings in Business and Economics,
https://doi.org/10.1007/978-3-030-17743-0_3

in which one entity is capable of meeting the entire demand in a more effective manner than in case when competition exists. A special tool for opening access to market is public procurement system, organised in compliance with the principles of fair competition, impartiality, objectivism, openness and equal treatment of contractors. According to the calculations made by the European Commission, EU member states purchase goods and services with the total value amounting to approximately 16% GDP of the European Union through the public procurement system. The primary goal of conducting tender proceedings is to guarantee reasonable and effective public spending. Among the group of entities obliged to apply public procurement law are public sector units, including public transport organisers.

2 Determinants of Competition in the Public Transport Market in Poland

The urban transport market plays a public interest role and is strongly connected with the self-government administration. This market is fragmentary in nature, and its basic specification is determined by [1]:

- transport services related to carriage of passengers as an object of exchange,
- urban transport companies as service providers and households as service consumers,
- local spatial range limited to the area of one or several cities forming an agglomeration together with the related suburban areas.

In consequence, when analysing the domestic urban transport market, we must take into consideration its dispersed structure, which consists of a number of local markets of various sizes. The volume of particular markets measured by the volume of operation work performed in these markets determines the type and intensity of competition. The urban transport market is characterised by the existence of natural monopoly. This structure is defined as a state in which the functioning of a higher number of companies is uneconomical due to the market conditions [2]. With limited demand, existence of high entry barriers and due to the economies of scale effect, one company is capable of satisfying the needs in a more economically effective manner than in case of competitive interaction. Demand volume is particularly limited in smaller towns, where transport network is relatively small, which results in a lower demand for operation work [3]. In case of urban transport, the existence of natural monopoly is strongly related to the market size. High market entry costs, which in case of transport are identified with a relatively high share of fixed costs, provide the primary supplier in a particular market with significant competitive advantage over potential competitors [4]. The extent and intensity of competition depend on the adopted model of market organisation.

In case of big cities or metropolitan areas, the volume of public transport services is large enough to support the application of competitive solutions in respect of selection of public transport operator. Two solutions are most popular in case of Polish cities: establishment of internal operator and competition regulated by the transport authority. Both forms of organisation are associated with operator selection. In this respect, the legislator has provided for the following options for contracting public transport services [5]: conclusion of a directly awarded contract (applied in case of procurement amounting up to EUR 1 million or 300,000 vehicle-kilometres per year), establishment of internal operator (based on EC Regulation no. 1370/2007), conduct of competitive procedure (pursuant to the Public Procurement Act). The organiser has the possibility to apply either non-competitive mode (internal operator), competitive mode or mixed solutions, in which a part of the market is reserved for the internal operator, whereas the other part is subject to competitive procedure. The privileged position of internal operator is subject to the following restrictions: it must be owned by self-government units (100% of shares); it may provide public transport services in the area covered only by one organiser, while at the same time other commercial activity is strongly limited.

The agreement concluded between the organiser and the internal operator must contain evaluation of the so-called compensation for provision of public services, which do not allow to obtain coverage for operating costs in specific conditions. The so-called 'reasonable profit' should be determined in compensation. Despite the statutory obligation, no regulation concerning the principles for determination of reasonable profit has been issued, whereas compensation is usually not determined in a clear manner. The internal operator is in a very comfortable market position, because it does not have to fear competition. This model is not oriented towards development of market competition, but its advantages include huge opportunities for introducing innovative and ecological solutions in the field of transport. The stability of conditions for agreement implementation and cooperation with the organiser, frequently being the owner of the internal operator, provides the opportunity to implement innovations, e.g. in respect of ecology (electric buses).

Two types of competitive interaction in the public transport market can be distinguished: competition for market entry and competition in the market. Competition for market entry assumes competition between carriers for a contract for transport services in a particular market or its segment. The object of competition is market entry, because further terms of service performance have been pre-defined by the organiser. Competition in the market takes places in a situation when carriers mutually compete for winning passengers; however, this model has not been widely applied in Poland.

The development of competition in the domestic urban transport market is based on the market competition assumptions. The most competitive form of organisation of public transport market in Polish conditions is the competition model regulated by the transport authority. It occurs when the functions of the carrier and the organiser are separated and the operator market is not monopolised. In this model, transport organiser awards the contract for execution of specific volume of operation work through competitive procedure. This solution brings economic profits resulting from the market pressure imposed on the operators. In order to maintain the market position, they aim to provide services that meet customer expectations. The customer is the transport organiser that is able to verify the level of operator costs and has the possibility to select the most favourable bid submitted in a tender procedure, thanks to the competition existing in the market [6]. A measurable effect of competition between operators is the reduction in expenses, demonstrated by lower costs of an operational work unit.

The most popular form of contracting urban transport services in Poland is gross cost agreement, which assumes that risk is divided between two entities. The industrial risk is borne by the carrier, whereas the commercial risk is incurred by the public authorities, i.e. transport organiser [7]. The subject matter of the agreement consists in implementation of specific volume of operation work expressed in vehicle-kilometres. Agreement is implemented under the terms strictly defined by the public party; in particular, this concerns the type and parameters of the rolling stock, routing, timetable, etc. Payment for a provided service is made on the basis of completed operation work and is not dependent to any extent on the number of passengers carried. This facilitates a far-reaching integration, because there are no problems with distribution of income from the tickets sold, which contributes to the durability of concluded agreements [8].

The opening and management of urban transport rests with the public party and is a long-term process that requires a change of approach to transport management and evolution of transport service market. Due to the existence of natural monopoly, the market-based urban transport strategies have been implemented mainly in agglomeration areas and major urban centres. The implementation of competition exacerbated certain problems that are difficult to solve: limitations in differentiating service prices from the marketing perspective because of the concessionary fares scheme, limited possibility of authority response to demand changes (necessity to renegotiate agreements with carriers), and lack of regular rolling stock replacement [9].

The regulated competition markets are dominated by oligopolistic structures. Such markets are characterised by high concentration level and domination of public operators. Despite the oligopolistic nature of the market, such entities cannot increase prices freely, e.g. through price fixing at the time of submission of tenders. Due to the relatively low market entry barriers, the operators in a particular market are afraid of potential market entries and tend to increase prices above their actual market level. Despite the concentrated structure, their behaviours are closer to a perfect competition market. Therefore, it can be concluded that a high competition level can be achieved with genuine political will to open the market, smooth functioning of the organiser and market attractiveness for carriers. The development of competition in the operator market will have the following effects [10]:

- reduction of direct involvement of municipalities in the functioning of municipal companies,
- higher effectiveness of service performance, resulting in lower costs or higher service quality.

The changes in the structure of public transport operators in Poland are presented in Table 1. According to the data of the Economic Chamber of Urban Transport, there has been a clear change in the operator structure since 1992. The budgetary units and enterprises, as well as state-owned enterprises, have been either transformed

Table 1 Organisational and legal forms of public transport entities in Poland [11, 12]

Organisational and legal form		1992		2000		2015	
		Number	Structure (%)	Number	Structure (%)	Number	Structure (%)
Limited company	Joint-stock company	12	10.3	4	2.6	3	2.2
	Limited liability company			84	55.6	99	73.9
Self-government budgetary enterprise/budgetary unit		70	59.8	59	39.1	12/20	9.00/14.9
Partnership		No data	No data	2	1.3	0	0.0
Natural person		2	1.7	1	0.7	0	0.0
State-owned enterprise and municipal company		33	28.2	1	0.7	0	0.0

or replaced by corporations. In 2015, limited liability companies constituted almost 74% of all operators. This change is a result of introducing market mechanisms in urban transport, as well as restrictions concerning the legal form of budgetary enterprises.

The transport services market develops together with the public transport market. The experiences of recent years have helped to identify the following tendencies in this respect:

- the activity of private units is increasing on a nationwide scale,
- the operators are willing to form consortia in order to win contracts for transportation effectively.

The systematic removal of barriers in access to urban transport market makes it more similar to the concept of contested market, characterised by low sunk costs [13]. Due to the well-developed market of bus providers and lessors, as well as subcontracting opportunities, the carrier's area of activity covers practically the whole country. The mobility of resources used is also increasing.

3 Public Procurement System in Poland

The legal basis for public procurement implementation in Poland is the Public Procurement Act of 29 January 2004 [14]. Within the group of entities obliged to apply its provisions (Art. 3), the legislator indicated, for example, public finance sector units, including public transport organisers. A public transport organiser always has the opportunity to apply the Public Transport Act, both in case of contracting transport services in parts and for the entire public transport network served.

The contractor selection should be based on the following criteria: equal treatment of contractors, fair competition, impartiality, objectivity, legality, openness, written nature of procedures and priority of tender modes. The evaluation and selection of the best bid is based on specific criteria. Initially, upon implementation of the Public Procurement Act, it was allowed to apply only the price criterion, as a result of which the lowest bid meeting specific requirements would win the tender proceedings. Over time, the legislator strove to reduce the role of price and increase the significance of non-price criteria. In accordance with the current wording of the Act, the share of price in tender evaluation process cannot exceed 60%. The other criteria with the total weight of 40% were not defined, whereas their selection and later verification were passed on to the contracting parties.

The contracting party selects the best bid among those that were not subject to exclusion on the basis of the evaluation criteria specified in the Terms of Reference. Tender evaluation can be performed on the basis of price and other criteria, in particular: quality of order processing expressed, for example, by technical parameters, environmental aspect, including energy efficiency of the object of contract, or experience of the individuals appointed for order processing.

The public procurement practice in Poland indicates that contracting parties underestimate the significance of non-price criteria for tender evaluation. The following criteria were among the most frequently applied in 2015: lead time, warranty terms and payment conditions [15].

The public procurement system is an element of public sector regulation, intended to guarantee reasonable, transparent and effective public spending. Similarly to any other regulation, the system generates specific costs and risk of negative and adverse impacts. The following risks to economic efficiency of public procurement can be specified: increase in administrative costs, prolongation of the purchasing process (due to submitted protests and appeals), increased risk of undertaking cooperation with an unknown and unverified (but the cheapest) contractor and loss of benefits resulting from long-term cooperation with one contractor [16]. Public Procurement Act is only a tool applied by specific market entities. Apart from the quality of legal regulations, their efficiency for the purpose of reasonable management depends on the contracting parties themselves.

4 Analysis of the Results of Tender Proceedings Based on the Example of KZK GOP

KZK GOP (Municipal Transport Union of the Upper Silesian Industrial District) is an inter-municipal association established in 1991 for the purpose of public transport organisation in the area of the associated municipalities. The operating area of KZK GOP currently covers 29 municipalities located in the central part of the Silesia Province. KZK GOP organises public transport on more than 330 bus lines and 28 tram lines. In case of tram transport, there is one monopolistic operator, whereas bus transport is provided on the basis of regulated competition. KZK GOP selects the transport operator on the basis of results of tender proceedings. Tenders are usually announced for one or several bus lines. There are both municipal transport companies (originating from former WKP, i.e. Provincial Municipal Transport Company in Katowice) and private companies in the bus operator market. The number of associated municipalities has changed since the establishment of KZK GOP. In particular, the first few years of KZK GOP activity were characterised by a dynamic increase in the number of associated municipalities. This resulted from the integration of public transport in KZK GOP area, which both increased its attractiveness to passengers and facilitated transport. The implementation of market mechanism in the process of public transport organisation required some time.

In 1999–2016, the volume of operation work commissioned by KZK GOP was relatively constant, within the range of 64–70 million vehicle-kilometres per year. The number of operators in the presented period ranged between 25 and 40 (including consortia). The number of entities is not directly correlated to the market size. Therefore, the appearance of new entities does not result in changes in the volume of operation work.

Until 2014, the KZK GOP market was characterised by a low level of market concentration. A sharp increase in concentration took place in 2014 and resulted from the fact that three largest bus operators formed a consortium that won a large tender for operating 181 bus lines. In 2016, the municipal companies had the largest share in the market—the total share of three largest municipal operators amounted to 67%.

The operating conditions of KZK GOP have a positive impact on the operator market; on the one hand, a significant number of vehicle-kilometres are executed, and on the other hand, the territorial scope of KZK GOP is very wide. Due to the geographical expansion combined with various levels of urbanisation in the operated areas, even the largest single entity is not capable of effective operation in the entire area, because due to the costs of travel to the initial bus stop on a particular route and return to the bus depot after finishing the service, such a bid is uncompetitive in terms of price in comparison with smaller local entities.

The analysis of tender proceedings at KZK GOP covers the period of full four years (2013–2016). The total number of contracts awarded during that period was 131. The subject of the tender was usually to operate between one and several transport lines, with the exception of the above-mentioned large tender for operating 181 bus lines. During the analysed period, two bids were placed per one tender on the average (Table 2). The number is not high, if we take into account how large the operator market in the KZK GOP area is. In a majority of cases, the bids were submitted by single entities; however, the number of bids placed by consortia is systematically increasing. Consortia are usually formed by private companies and less frequently by municipal companies, although there were cases when consortia were formed

Table 2 Characteristics of tenders at KZK GOP (own study)

Year	Number of proceedings	Number of submitted bids	Number of bids submitted by consortia
2013	28	53	9
2014	47	79	23
2015	51	88	22
2016	19	29	8

jointly by private and municipal entities. The formation of a consortium is particularly reasonable in case of applying for a large contract whose implementation requires obtaining an appropriate number of rolling stock units. However, the practice shows that consortia were formed both in case of multi-million contracts and orders at the level of PLN 200,000–300,000.

In the analysed tender proceedings, KZK GOP made selection of the best bid conditional on the price. Before implementation of the regulations resulting from the amendment of the Public Procurement Act of 2014, the price criterion constituted 100% of the tender evaluation. One non-price criterion (payment terms) began to be taken into consideration within the framework of this process at the end of 2014. The weight of this criterion was only 5%, and it could have been achieved upon settlement of the entire amount due for a particular month within 10 days from the date of receiving the invoice. In case of a shorter period of settling the amounts due, the condition was not fulfilled, and no points would be awarded. In 2016, the price criterion share was reduced in favour of payment terms, and the following system was adopted for the process: 60%—price and 40%—term of payment. The non-price criterion applied by KZK GOP has no impact on the quality of service performed and is easy to comply with. In a majority of the analysed documents concerning selection of the best bid, the bidders achieved the maximum number of points in this respect. Selection of the appropriate non-price criterion may be difficult, particularly in a situation when the expected service quality has been precisely described in the Terms of Reference. On the other hand, rigid and excessive requirements defined in documentation may constitute a barrier, e.g. in a situation when the operator does not have the appropriate number of vehicles of a certain age or with pre-defined emission of pollutants and is unable to obtain such vehicles within a relatively short period of time, for organisational reasons. This is likely one of the reasons why the tenders organised by KZK GOP are so frequently attended by consortia.

The market is clearly divided into municipal entities and numerous private enterprises performing different volumes of services. Due to the geographical location, municipal entities are basically not direct competitors of one another. In the analysed tenders, there were only three cases of mutual competition between municipal entities in tender proceedings.

During the analysed period of time, in 34 cases the bids were submitted by municipal and private entities at the same time. A vast majority of these proceedings were won by private operators that declared a lower service price. Municipal entities won only in 5 tenders. In all other proceedings, the price declared by the municipal operators was frequently over 20% higher than the best bid. Therefore, it can be concluded that despite the dominating position in the market, municipal entities do not apply an aggressive pricing policy and are not very active in winning new contracts. In the upcoming years, their stability of functioning will be ensured by operation work within a large tender for 181 bus lines, in which the bids were submitted only by the consortium of three largest municipal operators.

The public procurement system is undoubtedly a tool for market opening and rationalisation of public spending; however, the effectiveness in this respect depends on the willingness and efficiency of its application by the contracting parties. On the basis of the previous analyses, the following conclusions may be drawn:

- public procurement system enables new entities to enter the market,
- competition between the operators is based on prices,
- contracting parties underestimate the significance of non-price criteria in bid selection,
- private entities are able to declare a lower price for the services provided,
- competition between municipal entities is to a lesser degree based on prices.

Differences in the price levels between municipal and private operators result from many factors. The municipal entities have their own technical infrastructure which they have to maintain. They hire employees on the basis of employment contracts and frequently invest in development of young staff. Additionally, there are trade unions operating in these entities which exert constant pressure on improving working conditions, including the appropriate remuneration, social protection or protective clothing. On the other hand, private operators frequently reduce the operating costs by introducing flexible forms of employment. There is also a likelihood that in case of a very low price offered by private carriers, it may be presumed that the service will not be properly performed, e.g. certain services will be performed using inappropriate rolling stock (e.g. older or smaller than required), or the commissioned operational work will not be fully implemented. A dishonest contractor assumes that the efficiency of order processing inspection carried out by the contracting party is limited.

5 Conclusions

The application of public procurement system in the context of contracting transport services should be evaluated as positive. From the perspective of the market functioning, any sign indicating that competition has been introduced, even to a minimum extent, is better than maintaining a monopoly and processing orders without the market price verification. Public procurement is a tool that requires not only proper legislation, but also involvement on the part of the contracting party. Unfortunately, in case of regulated competition in Poland, there is a conflict of interest. A public transport organiser being a contracting party in tender proceedings is at the same time frequently the owner of the municipal operator that competes with other entities in the same tender. The problem does not stem from the structure of Public Procurement Law, but from the excessive involvement of state-owned companies in the public transport sector. It may seem that municipal operators will be partly protected by the contracting parties. Partial competition has been applied in many large cities in Poland—a municipal entity in the role of an internal operator has the dominating share in the market (frequently over 80%), whereas the remaining part of the market is exposed to competition. KZK GOP is the only large organiser fully applying the regulated competition. However, in this case there are also tendencies to increase market concentration by organising a large tender, whose terms can only be met by the largest municipal entities.

References

1. Polski rynek usług transportowych. Funkcjonowanie – przemiany - rozwój, Rucińska D. (ed.), p. 294. Polskie Wydawnictwo Ekonomiczne, Warszawa (2012)
2. Mas-Coell, A., Whinston, M.D., Green, J.R.: Microeconomics Theory, p. 570. Oxford University Press, New York (1995)
3. Kołodziejski, H.: Istota, formy i intensywność konkurencji w komunikacji miejskiej. In: Gospodarowanie w komunikacji miejskiej, Wyszomirski O. (eds.), p. 164. Wydawnictwo Uniwersytetu Gdańskiego, Gdańsk (2002)
4. Fiedor, B.: Regulacja państwowa a monopole naturalne. Prace Naukowe Akademii Ekonomicznej im. Oscara Langego we Wrocławiu, vol. 598, p. 35 (1991)
5. Act of December 16, 2010 on public collective transport. J. Laws, No. 5, item 13 (2011)
6. Dydkowski, G.: Integracja transportu miejskiego. Wydawnictwo Uniwersytetu Ekonomicznego w Katowicach. Katowice, pp. 187–188 (2009)
7. Dydkowski, G., Tomanek, R.: Finansowanie rozwoju transportu zbiorowego w warunkach liberalizacji. Transport Miejski i Regionalny **11**, 25 (2008)
8. Ceny transportu miejskiego w Europie. In: Tomanek, R. (ed.) Wydawnictwo Akademii Ekonomicznej w Katowicach, Katowice (2007)
9. Dydkowski, G., Tomanek, R.: Regulacyjne ograniczenia rynkowego zarządzania komunikacją miejską. Przegląd Komunikacyjny **7–8**, 24–25 (1999)
10. Hebel, K., Wyszomirski, O.: Liberalizacja i prywatyzacja lokalnego transportu zbiorowego na przykładzie Aglomeracji Gdańskiej. In: Liberalizacja transportu w warunkach transformacji gospodarczej, Dydkowski, G., Tomanek, R. (eds.), p. 99. Wydawnictwo Akademii Ekonomicznej w Katowicach, Katowice (2003)
11. Grzelec, K., Karolak, A., Radziewicz, C., Wolański, M.: Raport o stanie komunikacji miejskiej w Polsce w latach 2000–2012, p. 40. IGKM, Warszawa (2013)
12. Komunikacja miejska w liczbach, No. 2. IGKM, Warszawa (2016)
13. Tomanek, R.: Transport zbiorowy w świetle teorii rynków kontestowanych. Przegląd Komunikacyjny **6**, 25 (2005)
14. The Act of 29 January 2004 on Public Procurement Law, Dz. U. U. No. 19, item 177 (2004)
15. Sprawozdanie Prezesa Urzędu Zamówień Publicznych o funkcjonowaniu systemu zamówień publicznych w 2015 r., p. 34. Public Procurement Office, Warszawa (2016)
16. Prawo europejskie w systemie polskiej gospodarki. In: Królikowska, M. (ed.), p. 48. Difin, Warszawa (2005)

Management of Urban Mobility in Metropolitan Areas on the Example of the Upper Silesian Metropolis

Robert Tomanek

Abstract Mobility management in agglomerations requires an integrated approach. For this purpose, Metropolitan Association was established in the Upper Silesia Agglomeration, which is polycentric in nature, with over 2 million inhabitants living in several dozen cities here. Metropolitan Association began to operate on 1 January 2018, after several months of organizational work. It has its own separate (although small) budget. The first task is to carry out the tariff and ticket integration of public transport in the Agglomeration area. However, several redistributive actions have also been taken: implementation of free travel on public transport for individuals under the age of 16 and financial support for municipality investment within the framework of the so-called Solidarity Fund in the amount of almost one-third of the Metropolitan Association budget. The efficiency and effectiveness of the presented organizational solution can only be assessed after a longer period; however, the activity of Metropolitan Association should already be analysed, because it is regarded as an organizational solution for the future, serving as model for other agglomerations.

Keywords Metropolization · Urban mobility · Public transport integration

1 Introduction

Urbanization leads to formation of metropolises where municipal functions are cumulated, while the number of inhabitants and complexity of interconnections are growing. A metropolis can only be competitive in case of its effective (economic and social) functioning; therefore, it is important to introduce changes allowing to achieve the synergy effects. In Poland, the processes of cooperation between cities within an agglomeration occur slowly and spontaneously. The exception is the Upper Silesia Agglomeration, where on 1 July 2017 Metropolitan Association (hereinafter referred to as Metropolis) was established on an experimental basis. Following the

R. Tomanek (✉)
University of Economics in Katowice, Katowice, Poland
e-mail: robert.tomanek@ue.katowice.pl

M. Suchanek (ed.), *Challenges of Urban Mobility, Transport Companies and Systems*, Springer Proceedings in Business and Economics,
https://doi.org/10.1007/978-3-030-17743-0_4

organizational period, Metropolitan Association began to perform its statutory tasks, particularly related to public transport, on 1 January 2018.

The purpose of this chapter is to assess the role of Metropolis in sustaining mobility. Despite its short period of operation, it is already worth investigating the activity of the Metropolitan Association in the field of sustaining mobility, because the operating problems of transport in the Upper Silesia Agglomeration, and especially the insufficient scope of integration were the main reasons for the formation of Metropolis.

2 Contemporary Urban Tendencies

The number of global urban population is increasing, and therefore, the importance of cities in social and economic development of the world is growing simultaneously. In the twenty-first century, the number of city inhabitants exceeded the population in rural areas and further growth is forecasted. It is estimated that while 54% of global population lived in the cities in 2014, by 2050 this index will have increased to 66%. The level of global urbanization is diversified—the biggest number of urban population is in North America (82%), which is followed by Latin America and Caribbean (80%), and then Europe (72%). The forecasted level of urbanization in Europe for 2050 is 82%. In Poland, 61% of population lived in the cities in 2014, whereas by 2050 this number will have reached 70% [1].

In North America and Europe, the inhabitants want to participate actively in urban development processes, as evidenced by the increasingly strong citizens' initiatives, referred to as new urban movements or urban social movements (USM). In Poland, an example of the USM operation is Kongres Ruchów Miejskich (KRM, urban movement congress), which first took place in Poznań in 2011. The activists at the congress diagnosed the situation in the cities in the following way: 'spatial chaos, decline of public transport and housing, city gentrification instead of revitalization, metropolization problems, social stratification and lack of citizens' participation, financial crisis, decreasing financial envelope for social policy and culture' [2]. Among the 'Urban Theses' adopted by KRM, the problems of mobility and sustainable transport are of primary significance. They constitute an important element of sustainable urban development, which is based on treating citizens' needs as a priority and on the increase in social capital value. At the fifth congress, which took place in Dąbrowa Górnicza in May 2017, the manifesto entitled 'We, Passengers' was announced, including the following demands [3]:

- Increase in spatial availability of public transport ('we want no blank spots');
- Increase in availability of public transport for passengers with reduced mobility, including wheelchairs and prams;
- Tariff integration and simplification of ticket and information systems;

- Regulatory and financial support for municipal transport provided by central authorities;
- Determination of minimum quality standards for public transport.

3 Mobility Management in Metropolitan Areas

Metropolitan areas are aiming to increase the share of low-carbon and zero-emission transport (pedestrian and bicycle), as well as of public transport, in fulfilling transport needs. In the published reports, European Metropolitan Transport Authorities (EMTA) quoted the results of research conducted in 24 metropolitan areas in Europe and Montreal. As it turns out, in the examined capitals of urbanized areas in Europe, public transport currently fulfils between 11% (Cádiz) and 46.8% (Warsaw) of transport needs, whereas pedestrian and bicycle traffic are chosen by between 21% (Warsaw) and 67% (Amsterdam) of inhabitants; therefore, the share of cars and motorbikes is between approximately 13 and 66% [4]. While car and pedestrian traffic are regarded as means of fulfilling daily mobility needs, public transport and bicycles are to a larger extent of an occasional nature. This means that despite their seemingly high share in modal split, for a majority of inhabitants public transport and bicycles are not the primary means for fulfilling mobility needs [5].

The basis of effective mobility management is mobility planning, which adopts the form of systemic approach leading to development of Sustainable Urban Mobility Plans (SUMPs) for the entire urbanized area. SUMP is developed with participation of many stakeholders, and the participatory approach is an important element of this concept [6]. SUMPs promoted by EU are a new trend; however, they result directly from the European Transport Policy, including the White Paper on Transport of 2011 [7]. The evaluation and improvement of urban practice in the field of SUMP is supported by benchmarking and audits. A special SUMP audit was developed and executed under ADVANCE project [8]. In line with the ADVANCE project assumptions, over 500 cities in Europe should have SUMPs in 2020. The process of developing SUMPs to replace Sustainable Urban Transport Plans (SUTPs), which must be prepared by a part of municipalities and intermunicipal associations, is also in progress in Poland. SUTP concerns only public transport, while its content is governed by the legal regulations. Dissemination of the practice of forming SUMPs in Poland is connected with EU-funded investments and adaptation of good practices within the framework of CIVITAS programme and other EU initiatives related to urban development.

Bicycles are of particular significance among the zero-emission forms of fulfilling transport needs. Municipal authorities in Poland (similarly as in other countries) are developing bicycle systems, and two basic directions can be distinguished in this respect: development of cycle paths and bicycle routes, and development of public bicycle systems (bikesharing). The examples of good practices in Europe are quite numerous, including the adaptation of inoperative railway infrastructure for bicycle routes [9] and development of bikesharing. Bikesharing was originated in Amsterdam in mid-1960s; however, it did not flourish until the twenty-first century [10]. Such systems are developing very rapidly around the world—the increase in the number of journeys in the USA in 2012–2016 was from 320,000 to over 28 million (55 systems, 42,000 bicycles) [11]. In Poland, the first system was installed in Kraków in 2008, whereas the largest system is Veturilo in Warsaw (351 stations, 5200 bicycles). The systems are administered by 5 operators [12].

4 Sustaining Mobility in the Upper Silesia Agglomeration

4.1 Conditions for Urban Mobility Management in the Agglomeration

The urbanization rates in Poland are lower than the EU average, but the spatial development is characterized by one of the greatest levels of polycentralization of urbanization [13]. The Upper Silesia Agglomeration, which is inhabited by approximately 2–2.5 million people (according to various methods of metropolitan area delimitation), dominates in the Silesia Province in terms of population and economy—69% of the province population live here, and this is the centre of economic activities and urban functions [14]. Public transport, which is key for urban mobility, is managed by 5 organizers with different financing systems and services provided (quality and price). Three local transport organizers are of key significance for the Agglomeration (KZK GOP—Municipal Transport Union of the Upper Silesian Industrial Region in Katowice, MZKP—Intermunicipal Public Transport Association in Tarnowskie Góry and the city of Tychy with the surrounding communes, where the Metropolitan Transportation Authority performs the function of organizer on behalf of the city). Local transport is also provided by a regional railway carrier—Koleje Śląskie SA, organized by the Self-Government of the Silesia Province. KZK GOP and MZKP have had a common tariff and ticket system since 1994. In the past, common tickets were introduced and withdrawn between KZK GOP/MZKP and Tychy, as well as between KZK GOP/MZKP and Koleje Śląskie. The fifth organizer is the city of Jaworzno; although buses from Jaworzno reach the very centre of the Agglomeration, Jaworzno decided not to join Metropolitan Association.

The first stage of broad integration of local governments in the Agglomeration area consisted in establishing an association called the Association of Municipalities and Districts of the Central Subregion of the Silesia Province by 81 municipalities and districts of the Upper Silesia Agglomeration and its surroundings in 2013 for the purpose of management of the ITI (Integrated Territorial Investments) financial instrument [15]. The establishment of this association was the condition for obtaining funds for the implementation of ITI. The Subregion area is inhabited by approximately 2.8 million people and includes Jaworzno. In terms of mobility, the Subregion has been implementing the Sustainable Urban Mobility Plan (SUMP) adopted in 2016 [16]. Half of EUR 793 million from ITI for the years 2014 to 2021 has been

Table 1 SUMP result indicators of the Central Subregion in terms of low emission mobility

Strategic goal	Result indicator	Target/intermediate value 2018	Data source
Increase in competitiveness of sustainable transport	Number of purchased rolling stock units in public transport (items)	165/51	Public transport operators
	Capacity of purchased rolling stock in public transport (people)	8,250/-	Public transport organizers and operators
	Length of built bicycle paths (km)	1,113/-	JST
	Length of rebuilt bicycle paths (km)	800/-	JST
	Length of designed bicycle paths (km)	337/-	JST
	Number of completed integrated transfer nodes	53/6	JST
	Number of completed park and ride facilities	53/-	JST
Transport integration	Number of parking spaces in completed park and ride facilities (items)	3,225/-	JST
	Number of parking spaces for the disabled in completed park and ride facilities (items)	700/-	JST
	Number of completed park and ride facilities	300/-	JST
	Number of installed smart transport systems (items)	8/2	Municipal roads authorities
	Length of roads covered with ITS systems activity (km)	1,200/-	Municipal roads authorities

dedicated to the implementation of investments for sustaining mobility—see the list in Table 1 (drawn up on the basis of source data of the Association of the Central Subregion of the Silesia Province).

4.2 *Metropolitan Association—Assumptions and Their Implementation*

The problems with public transport integration in the Upper Silesia Agglomeration area were the basic reason why Metropolitan Association (commonly referred to as Metropolis) [17] was established in this area on 1 July 2017. The name of the association is Silesia Metropolis, and it was formed by 41 municipalities inhabited by over 2.2 million people. In order to fulfil the statutory criterion of 2 million inhabitants, not only big cities of the Upper Silesia Agglomeration with high density of population (over 1,000 people per km^2), but also small rural communes surrounding the agglomeration (the density of population in 9 rural communes is less than 200 people per km^2) were included in Metropolitan Association. The extensity and spatial diversity of the Metropolis area (2,553 km^2) will probably be a source of mobility-related problems. Moreover, the principle of double-majority decision-making by the Association Assembly (majority of individuals and majority of inhabitants) is applied in Metropolis, which created problems with selection of the Association Board already at the beginning of its operation.

The following factors were analysed in order to verify the purpose of Metropolis formation:

- Assumptions of the motion for the establishment of Metropolitan Association;
- Silesian Metropolis budget for 2018, as well as executive resolutions of the Assembly and of the Board (until 20 February 2018).

The following aspects were characterized on this basis:

- Volume of Silesian Metropolis share in financing sustainable mobility of the Agglomeration;
- Actual range of mobility-related operations in Metropolis in 2018;
- Role of Silesian Metropolis in public transport integration in the Agglomeration;
- Range of activities of image-building, frequently redistributive nature.

The problems of transport and mobility dominate on the list of statutory tasks performed by Metropolitan Association [18]. The municipalities requesting to establish Metropolis emphasize the fact that the association would allow them to integrate sustainable management development in the metropolitan area. The potentially planned tasks in the field of transport and mobility include, for example [19]:

- Preparation of the Public Transport Development Plan, which would be a study of opportunities for sustaining the transport system of the Metropolitan Association through construction of integrated railway and bicycle traffic systems, as well as other planning projects;
- Introduction of preferences and incentives contributing to a change in the existing behaviours towards the use of public transport;
- 'Common Ticket', i.e. ticket integration of passenger railway transport with other public transport means;

- Creating conditions for the development of electromobility (easier identification of electric vehicle charging stations, purchase of electric buses) and support for the development of low-carbon public transport in the cities;
- Activities intended to improve the road traffic safety (Intelligent Transport Systems (ITS) in the cities and their functional areas);
- Shared maintenance of Drogowa Trasa Średnicowa (DTŚ);
- Metropolitan system of public bike rental;
- Metropolitan system of bicycle routes.

However, the funding of transport and mobility by the Silesian Metropolis planned at the stage of establishing the Association is modest in comparison with other sources of funding for these tasks in the Agglomeration. Table 2 shows the basic sources of funding for public transport and mobility in the metropolitan area identified as a result of investigating financial plans for 2017/2018—excluding the revenue obtained from the market (tickets, other services, which in case of transport organizers means that the actual financial resources are approximately 30–35% higher) and funds intended for other purposes, however, resulting in investments in the transport system, also in the field of low-carbon mobility (e.g. implemented within the framework of private and public projects—national road investments and accompanying investments related to pedestrian and bicycle traffic). Although the presented initial calculations are highly simplified (the particular operating areas of the examined organizational units do not match), they show the approximate scale of Silesian Metropolis activity in mobility management—including not more than 15% of income from sales).

However, the actual financial scale of activity of the Silesian Metropolis in sustaining mobility in 2018 is lower than the project assumptions. The budget adopted on 19 December 2017 provides for a revenue at the level of EUR 86.8 million (more than assumed in 2017), out of which 90% is own revenue assumed in the Act, originating from the personal income tax collected in the Metropolis area. Over 47% of revenue in the budget is the budgetary surplus towards expenditure in the upcoming years. As much as EUR 22 million (48% of expenditure for 2018) has already been dedicated to the so-called Metropolitan Solidarity Fund—for various investment projects implemented by the municipalities (the plan for 2018 forecasts expenditure at the level of EUR 24 million) [20]. It is assumed that the fund will totally amount to EUR 38.4 million until 2023; however, judging by the amendments in the planned expenditure of Silesian Metropolis introduced several months after establishing the Association, these assumptions may change. The review of the projects co-financed in 2018 has demonstrated that they concern sustainable mobility only to a small extent—a vast majority of them are related to construction of roads and car parks, usually of local and redistributive significance [21]. Formally, the planned expenditure remains within the assumed allocation of funds for transport tasks; however, the Solidarity Fund projects include tasks such as thermo-modernization of the administrative building and revitalization of historical monuments. The expenditure included in the above-mentioned fund appears to be a compensation for the municipalities for joining the Association.

Table 2 Estimated financing of transport and mobility in the metropolitan area in 2017 (with the average EUR exchange rate of 26 February 2018 at the level of EUR 1 = PLN 4.17)

Unit/instrument	Total annual figures in EUR million	Including: separate investments in EUR million	Share in per cent (%)
Silesian province (budget for 2017—the subsidies concern not only the metropolitan area)	59.2	11.3	21.30
Organizers (budgets 2017), including	109.9		39.50
– KZK GOP	97.6		
– MZKP Tarnowskie Góry	5.2		
– Tychy (MZK)	7.1		
ITI (it was assumed that half of the subsidy will be divided equally for the years 2015–2020)	67.0	67.0	24.0
GZM (estimated 'transport' budget for 2018)	42.4		15.20
Total	278.4		100

With reference to integration, Metropolis began to implement a common tariff and ticket system since 1 January 2018, which means that the transport association system (KZK GOP and MZKP) will be combined with the transport system in Tychy. The common systems are to be administered by Metropolitan Transport Authority (a unit of Silesian Metropolis) formed in February 2018, which is intended to replace KZK GOP, MZKP and ZKM Tychy. The diversification of tariffs caused the necessity to change the prices by introducing generally minor price reductions in the dominant metropolitan area and price increases in the area where transport is organized by the city of Tychy. In 2018, approximately EUR 7.7 million was dedicated to this purpose. The integration concerns three organizers; however, due to the existence of a common tariff and ticket system for KZK GOP and MZKP, the essence of the project is to combine KZK GOP/MZKP and Tychy systems on the basis of the agreement between Silesian Metropolis and the organizers of 19 December 2017. The project is implemented in stages—from 1 January 2018 in respect of single and short-term tickets (uniform system), and from 1 April 2018 in respect of long-term tickets. (Initially, MZK Tychy tickets will be accepted by KZK GOP/MZKP, and upon expansion of the electronic card system, a common ticket system will also be introduced.) Since 1 April 2018, the share of MZK Tychy in the common ticket system was determined on the basis of historical data on the sale of tickets as 8.04%, whereas the rest of funds remain the responsibility of KZK GOP and MZKP. (The share of MZKP constitutes 4.92% of the total funds.) The question arises about the actual significance of the problem of integration—since the existing KZK GOP/MZKP system covered such a large part of revenue (and thus of transport), it would seem reasonable and probably less expensive to expand KZK GOP (instead of forming a separate Silesian Metropolis structure).

Silesian Metropolis demonstrates high activity in image-building. It was for this purpose that a redistributive instrument (introduction of free transport for the youth under 16) was applied—the co-financing of Silesian Metropolis in 2018 was estimated in the amount of EUR 2.16 million, which seems to be a lowered estimate. The effectiveness of public funds distributed in this manner also raises serious doubts [22]. Moreover, in the field of sustaining mobility, Silesian Metropolis has undertaken many additional actions, which so far have been important only from the PR point of view:

- Announcement of implementation of the project for purchasing up to 300 electric buses in 2019–2023 funded by the National Centre for Research and Development within the framework of the support programme for development of electromobility (up to 1,000 electric buses for the entire country in total) [23];
- Announcement of designed network of bicycle routes in the Metropolis area.

5 Summary

The metropolitan development in Poland is less advanced than in Western Europe; however, the problems with effective and sustainable mobility management are already visible in Warsaw, Upper Silesia and Gdańsk agglomerations. The experiment consisting in establishment of a metropolitan association by way of legislation in the Upper Silesia Agglomeration, which has a complex and polycentric spatial arrangement, may be used for drawing conclusions, both theoretical (about the efficiency of regulatory instruments) and practical (concerning the organization of cooperation between administrative units forming the agglomeration). Metropolitan Association in the Upper Silesia Agglomeration began its activity on 1 January 2018. Based on the analysis of the available documents, especially the budget, it may be noticed that in the first year of the Metropolitan Association operation:

- The budget of Silesia Metropolis was dominated by redistribution expenditure, especially on projects related to road transport and free public transport.
- The activities related to providing ongoing support for sustainable mobility are restricted to financing free transport on smog days (approximately EUR 0.24 million was allocated for this purpose) and preparatory works related to the purchase of electric buses;
- The activities related to public transport integration were undertaken; Metropolitan Transport Authority (intended to take over the tasks performed by the organizers in Metropolis) was appointed; tariff and ticket integration was introduced.

In the next few years, the assumptions of Metropolis will be verified in practice—in particular, whether Metropolis creates added value and produces synergy effects, or it confirms the rule referred to as 'Ockham's razor' and is a 'being beyond necessity'—a formation serving primarily for redistribution of tax revenue.

References

1. World Urbanization Prospects: The 2014 Revision, Highlights, UN (2014)
2. Kongres Ruchów Miejskich: https://kongresruchowmiejskich.pl/o-nas/
3. Manifest "My pasażerowie", http://kongresruchowmiejskich.pl/wp-content/uploads/2017/05/Manifest_My_Pasazerowie.pdf
4. 2015 EMTA Barometer, 11th edn. CRTM Madrid, Paris (2017)
5. Attitudes of Europeans Towards Urban Mobility. Special Eurobarometer 406. EU, Brussels (2013)
6. Guidelines. Developing and implementing a sustainable urban mobility plan. EU, Brussels (2013)
7. WHITE PAPER Roadmap to a Single European Transport Area—Towards a competitive and resource efficient transport system. COM, 144 (2011)
8. Duportail, V., Meerschaert, V.: Final ADVANCE audit scheme and guidelines. ELTIS (2013)
9. Beim, M.: Polityka rowerowa Bolzano. Transport Miejski i Regionalny **11**, 11–17 (2011)
10. Shaheen, S., Guzman, S., Zhang, H.: Bikesharing in Europe, the Americas, and Asia: past, present, and future. J. Transp. Res. Board **2143**, 159–167 (2010)
11. National Association of City Transportation Officials: https://nacto.org/bike-share-statistics-2016/, 11.11.2017
12. Wikipedia: https://pl.wikipedia.org/wiki/Lista_systemów_rowerów_publicznych_w_Polsce, 11.11.2017
13. ESPON 1.1.1: Potentials for polycentric development in Europe. Project report, NORDREGIO (2004)
14. Korzeniak, G., Gorczyca, K.: Policentryczność rozwoju systemu osadniczego z udziałem miast małych i średnich w kontekście procesów metropolizacji In: Korzeniak, G. (ed.) Małe i średnie miasta w policentrycznym rozwoju Polski. Instytut Rozwoju Miast, pp. 127–156, Warszawa (2014)
15. Tomanek, R.: Telematics in the new EU cohesion policy on the example of integrated territorial investments strategy. In: Mikulski, J. (ed.) Telematics-Support for Transport, pp. 434–440. Springer, Heidelberg (2014)
16. Plan zrównoważonej mobilności miejskiej Subregionu Centralnego Województwa Śląskiego. Uniwersytet Ekonomicznyw Katowicach, Katowice (2016)
17. Rozporządzenie Rady Ministrów z 26 czerwca 2017 r. w sprawie utworzenia w województwie śląskim związku metropolitalnego pod nazwą „Górnośląsko-Zagłębiowska Metropolia" (Dz.U. 2017 poz 1290)
18. Ustawa z 9 marca 2017 r. o związku metropolitalnym w województwie śląskim (Dz.U. 2017, poz. 730)
19. Wniosek o utworzenie związku metropolitalnego w województwie śląskim, Katowice 2017 (załącznik do uchwały Rady Nr XLII/799/17 Miasta Katowice z 29 maja 2017 roku w sprawie złożenia wniosku o utworzenie związku metropolitalnego w województwie śląskim pod nazwą Górnośląsko-Zagłębiowska Metropolia)
20. GZM: https://fs.siteor.com/metropoliagzm/article_attachments/attachments/165109/original/Uchwała_nr_IV_24_2017.pdf?1517404994
21. GZM: https://fs.siteor.com/metropoliagzm/article_attachments/attachments/165767/original/Załącznik_do_Uchwały_Zarządu_nr_23_2018.pdf?1518529020

22. Tomanek, R.: Free-fare public transport in the concept of sustainable urban mobility. Transp. Probl. **12** (special edition), 95–105 (2017)
23. NCBiR: http://www.ncbr.gov.pl/aktualnosci/art,5821,2–2-mld-zl-na-nowoczesne-bezemisyjne-autobusy-dla-polskich-miast.html

Commuting to Places of Work and Study in the Light of Marketing Research Results with Particular Emphasis on Gdynia in Poland

Katarzyna Hebel and Olgierd Wyszomirski

Abstract As part of research into the demand for urban transport services, special attention is paid to the demand resulting from the implementation of transport needs of an obligatory nature, such as commuting to places of work and study. The research on commuting to places of work and study are carried on in different countries. Noteworthy is the example of Great Britain and Poland. The article presents the methodology of research on commuting to places of work and study as well as the results of research from 2000, 2008 and 2015 carried on in Gdynia. These changes in commuting to places of work and study are significant, since the travel time both to places of work and study has increased. The main mode of travel to work has become the use of own car, instead of travelling by public transport. This is one of the basic conclusions from the analysis carried on in the article.

Keywords Urban transport · Travel behaviour · Marketing research

1 Introduction

The demand for public transport services results from the fulfilment of the inhabitants' transport needs. Great importance is attached to the needs of an obligatory nature, such as commuting to places of work and study since they require regular use of public or individual transport. To provide attractive, from the perspective of inhabitants, public transport, first of all we need to learn the specificity of daily travels, which has been the subject of research of different nature and scope. A country with considerable experience in studying the travel behaviour, including commuting to places of work and study, is Great Britain. Research on commuting to places of work and study has also been developed in Poland. This area of research holds an

K. Hebel (✉) · O. Wyszomirski
University of Gdansk, Sopot, Poland
e-mail: khzh@wp.pl

O. Wyszomirski
e-mail: o.wyszomirski@wp.pl

M. Suchanek (ed.), *Challenges of Urban Mobility, Transport Companies and Systems*, Springer Proceedings in Business and Economics,
https://doi.org/10.1007/978-3-030-17743-0_5

important place in sociology. The city in Poland where research on commuting to places of work and study is conducted on a regular basis is Gdynia. This article aims to present and analyse the results of research on commuting to places of work and study with particular reference to Gdynia.

2 Research on Commuting to Places of Work and Study at the Example of Great Britain

The National Travel Survey—NTS—is conducted in Great Britain on a regular basis. The first studies were performed in 1960; then, they were not conducted regularly, and from 1988, the Department for Transport performed regular research with the final round conducted in 2016. These are surveys regarding individual travels of the inhabitants in Great Britain. The data is collected during interviews and supplemented with information from "travel diary" kept by the respondent during the week preceding the survey [1]. There are the following purposes of travel:

- commuting to and from work,
- business errands,
- study (including transporting others),
- shopping,
- personal errands (e.g. medical ones) and transporting for these purposes,
- visiting friends,
- spending free time (mainly entertainment, sport, holidays, and short one-day trips).

Transporting others and accompanying others refer to the main purposes (main activities), i.e. study, if it involves driving children to school, and personal errands, if it involves transporting others, e.g. to the doctor's (Table 1).

Commuting to places of work constitutes 15% of travels of the inhabitants of Great Britain, and commuting to places of study—12%. The travel distance to places of

Table 1 Share of particular purposes of travel in the average number of travels and distances in Great Britain

Purpose of travel	Share in average number of travels (%)	Share in average distance of travel (%)
Shopping	19	11
Personal errands	18	14
Spending free time	17	21
Commuting to and from work	15	20
Visiting friends	15	20
Study	12	5
Business errands	3	10

Source [1]

work constitutes 1/5 of average travel distance, whereas the travel distance to places of study constitutes only 1/20. On average, an inhabitant of Great Britain has 144 travels to work per year at a distance of 1296 mil (2085 km), which takes 30 min per day. 64% of travels to work are performed with the use of passenger car, 11% on foot and 7% by rail. The average number of travels to places of study per year amounts to 117, at the total distance of 321 mil (516 km), which takes 17 min per day. Children aged 5–10, in 94%, are accompanied by a minder, whereas elder children, i.e. aged 11–16, are accompanied by a minder in 56% of travels. Travel to primary school is done in 51% on foot. Travel to secondary school in 27% is done by bus.

3 Commuting to Places of Work and Study as a Research Problem in Poland

The issues related to commuting to places of work were analysed in Poland from 1960s [2–7]. During the second census in 1960, extensive material was collected, related to commuting to work, but due to technical reasons, the material was difficult to use. In 1964, one-off survey was conducted on commuting to work [5] with supplementary personnel register of the same year.

Together with the development of statistics, censuses, and registers of personnel commuting to work, kept by particular state companies, the knowledge on travels to places of work in Poland was improved. There were also studies based on surveys and interviews, conducted in places of work [4, 8] which analysed the structure of travels to work, modes of transport, costs and reasons for travel.

The census conducted in 1978 analysed travels to work on a representative sample of 10% housings [9]. In subsequent years, the topic became much less interesting, and even the data on travels to work collected during the census in 1988 was not analysed in detail and published.

Studies conducted by economic entities were also limited, due to difficulties related to selecting samples, which resulted from regulations on personal data protection [10]. Political transformations in Poland which involved restructuring and privatization of state companies had significant impact on commuting to work. Nevertheless, during the census in 2002, the question related to places of work was omitted, which led to creating a knowledge gap on changes in commuting to work.

After years of no information, for the first time GUS (Central Statistical Office) published detailed data on commuting to work in Poland in 2006 [11].

Under research results entitled: "*Przepływy ludności związane z zatrudnieniem w 2006 r.*" [Movement of population in relation to employment in 2006] conducted in compliance with data from tax registers, supplemented with information from other entities (ZUS—Social Insurance Institution, tax offices), new methodology of research on commuting to work was developed, used during the National Census of Population and Housing 2011 (NSP 2011). The data therefrom allowed to provide characteristics of commuters by gender and age. In particular, it was possible to identify in detail the places of work by territory. In combination with the actual place of residence of the surveyed, it was possible to determine the directions of movement related to work. However, the data collected in such manner fails to define the mode of transport, frequency of travel service and travel time of commuters. Such information on a national basis and in relation to voivodeships was provided by representative sample within NSP 2011. We need to realize that the results of research on commuting to work under administrative registers are not comparable with data obtained under representative sample, mainly due to differences in the sources of data and methodological differences [12].

Within representative sample, following the definition providing that a commuter is a person whose place of work is located outside their town of residence—in 2011 in Poland there were 4,462,500 commuters. Then, the group was limited by selecting only employees whose source of income resulted from remuneration payable under employment relationship. The group comprising 3,130,600 commuters was analysed as per various structural and spatial cross-sections [13]. It was found that employment outside the place of residence was rather common in Poland since 3.1 million workers with employment relationship were commuters, which constituted 32.5% of the group. The most "mobile" were the inhabitants of Śląskie and Wielkopolskie voivodeships with over 487,000 and 363,000 commuters, respectively. Subsequent voivodeships as per the number of inhabitants who worked outside their municipalities were: Mazowieckie Voivodeship (357,000) and Małopolskie Voivodeship (311,000). Significant mobility of workers living within these voivodeships resulted mainly from the vicinity of large urban municipalities attracting workers from neighbouring towns. The least mobile were the inhabitants of Podlaskie Voivodeship (53,000) and Lubuskie Voivodeship (83,000). In Pomorskie Voivodeship there were 180,000 people who commuted to work outside their municipality.

The situation is slightly different when we refer the percentage share of commuters to the total number of the employed in a particular voivodeship. The highest share of commuters among the employed under employment relationship was observed in Podkarpackie Voivodeship (43.2%). Subsequent ones were Opolskie and Śląskie Voivodeships with ca. 38%. Whereas, on the other side, we had Podlaskie Voivodeship with only 21.5% of the employed who commuted to work outside their municipality. In Pomorskie Voivodeship, the share of people commuting to work outside their municipality constituted 31% of all people employed.

The spatial mobility is not equal among the Polish workers (Table 2). The key factors affecting the diversity of mobility are age and gender. Travels to places of work refer mainly to people aged 25–34. The share of this age group among all the commuting Poles in 2011 amounted to 32.5%. The share of other groups was much lower. The smallest share in the structure of commuters referred to the elderly aged over 65 (0.5%). Men constituted nearly 60% of all commuters.

The issue of commuting to places of work was referred to travels for other purposes, with travels to places of study included as late as in 2015 within the pilot study on travel behaviour in Poland conducted by GUS. In the study, the following research methods were applied: CAII (questionnaires filled in by respondents via Internet application), CAPI (computer-assisted personal interviewing conducted by

Table 2 Commuters in Poland in 2011 by gender and age

Characteristic	Total	Percentage
Age		
15–24	330,194	10.5
25–34	1,017,364	32.5
35–44	818,614	26.1
45–54	677,276	21.6
55–64	272,081	8.7
65 and more	15,085	0.5
Gender		
Woman	1,317,899	57.9
Man	1,812,715	42.1

Source Own elaboration based on [12]

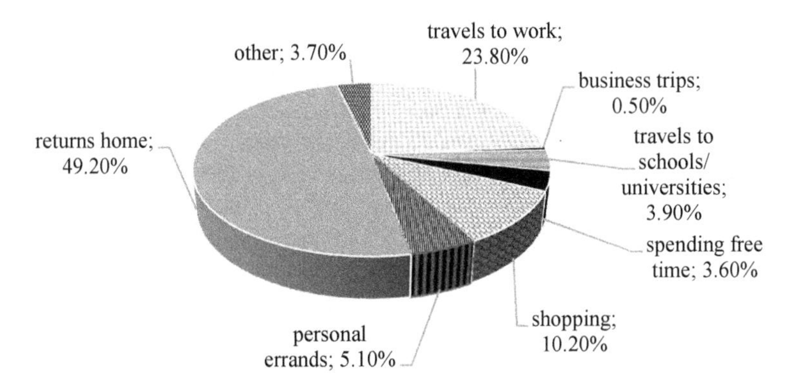

Fig. 1 Structure of travels by purpose in Poland in 2015 *Source* Own elaboration based on [14]

pollsters) and combination of PAPI and CAPI (computer-assisted personal interviewing conducted with the use of paper questionnaire filled in by pollster followed by the introduction of obtained data to CAPI-based application) on the sample of 18,000 households in the whole country (interviews conducted in 13,551 households). The respondents were people aged 16 and more living in selected households (Fig. 1).

Travels to work constitute 24% of all travels and travels to schools and universities 4%. People aged 16 and more make ca. 13 billion travels. There are 422 travels per person on a yearly basis. Travels to work constitute over 100 of these travels, and only 17 refer to travels to schools and universities. Travels to work take on average ca. 24 min, and travels to schools and universities—30 min. The average distance of 1 person travelling by car in Poland is 17.9 km and by public transport in cities 9.5 km. The majority of travels in the country are travels by car; their share amounts to 54.5%. Most frequently, passenger cars are used to commute to work (27.2%) [14].

There are plans to continue passenger travel behaviour studies in Poland every 3 years and the methodology of research will be adjusted to Eurostat requirements.

4 Commuting to Places of Work in the Light of Results of Sociological Study

Research on commuting to work also constitutes a part of sociological study. An example of such research involves "Commute survey" of global recruitment company, Michael Page, operating in 36 countries, which examined modes of commuting to work and the related stress of employees via Internet questionnaire. In the study, there were 12,500 participants from 11 European countries, including nearly 1,100 from Poland [15, 16].

The results of survey of 2017 prove that as many as two out of three Poles commute to work by car. The largest share, i.e. 80%, refers to people who drive children to kindergarten or school on the way to work. The average travel time to work amounts to 41 min one way (on average in Europe—42 min, the longest in Turkey—48 min, the shortest in Portugal—slightly over 30 min). The main reasons for choosing the car include: shorter travel time and independence from public transport timetables. According to the respondents, the advantages of public transport related to commuting to work include: relatively low costs (60% of the respondents), short travel time (44% of the respondents), in particular when it is possible to use underground or fast train, and no problems with parking the car (40% of the respondents). In Poland, 88% of commuters are satisfied with public transport services. Among the inhabitants and workers of the same town, the percentage of positive opinion totals 92%, whereas among those who travel to work less than half an hour 96% are satisfied with public transport services. Such results rank Poland third among the analysed countries, after Austria and Switzerland (with satisfied passengers amounting to more than 90%). The European average is decreased by the Italians; only 54% of workers travelling by public transport value their services.

It was also found that despite traffic congestions in the European cities travelling by car causes less stress than travelling by public transport. Crowded buses, trains and underground causes stress for 38% of people travelling by public transport (in Poland 35%). Meanwhile, travel by car causes stress for 34% of the Europeans, including 23% of the Poles (as per report by Deloitte of 2016 in large cities in Poland, i.e. Warszaw, Poznań, Kraków and Wrocław, drivers spend 8–9 h a month in traffic jams) [17, 18].

The survey conducted at portal, *Praca.pl* by independent research institute, *Kantar Public* commissioned by *Crossover* in September 2017, under CATI method (computer-assisted telephone interviewing), at the sample of 1000 Poles aged 18 and more, proved that for six out of nine Poles the job offer is attractive, if the travel time to work does not exceed 30 min, whereas every fifth would like to travel max. 15 min. According to M. Filipkiewicz, an expert at portal, *Praca.pl*, the less time spent on travelling to work, the better for mental health of employees. Therefore, long travel time to work disturbs the work-life balance, namely balance between the private and professional life [19].

Similar conclusions can be drawn under yearly research conducted among Londoners by portal "Time Out" on happiness and satisfaction with life at the sample of 60,000 people. The respondents are asked about the travel time to work, feeling of happiness or irritation and general feeling of satisfaction related to the preceding day. The majority of the surveyed would like to travel to work fast. Even several minutes make significant difference. Travelling to work every day makes the British more anxious and unhappy [20].

These conclusions are confirmed by research conducted by the University of Waterloo, published in 2012 in the American Journal of Preventative Medicine, which prove that travel time to work is related to the general well-being. People who spend more time travelling to work declare lower satisfaction with life.

5 Commuting to Places of Work and Study of the Inhabitants of Gdynia

5.1 Methodology of Research on Commuting to Places of Work and Study in Gdynia

Research on commuting to places of work and study has been conducted in Gdynia since 1985 as part of research into the inhabitants' travel behaviour. The analysis of three research rounds carried out in 2000, 2008 and 2015 allows to identify the trends of changes taking place in commuting to places of work and study [21–23].

The representative research conducted on a sample of 1% of inhabitants aged from 16 to 75, was held by the Chair of Transportation Market at the University of Gdańsk, Poland, in cooperation with the Transport Board of Gdynia. The method adopted for this research consisted of in-person survey in the form of a questionnaire which was carried out in the respondents' homes. In subsequent rounds, certain aspects were deemed less important and were removed from the questionnaire, whilst others increased in significance and were researched more thoroughly [24].

5.2 Commuting to Places of Work and Study as Purpose of Travel for the Inhabitants of Gdynia

The inhabitants of Gdynia travel for various reasons. In the studies conducted in three analysed research rounds, the fundamental purposes of travel involved home, work, study, personal errands, shopping, professional and business errands, social life and recreation (Table 3). It was stated that the most important purpose of travel

Table 3 Commuting to places of work and study among travel purposes of the inhabitants of Gdynia between 2000 and 2015 (% of answers)

Purpose of travel	Year		
	2000	2008	2015
Home	45.02	45.66	43.77
Work	18.57	22.42	22.91
Study	5.63	4.73	3.40
Personal errands	13.54	14.84	11.62
Shopping	7.51	7.55	7.94
Professional and business errands	5.57	2.41	1.84
Social life	2.08	1.38	2.97
Recreation	2.08	1.00	2.54
Transporting others	–	–	3.30

Source Own elaboration based on [21–23]

was home, which was obvious since regardless of the basic reason for travelling, the travellers returned to their place of residence. Lower than 50% share of such travels means that some inhabitants travel not only from home and back home. The largest share in the structure of travels among the inhabitants of Gdynia refers to compulsory travels, i.e. travels to places of work and study (every time these are in total ca. 1/4 of all travels). The analysis of data specified in Table 3 indicates that travels to work constitute increasingly higher share in all travels, and travels to places of study—increasingly lower.

Relatively large share belongs to travels for personal reasons (12–15%). Significant changes were observed in relation to professional and business trips. Their share decreased from nearly 6% in 2000 to 2% in 2015. In the final round of research, new purpose of travel was included, namely transporting others to various places. It has become more frequent purpose of travel than professional and business errands, social life and recreation.

5.2.1 Mode of Commuting to Places of Work and Study of the Inhabitants of Gdynia

The declared mode of urban travels is determined under an answer to the question: "How do you travel in the city"; among possible answers there are 6 categories [25]:

- always by public transport,
- usually by public transport,
- equally by public transport and private car,
- always by private car,
- usually by private car,
- other.

In the light of data presented in Table 4, the share of inhabitants who declare that they always and usually travel by public transport has been decreasing. In 2000,

Table 4 Declared mode of urban travels by the inhabitants of Gdynia between 2000 and 2015 (% of answers)

Mode of travel	Year		
	2000	2008	2015
Always by public transport	34.48	27.62	25.15
Usually by public transport	27.29	25.67	22.00
Equally by public transport and passenger car	12.54	11.06	11.05
Usually by passenger car	13.79	16.31	18.15
Always by private car	10.96	18.93	22.50
Other	0.91	0.41	1.15

Source Own elaboration based on [21–23]

Table 5 Mode of travel to places of work and study among the inhabitants of Gdynia between 2000 and 2015 determined under travels made on the day preceding the survey (as per used modes of transport)[a]

Mode of travel	To the place of work			To the place of study		
	Year					
	2000	2008	2015	2000	2008	2015
By bus	40.12	30.59	20.64	68.69	45.70	40.48
By trolleybus	14.24	11.31	6.13	14.65	21.85	15.08
By rapid railway or rail	5.97	5.73	5.31	11.11	19.21	19.84
By private car	38.74	51.67	58.60	4.55	11.92	11.12
By bicycle	0.46	0.70	1.30	–	1.32	–
Other	0.46	–	8.02	1.01	–	13.48

[a]In the travel the proportional share of the mode of transport was included in case of a change to other mode of transport, and round figures were applied
Source Own elaboration based on [21–23]

there were over 60% of such people, when in 2015 only nearly 50%. Whereas the share of people who declare that they always or usually travel by private car has been increasing. Within 15 years, their share grew from 25% to slightly more than 40%. The share of inhabitants who declare that they equally travel by public transport and private cars remains more or less on the same level. Their share in the first round of research amounted to 13%; in the last two—11%. The main type of travel includes travels to places of work and study. The dominant mode of transport here is the private car. Its share increased between 2000 and 2015 from 39 to 59%. Whereas, among travels to places of study, the most important is public transport. However, its total share in such travels decreased between 2000 and 2015 from 94 to 75%. The share of private car in travels to places of study is relatively small; in the analysed period, it grew from nearly 5 to 11% (Table 5).

Other modes of transport included, e.g. travels on foot.

5.2.2 Travel Time to Places of Work and Study of Gdynia Inhabitants

As in the case of modes of travel, first the declared travel time to places of work and study was analysed (Table 6), and then, it was compared with real travel time as on the day preceding the survey (Table 7).

The travel time to places of work and study by both public transport and private car was declared by all the surveyed inhabitants, regardless of their mode of travel, as self-perceived travel time. In relation to public transport in 2015, the highest percentage of people declared for both purposes of travel the travel time longer than 30 min. Such travel time was declared by ca. 50% of the surveyed. Whereas, in relation to private car, similar number of people, ca. 20% declared the travel time in four categories from 6 to 30 min.

Table 6 Declared travel time to places of work and study as per "door-to-door" system by public transport and private car of Gdynia inhabitants between 2000 and 2015 (% of answers)

Travel time (min)	Year	To the place of work		To the place of study	
		By public transport	By private car	By public transport	By private car
To 5	2000	1.31	9.09	0.66	6.91
	2008	0.43	4.64	1.54	4.63
	2015	2.33	6.88	2.90	7.73
6–10	2000	5.95	23.48	3.96	20.07
	2008	2.27	18.04	4.25	18.15
	2015	5.29	19.79	5.31	17.87
11–15	2000	10.95	24.08	10.23	20.38
	2008	5.72	24.62	5.02	20.08
	2015	7.20	22.01	6.76	19.81
16–20	2000	13.45	14.30	9.90	13.49
	2008	8.21	16.09	8.11	16.60
	2015	10.47	20.00	11.11	16.91
21–30	2000	25.60	13.71	24.42	12.83
	2008	26.35	18.90	25.48	18.53
	2015	21.48	17.25	24.64	21.74
Over 30	2000	35.00	7.15	45.55	13.49
	2008	52.81	15.12	54.44	18.53
	2015	47.83	11.00	46.86	10.63
I don't know	2000	7.74	8.22	5.28	12.83
	2008	4.21	2.59	1.16	3.48
	2015	5.4	3.07	2.42	5.31

Source Own elaboration based on [21–23]

Table 7 Average travel time to places of work and study as per "door-to-door" system by public transport and private car of Gdynia inhabitants between 2000 and 2015 (min)

Years	To the place of work		To the place of study	
	By public transport	By private car	By public transport	By private car
2000	32	17	41	20
2008	41	22	40	23
2015	39	20	36	21

Source Own elaboration based on [21–23]

The average declared travel time to places of work and study of Gdynia inhabitants was longer by public transport than by private car (Table 7) from 71 to 105% depending on the year of survey and purpose of travel. More dissimilarity in the travel time among people travelling by public transport and private car was observed during the analysed years in relation to travels to places of study with extreme values referring to those travels. The dissimilarity in the travel time by both modes of transport during the analysed period was lower for travels to work, falling within 86 to 95%. We can specify in general that the travel time to places of work and study by public transport is perceived as nearly twice as long. Such situation determines the advantage of private car over public transport in commuting to places of work and study.

Table 8 presents the average time of particular stages of travel to places of work and study with walk to the bust stop from the starting point, waiting for the vehicle, travel and walk from the bus stop to the place of destination. In case of travels with changes, the travel time was presented separately for each mode of transport used during the travel. Additionally, such travels also include time for change and waiting for another mode of transport. The data in this table shows that the highest share in the structure of travel time belongs to the ride itself, constituting ca. 60% of travels without any change and travels with one and two changes, which means that its share in the total travel time is independent of the number of changes. In general, the average travel time with two changes was longer than travel with one change, and the latter was longer than the travel without any changes. The travel with one change took nearly twice as much time as the travel without any change. The increase of time in case of the second change was not so high, amounting to ca. 50%. The main factor which determines longer travel time with changes is the distance of travel. It is reflected in the longer combined duration of ride in such travels. However, another element involves time for changes and waiting which takes at the first and second change ca. 6 min and constitutes ca. 10% of travel time in case of one change and 15% in case of another change. The longer, on average, travel time with changes with increasing losses of time needed for the change proves primacy of direct connections among other demands submitted by the surveyed inhabitants.

Table 8 Average time of particular stages of travel by public transport to places of work and study of Gdynia inhabitants between 2000 and 2015 (min)

Stage of travel	Year		
	2000	2008	2015
Walk to the bus stop	–	5.21	4.56
Waiting for transport	–	3.19	2.29
Travel time	–	18.04	18.13
Walk from the bus stop	–	6.46	6.09
Travel time without a change	–	28.30	31.47
Change and waiting	–	6.01	5.45
Travel time by the second vehicle	–	16.26	19.32
Travel time with one change	–	54.57	57.04
Change and waiting	–	5.58	5.55
Travel time by the third vehicle	–	10.44	11.50
Travel time with two changes	–	74.39	74.49

Source Own elaboration based on [21–23]

6 Conclusions

In studies on travel behaviour, special attention is paid to the issue of commuting to places of work and study. It is proven by the examples from Great Britain and Poland.

The issues of commuting to places of work and study also constitute a part of sociological study. Such situation occurred in 11 European countries, including Poland.

For over 30 years, commuting to places of work and study has been the subject of research on travel behaviour among the inhabitants in Gdynia. The results of research from 2000, 2008 and 2015 prove that:

- travels to places of work and study considered collectively have the highest share in the structure of travel and constitute ca. 1/4 of travels to all places together with a home and nearly half of travels to all places excluding a home;
- in travels to places of work private car is dominant—its share in the analysed period increased from 39 to 59%;
- in travels to places of study modes of public transport are dominant, although their share in the analysed period decreased from 94 to 75%;
- travel time to places of work and study by public transport is perceived approximately twice as long as travel by car;
- average travel by public transport with changes takes at least twice as much time as direct connection, due to larger travel distance with two and more modes of transport and due to the activity of changing vehicles combined with waiting for another vehicle.

References

1. National Travel Survey: Statistical Release, 27 July 2017. Department for Transport, England (2016)
2. Herma, J.: Dojazdy do pracy w województwie krakowskim, Rocznik Naukowo-Dydaktyczny Wyższej Szkoły Pedagogicznej, Prace Geograficzne, 10, Kraków, 129–141 (1962)
3. Herma, J.: Kwalifikacje zawodowe pracowników dojeżdżających do przemysłu w ośrodkach w województwa krakowskiego, Rocznik Naukowo-Dydaktyczny Wyższej Szkoły Pedagogicznej, Prace Geograficzne, 22, Kraków, strony (1964)
4. Herma, J.: Dojazdy do pracy w Polsce Południowej (województwo katowickie, kieleckie, krakowskie, opolskie, rzeszowskie) 1958–1961, Wydawnictwo Wyższa Szkoła Pedagogiczna w Krakowie, Kraków (1966)
5. Lijewski, T.: Dojazdy do pracy w Polsce, Studia Komitetu Przestrzennego Zagospodarowania Kraju Polska Akademia Nauk, t. 14, Warszawa, strony (1967)
6. Cegielski, J.: Dalekie dojazdy pracownicze do Warszawy, Kronika Warszawy, 1, 1, 7, 45–56 (1974)
7. Gawryszewski, A.: Związki przestrzenne między migracjami stałymi a stałymi dojazdami do pracy oraz czynniki przemieszczeń ludności, Ossolineum, Instytut Geografii Polska Akademia Nauk, Prace Geograficzne, 109, Wrocław – Warszawa – Kraków – Gdańsk (1974)
8. Olędzki, M.: Dojazdy do pracy. Zagadnienia społeczno-ekonomiczne na przykładzie rejonu płockiego. Książka i Wiedza, Warszawa (1967)
9. Gawryszewski, A.: Ludność Polski w XX wieku, Monografie, 5, Instytut Geografii i Przestrzennego Zagospodarowania im. S. Leszczyckiego PAN, Warszawa (2005)
10. Matykowski, R., Tobolska, A.: Funkcjonowanie zakładów przemysłowych XXI wieku na przykładzie Swedwood Poland i Volkswagen Motor Polska Sp. z o.o. Analiza dojazdów do pracy, Prace Komisji Geografii Przemysłu, 14, Warszawa-Kraków, 65–75 (2009)
11. Kruszka, K. (ed.): Dojazdy do pracy w Polsce. Terytorialna identyfikacja przepływów ludności związanych z zatrudnieniem, Główny Urząd Statystyczny, Urząd Statystyczny w Poznaniu, Poznań (2010)
12. Raport z wyników Narodowy Spis Powszechny Ludności i Mieszkań (2011). https://stat.gov.pl/cps/rde/xbcr/gus/lud_raport_z_wynikow_NSP2011.pdf
13. Dojazdy do pracy NSP, Główny Urząd Statystyczny, Warszawa (2014)
14. Badanie pilotażowe zachowań komunikacyjnych ludności w Polsce. Główny Urząd Statystyczny, Warszawa (2017) https://stat.gov.pl/files/gfx/portalinformacyjny/pl/defaultstronaopisowa/5851/1/1/streszczenie_raportu_badanie_pilotazowe_zachowan_komunikacyjnych.pdf
15. Tylko 23% Polaków odczuwa stres, jeżdżąc do pracy samochodem. https://www.michaelpage.pl/dla-medi%C3%B3w/tylko-23-polak%C3%B3w-odczuwa-stres-je%C5%BCd%C5%BC%C4%85c-do-pracy-samochodem
16. Transport and Commute The European Commute: calm and relaxed or stressful and inefficient? https://www.pagepersonnel.nl/en/news-insights/studies/transport-and-commute
17. Polak docenia transport publiczny, ale woli dojeżdżać autem. http://www.rp.pl/Praca/312049982-Dojazdy-do-pracy-Polak-docenia-transport-publiczny-ale-woli-dojechac-autem.html
18. Kierowcy w siedmiu największych polskich miastach tracą już 3,8 mld zł rocznie przez korki. https://www2.deloitte.com/pl/pl/pages/press-releases/articles/korki-w-polskich-miastach-2016.html
19. Dojazd do pracy: Pracownicy tracą 40 godzin miesięcznie w korkach. Rozwiązaniem praca zdalna. http://www.pulshr.pl/zarzadzanie/dojazd-do-pracy-pracownicy-traca-40-godzin-miesiecznie-w-korkach-rozwiazaniem-praca-zdalna,49155.html
20. Rojek-Kiełbasa A., Czy przez dłuższy dojazd do pracy jesteśmy mniej szczęśliwi? Brytyjczycy owszem są. http://natemat.pl/91921,czy-przez-dluzszy-dojazd-do-pracy-jestesmy-mniej-szczesliwi-brytyjczycy-owszem-sa

21. Badania pierwotne pt. "Preferencje i zachowania transportowe mieszkańców Gdyni", Katedra Rynku Transportowego Uniwersytetu Gdańskiego i Zarządu Komunikacji Miejskiej w Gdyni, Gdynia (2000)
22. Badania pierwotne pt. "Preferencje i zachowania transportowe mieszkańców Gdyni", Katedra Rynku Transportowego Uniwersytetu Gdańskiego i Zarządu Komunikacji Miejskiej w Gdyni, Gdynia (2008)
23. Badania pierwotne pt. "Preferencje i zachowania transportowe mieszkańców Gdyni", Katedra Rynku Transportowego Uniwersytetu Gdańskiego i Zarządu Komunikacji Miejskiej w Gdyni, Gdynia (2015)
24. Hebel, K.: Zachowania transportowe mieszkańców w kształtowaniu transportu miejskiego. Fundacja Rozwoju Uniwersytetu Gdańskiego, Gdańsk (2013)
25. Hebel, K.: Zmiany preferencji i zachowań transportowych mieszkańców Gdyni w latach 1996–2013. Transport Miejski i Regionalny, Nr 4, 10–15 (2014)

The Analysis of Roundabouts Perception by Drivers, Cyclists and Pedestrians

Elżbieta Macioszek and Damian Lach

Abstract Statistics show that roundabouts are safe road solutions. The main objective of presented paper is the analysis of roundabouts perception by drivers, cyclists and pedestrians. For this purpose, this paper attempts to check and verify the hypothesis that roundabout users in Poland—especially single-lane roundabout users—generally perceived roundabouts positively. In connection with the above, the use of the preference study determined the perception of the roundabouts of their users. Survey research was conducted among the inhabitants of the Silesian Voivodeship. The results of the research allowed to state that roundabout users express a general positive attitude to the reconstruction of other types of intersections at the crossroads of the roundabout. The perception of roundabouts by cyclists and pedestrians coincides with driver ratings. In terms of traffic safety, among all types of roundabouts, users are best perceived as single-lane roundabouts and the spiral roundabouts are the worst.

Keywords Roundabouts · Car drivers at roundabouts · Cyclists at roundabouts · Pedestrians at roundabouts

1 Introduction

Transport, ensuring the possibility of efficient and quick displacement of people and cargo, is one of the most important branches of the national economy. From year to year, the demand for transport services is systematically increasing, which in turn causes a dynamic increase in traffic volumes on the transport network. Technological progress is trying to keep up with demand, contributing simultaneously to ensuring

E. Macioszek (✉) · D. Lach
Transport Systems and Traffic Engineering Department, Faculty of Transport,
Silesian University of Technology, Katowice, Poland
e-mail: elzbieta.macioszek@polsl.pl

D. Lach
e-mail: damian.lach@polsl.pl

M. Suchanek (ed.), *Challenges of Urban Mobility, Transport Companies and Systems*, Springer Proceedings in Business and Economics,
https://doi.org/10.1007/978-3-030-17743-0_6

an adequate level of road safety and reducing the negative impact of transport on the natural environment. Despite this, road accidents are one of the basic causes of disability and death of people both in Poland and in the world. Ensuring efficient, safe and low-polluting transport has become the main goal of modern traffic engineering.

In Poland since the 1990s of the twentieth century, the number of registered vehicles using public roads is constantly increasing. According to statistics from the Central Statistical Office, in 2016 this value amounted to less than PLN 29 million [1]. In addition, it should be noted that Poland is located at the intersection of several European transport routes connecting east to west, and north to south which further generates significant transit traffic. Increased traffic impacts the number of accidents and collisions. Recent years have brought a decisive improvement in the field of road safety in Poland, which is probably the effect of constantly running government programs aimed at improving road safety conditions of various road users. In 2007, there were 49,536 accidents, in which 5583 people were killed and 63,224 were injured. In 2008–2010, the number of accidents and their victims decreased, until 2011, when they again increased. Since 2012, the downward trend in the number of road accidents and their victims has been noticeable again (except for 2016, where these values have risen again) [2, 3].

Statistics show that roundabouts are safe road solutions. This is confirmed by many results of research presented in national works (including: [4–9]) as well as foreign ones (including: [10–13]). Of all types of roundabouts, single-lane roundabouts are considered by users to be the safest solution. The main objective of the presented paper is the analysis of roundabouts perception by drivers, cyclists and pedestrians. For this purpose, this paper attempts to check and verify the hypothesis that roundabout users in Poland—especially single-lane roundabout users—generally perceived roundabouts positively. In order to verify the aforementioned hypothesis, surveys were carried out. In connection with the above, the use of the preference study determined the perception of the roundabouts of their users. Survey research was conducted among the inhabitants of the Silesian Voivodeship.

2 Characteristics of the Research Area

Silesian Voivodeship is characterized by the highest indicator of urbanization in the country. The existing infrastructure is well developed in comparison with other voivodships, and there is a dense road and street network. The road system of this area, in particular the Katowice agglomeration, is characterized as one of the largest traffic loads in the country for freight and passenger transport. All these factors determine the level of security and often contribute to the failure to meet the expectations of some road users.

According to statistical data from the Municipal Police [3], in 2016 the Silesian Voivodeship was characterized by the second largest accident density indicator for 100 km of public roads in Poland. There were 3650 accidents, in which 257 people were killed, and another 4347 were injured. About 30% of these types of traffic

Table 1 Traffic incidents at intersections and their effects on the Silesian Voivodeship in the years 2010–2016

Types of incidents/victims	Intersection with circular traffic	Equivalent intersections	Intersection with the road with the priority of passing	Sum
Accidents	380	182	11,266	11,828
Collisions	7902	2048	98,614	108,564
Injured	442	224	14,169	14,835
Kills	2	2	506	510

Source Own on the basis of statistical data received from the provincial police headquarters in Katowice [14]

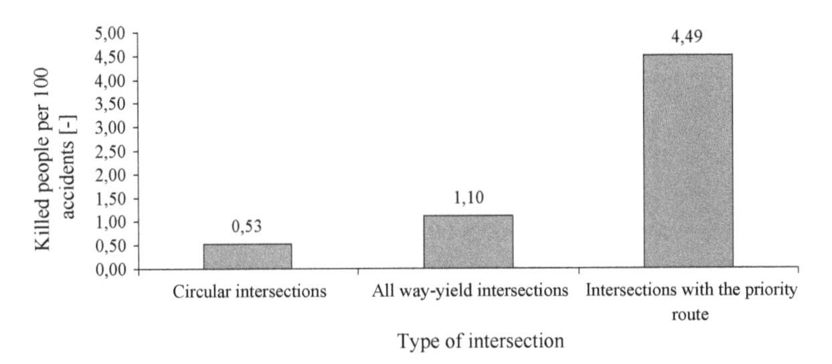

Fig. 1 Killed people per 100 accidents. *Source* Own on the basis of data [14]

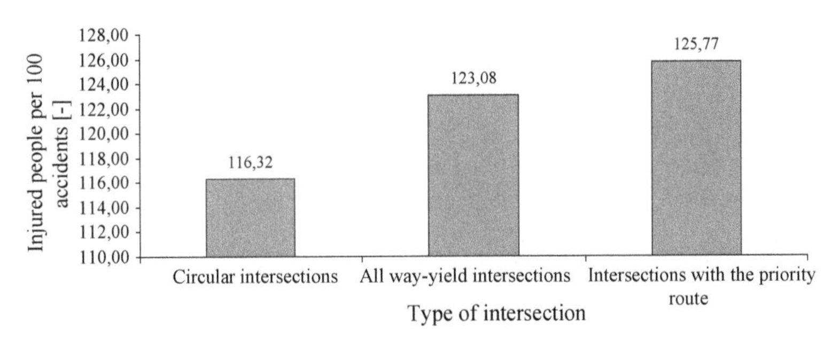

Fig. 2 Injured people per 100 accidents. *Source* Own on the basis of data [14]

incidents occurred at road crossings. The number of traffic incidents at intersections and their effects, in 2010–2016 in the Silesian Voivodeship, are presented in Table 1. The data presented in Table 1 do not take into account the number of intersections of particular types located in the Silesian Voivodeship, and therefore, it is not possible to directly formulate conclusions based on them on the general level of road traffic safety occurring at the types of intersections shown. Hence, in Figs. 1 and 2, the number of people killed and injured per 100 traffic accidents is shown.

The data shown in Figs. 1 and 2 indicate that the fewest persons are killed and injured in accidents that occur at intersections with circular traffic (including roundabouts).

Roundabouts are often and willingly designed on the transport network mainly due to the fact that they provide their users with a high level of road safety resulting mainly from:

- A smaller number of collision points compared to other types of intersections,
- Lower speed of crossing the intersection,
- Higher pedestrian protection—thanks to the separation of traffic flows at the inlets from the traffic flows at the outlets (through the use of separating islands at the inlets, which means that pedestrians have the option of passing the inlet and outlet separately),
- Less time wasted compared to other types of intersections, which also entails lower environmental pollution, lower fuel consumption and lower travel costs through the intersection,
- Increase in flowability compared to other types of intersections.

In order to make an inventory of the number of roundabouts in the Śląskie Voivodship, the identification of these types of intersections in all 19 cities with powiat rights was carried out first of all. The area covered by the analysis is over 1800 km^2, which is 14.6% of the total area of the Silesian Voivodship. The data were obtained mainly from relevant public institutions in individual cities as well as from any other available sources. Inventory included in its scope:

- The location of the roundabout, the name of the roundabout, the names of intersecting streets,
- Location (built-up area, outside built-up area, outskirts of built-up area),
- Outside diameter of the roundabout (in the case of turbine roundabouts—first radius of the R_1 roundabout),
- Roundabout type (single-lane, multi-lanes (with distinguishing the number of lanes on the roundabout and on individual inlets), turbine, spiral),
- Development of the central island (low, medium, high),
- Form of development of the central island (greenery, elements of small architecture),
- Comments.

In order to obtain a full list of roundabouts in the Silesia Province, further inventory works are carried out in other cities of the voivodship. A detailed inventory of roundabouts in the Silesian Voivodeship along with their characteristics can be found at work [15]. In total, 304 roundabouts intersections have been distinguished in all cities with powiat rights in the Silesian Voivodship (Fig. 3):

- 29 mini-roundabouts,
- 189 single-lane roundabouts,
- 32 multi-lane roundabouts,
- 6 semi-two-lane roundabouts,

Fig. 3 Distribution of the number of roundabouts in cities with district right in the Silesian Voivodeship

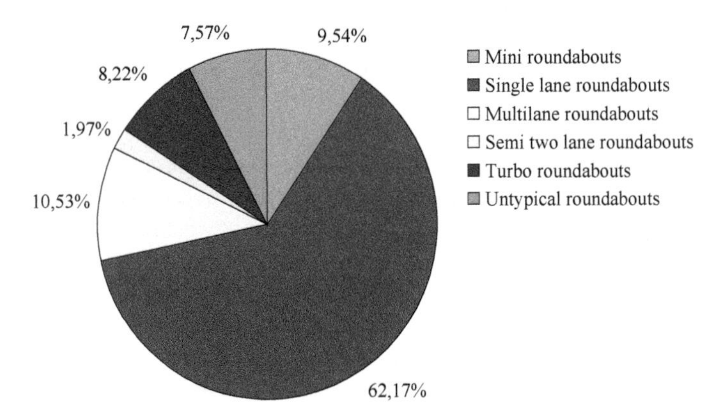

Fig. 4 Percentage of individual types of roundabouts in the total number of roundabouts in cities with district right in the Silesian Voivodeship

- 25 turbo roundabouts,
- 23 non-standard roundabouts.

The largest number of roundabouts was recorded in Rybnik (38). On many of them, the central island is developed using elements of small architecture. The smallest roundabouts are located in Piekary Śląskie and Świętochłowice (4).Figure 4 shows the percentage of particular types of roundabouts in the total number of roundabouts in the Silesian Voivodeship.

On the basis of data presented on Fig. 4, it can be stated that the largest share in the total number of roundabouts in all cities with powiat rights in the Silesian Voivodeship has single-lane roundabouts (62.17%). The second in terms of frequency is multi-lane roundabouts (most often double-lane), whose share is 10.53%. The least common is semi-two-lane roundabouts (1.97%), of which the vast majority are located in Rybnik.

3 Analysis of the Results

In order to learn about the perception of a roundabout crossroads by drivers, cyclists and pedestrians traveling on roundabouts, surveys have been carried out. The research was carried out online using the "ProfiTest" Internet application from April to July 2017. The survey was addressed only to the inhabitants of the Silesian Voivodeship. During the study, 351 respondents began to provide answers, of which 300 of them completed the survey correctly and in their entirety. Due to the selection of respondents method, the conducted surveys were of quasi-representative character, as the surveyed group of respondents only in some respects met the requirements of the representative method. In the initial part of the survey, a short theoretical introduction was added, which aimed to familiarize students with the issues related to crossings such as the roundabout including definitions of geometric elements of roundabouts and roundabouts classification. The next part of the survey was concerned with general questions allowing to characterize the respondents. In the main part of the questionnaire, respondents made assessments regarding the perception of roundabouts. This part of the survey consisted of detailed questions related to specific road solutions, their development and eventual events in which the respondents participated. In the questions in which the respondents were asked to sort the answers in ascending order or to indicate the best solution they considered, ranking was applied. In all questions, the best—according to the respondents—solution received the rank equal to 1.00, while the worst rank equaled 3.00 or 6.00 (depending on the number of possible answers). On this basis, the average rank value was calculated for each answer. The questionnaire contained closed ones and multiple choice questions and ranking questions. 49.33% of the respondents were men and 50.67% women. Due to the limitations of the volume of the article, selected survey results are presented as follows.

Figure 5 presents age structure of respondents. The most numerous group among all respondents were people aged <40–45, and they constituted 22.33% of all respondents. In the second place, there were people from the age range <35–40. Their share in the study was 17.67%. The least represented age-group among all drivers participating in the study was people over 60 years of age. Their participation in the study was 2.00%.

In addition, the largest group of respondents were drivers who already have considerable experience in driving a vehicle. They were drivers who have the right to drive a vehicle in the range of <25–30 years—which accounted for 31.33% of all

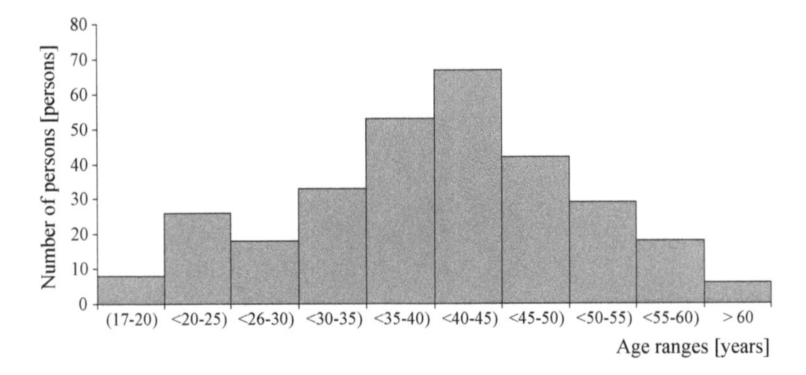

Fig. 5 The age structure of responders

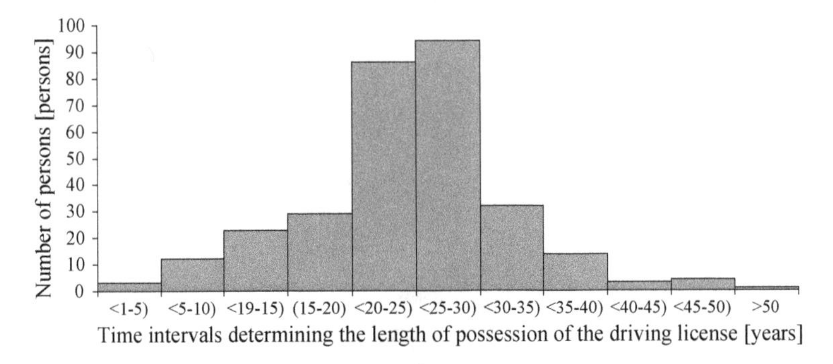

Fig. 6 The structure of the length of possession of the driving license by the responders

respondents and drivers from the group <20–25—which in turn accounted for 28.67% of respondents (Fig. 6). The least represented group were the respondents who have the right to drive a car vehicle for more than 50 years (0.33% of respondents) and beginner drivers who have a driving license for no more than 5 years (1.00%).

All people who completed the survey at least once passed through a roundabout (Fig. 7). Almost half of all drivers (49%) pass through these types of crossroads every day, while 37% of respondents answered that they cross this type of intersection several times a week. Seven percent of all respondents pass through the roundabout once a week. The same value was obtained for the “Once a month” answer. None of the respondents while driving a vehicle do not avoid passing through this type of intersection.

In order to ensure positive reception of new engineering solutions among the residents, especially those that in the initial phase of functioning may cause difficulties in moving or raise doubts as to the legitimacy of their use, such as like a mini-roundabout or a turbine roundabout, there should be use of, for example, advertising campaigns on the TV radio, billboards, leaflets or meetings with local people to familiarize drivers with how to navigate these types of objects. Such information

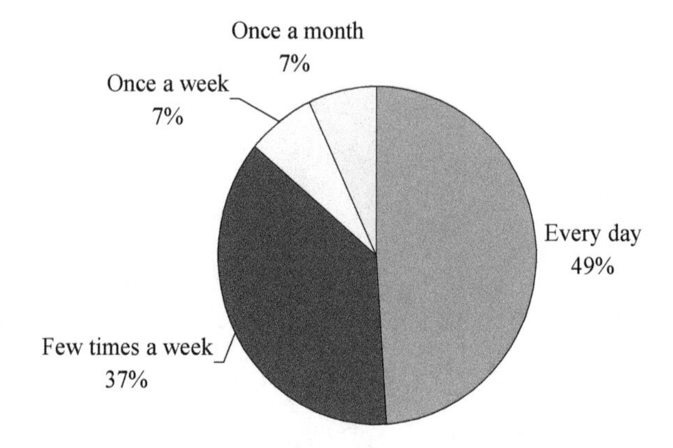

Fig. 7 The structure of the frequency passing through the roundabouts

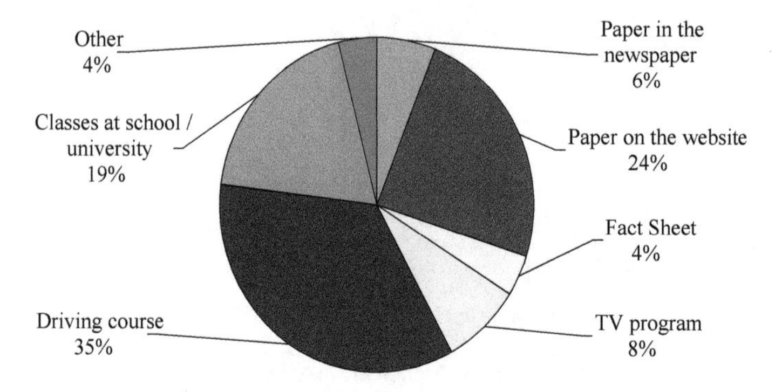

Fig. 8 The structure of the sources of information about roundabouts

will help to eliminate mistakes and mistakes in the behavior of drivers while navigating new elements of transport infrastructure. The results of the conducted surveys allowed to state that 57% of the respondents met with educational information on the roundabouts. As their source, they mainly exchanged driving licenses, articles in the newspaper, articles on websites, information brochures, TV programs, classes at school and/or at the university. However, as many as 43% of respondents stated that they never had to deal with these types of materials. It should be noted, however, that most of them probably did not consider the driving course at all, or they decided that the knowledge given to them at that time was not sufficient. The structure of sources of information about the roundabouts is shown in Fig. 8.

The vast majority of drivers participating in the study found that riding on roundabouts did not cause them any difficulties (97%). Only 3% of respondents said that driving roundabouts such as: spiral roundabouts (rank 4.67), semi-two-lane roundabouts (rank 4.00), turbo roundabouts (rank 3.00) and two-lane roundabouts (rank 2.00), makes them feel difficult. All respondents have agreed that driving on single-edged roundabouts does not cause them any difficulties (rank 1.00).

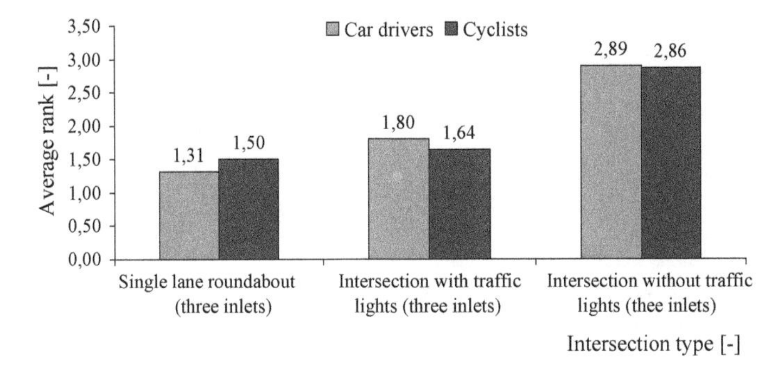

Fig. 9 Perception of drivers and cyclists of various types of three-inlet intersections in the aspect of road traffic safety

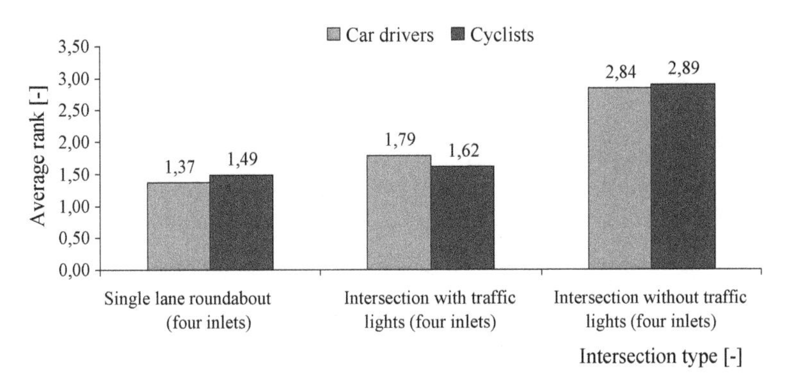

Fig. 10 Perception of drivers and cyclists of various types of four-inlet intersections in the aspect of road safety

Respondents also answered questions that placed them in the role of drivers, cyclists and pedestrians. This exercise was aimed at examining the impact of modifying the mode of transport on the perception of individual road solutions. Among drivers and cyclists, both in the case of three- and four-inlet junctions, the single-lane roundabout is considered the safest solution. Next, there was successively an intersection with traffic lights and an intersection with no traffic lights (Figs. 9 and 10).

The perception of various types of roundabouts by vehicle drivers and cyclists as safe road solutions is shown in Fig. 11. On the basis of Fig. 11, it can be stated that the safest type of roundabout is a single-lane roundabout. In the second place in the respondents opinion, there are turbo roundabouts with lane separators raised above the road surface, and on the third—a turbo roundabout with lane separators in the form of a continuous P-2 line. The fourth place was taken by semi-two-lane roundabouts. The penultimate place was occupied by two-lane roundabouts commonly found on Polish roads. The worst respondents assessed the spiral roundabouts. In turn, the biggest difference in the perception of roundabouts by vehicle and bicycle drivers

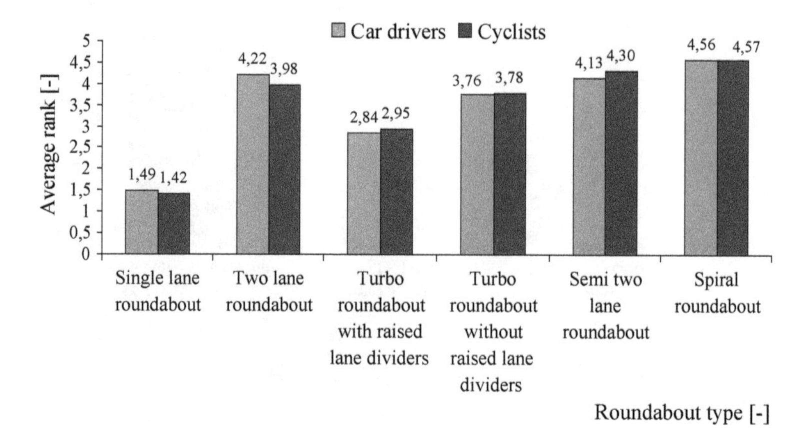

Fig. 11 The perception of various types of roundabouts by drivers of vehicles and cyclists as safe road solutions

was recorded in the case of two-lane roundabouts (0.24). This means that cyclists, in contrast to vehicle drivers, better rated two-lane roundabouts in the aspect of road safety. Differences in vehicle and cyclist driver ratings are low for other types of roundabouts.

Respondents—pedestrians in assessing the perception of roundabouts in the aspect of road safety, pointed to the need to use asylum islands on the inlets, whose presence is a significant facilitation in crossing the road and thus increases the level of road safety.

The survey also allowed to the identification of those respondents who participated in the traffic incident at the roundabout. Those respondents were asked to indicate the types of participants in the event in which they participated (i.e., vehicle—vehicle, vehicle—pedestrian, vehicle—cyclist, vehicle—infrastructure element), type of incident and its causes. Respondents participated in such types of road incidents on roundabouts as:

- Side collision,
- Back collision,
- Hovering over an element of infrastructure.

The reasons for these events were:

- Not giving the right of way,
- Non-adjustment of speed to prevailing traffic conditions,
- Non-observance of signs and other signals,
- Failure to maintain a safe distance between vehicles,
- Bad state of the road.

The types of incidents indicated by the respondents and their causes are characteristic of the roundabouts.

4 Selected Results of Statistical Inference

On the basis of the descriptive statistics presented above, some statistical inference such as cross-analysis was performed. The analyses were performed using IBM SPSS Statistics Software. Cross-analysis is used to present data, where the main measure is the number of responses and the percentage of the group. In subsequent cross-analysis, it was checked whether in the surveyed group of respondents, women differed from men in terms of:

- The structure of the frequency passing through the roundabouts,
- The age structure of responders,
- The structure of the length of possession of the driving license by the responders,
- The structure of the sources of information about roundabouts,
- Perception by car drivers of various types of three-inlet intersections in the aspect of road traffic safety,
- Perception by cyclists of various types of three-inlet intersections in the aspect of road traffic safety,
- Perception by car drivers of various types of four-inlet intersections in the aspect of road traffic safety,
- Perception by cyclists of various types of four-inlet intersections in the aspect of road traffic safety,
- The perception of various types of roundabouts by drivers of vehicles as safe road solutions,
- The perception of various types of roundabouts by cyclists as safe road solutions.

Table 2 The structure of the sources of information about roundabouts

Sources of information about roundabouts	Gender					
	Woman		Men		Total	
	Number	Percentage (%)	Number	Percentage (%)	Number	Percentage (%)
Driving Course	54	35.53	51	34.46	105	35
Paper on the website	25	16.45	47	31.76	72	24
Classes at school/university	39	25.66	18	12.16	57	19
TV program	15	9.87	9	6.08	24	8
Paper in the newspaper	6	3.95	12	8.11	18	6
Fact Sheet	5	3.29	7	4.73	12	4
Other	8	5.26	4	2.70	12	4
Sum	152	100.00	148	100.00	300	100

Table 3 Perception of car drivers of various types of three-inlet intersections in the aspect of road traffic safety

Perception by car drivers of various types of three-inlet intersections as safe road solutions	Gender					
	Woman		Men		Total	
	Number	Percentage (%)	Number	Percentage (%)	Number	Percentage (%)
Single-lane roundabouts (three inlets)	84	55.26	103	69.59	187	62.33
Intersections with traffic lights (three inlets)	43	28.29	26	17.57	69	23.00
Intersections without traffic lights (three inlets)	25	16.45	19	12.84	44	14.67
Sum	152	100.00	148	100.00	300	100.00

In each case, the analyzed variables are nominal variables. Due to the fact that nominal statistics cannot be used to calculate other statistics such as the mean and standard deviation, the best way to present the results is to present the number (number of people) for given answers. This form of data presentation provides full insight into obtained results. The tables show the number of response categories for one and the other variables and how the results of people for these categories were distributed. Due to the limitations related to the paper volume in the following cross-analyses (Tables 2, 3, 4 and 5), selected results of the conducted research were presented.

Based on the data presented in Table 2, it can be concluded that the basic source of knowledge about proper behaviors at the roundabouts is driving courses (both for men and for women) and information in the website (more often for men than for woman). During the analysis of partial results, it can be stated among others that women's basic knowledge about roundabouts was much more likely than men to gain from classes at school/universities.

Based on the data presented in Table 3, it can be concluded that from among various road infrastructure solutions, the safest road solutions are perceived to have three-inlet single-lane roundabouts and successively three-inlet intersections with traffic lights. Analyzing the partial results, among others it can be stated that women more often than man perceive three-inlet intersections with traffic lights as safe road solutions.

A similar situation as in the case of perception of three-inlet intersections takes place in the case of perception of four-inlet intersections (Table 4). Among vari-

Table 4 Perception of car drivers of various types of four-inlet intersections in the aspect of road traffic safety

Perception by car drivers of various types of four-inlet intersections as safe road solutions	Gender					
	Woman		Men		Total	
	Number	Percentage (%)	Number	Percentage (%)	Number	Percentage (%)
Single-lane roundabouts (four inlets)	73	48.03	83	56.08	156	52.00
Intersections with traffic lights (four inlets)	53	34.87	35	23.65	88	29.33
Intersections without traffic lights (four inlets)	26	17.11	30	20.27	56	18.67
Sum	152	100.00	148	100.00	300	100.00

ous solutions, the safest road solutions are perceived to have four-inlet single-lane roundabouts and next four-inlet intersections with traffic lights.

In turn, among all types of roundabouts, also single-lane roundabouts are considered as one of the safest solutions (both by men and by woman) followed by the turbo roundabouts with raised lane dividers. Detailed analysis shows that women more often than men perceive turbo roundabouts with raised lane dividers as safe road solutions. The results of cross-analysis are presented in Table 5.

5 Conclusions

In Poland, roundabouts in the last twenty years have become a popular solution for crossing roads. This is mainly due to their many advantages. Statistics show that transforming the intersection into a roundabout improves traffic conditions at a given intersection, reduces the time losses incurred by vehicle drivers and increases the level of road safety. In order to ensure the highest level of traffic safety at the intersection, it is also important to know how users perceive it. The research results presented in the article do not provide complete information on the way roundabouts are perceived by their users, but this knowledge can be used as an auxiliary element in the geometric design of roundabouts. Based on the analysis presented in the article, the following conclusions were drawn:

Table 5 The perception of various types of roundabouts by drivers of vehicles as safe road solutions

Perception of various types of roundabouts by drivers of vehicles as safe road solutions	Gender					
	Woman		Men		Total	
	Number	Percentage (%)	Number	Percentage (%)	Number	Percentage (%)
Single-lane roundabouts	57	37.50	90	60.81	147	49.00
Two-lane roundabouts	6	3.95	15	10.14	21	7.00
Turbo roundabout with raised lane dividers	48	31.58	19	12.84	67	22.33
Turbo roundabout without raised lane dividers	21	13.82	9	6.08	30	10.00
Semi-two-lane roundabouts	17	11.18	4	2.70	21	7.00
Spiral roundabouts	3	1.97	11	7.43	14	4.67
Sum	152	100.00	148	100.00	300	100

- The surveyed population of roundabout users generates a general positive attitude to the reconstruction of other types of intersections at the roundabouts. Ninety-seven percent of people participating in the study found that riding around the roundabout did not cause them any difficulties,
- Respondents (as drivers of vehicles, cyclists and pedestrians) perceive them as safe road solutions. The vast majority of respondents, as many as 95%, think that this type of intersection has a positive effect on the level of road safety,
- The cyclists and pedestrians responses were in line with the drivers assessments,
- In the aspect of road safety, roundabout users are best perceived among roundabouts by single-lane roundabouts, and the worst roundabout is the spiral roundabout,
- Public information and education in the use of various types of roundabouts remain an important problem in improving road safety. The results of the research allowed to state that 57% of respondents received educational information about roundabouts, and the most common source of this knowledge was indicated by driving courses and articles in newspapers (59%),

- The results of cross-analysis confirmed the hypothesis put at the beginning of this paper that roundabout users in Poland—especially single-lane roundabouts users—generally perceived roundabouts positively.

References

1. Statistical Poland Information Portal. http://stat.gov.pl/obszary-tematyczne/transport-i-lacznosc/transport/transport-wyniki-dzialalnosci-w-2016-r-,9,16.html
2. National Road Safety Council: State of Road Traffic Safety and Works in this Area in 2016. http://statystyka.policja.pl/st/ruchdrogowy/76562,Wypadki-drogoweraporty-roczne.html
3. Police Headquarters: Road Accidents in Poland in 2016. http://statystyka.policja.pl/st/ruch-drogowy/76562,Wypadki-drogowe-raporty-roczne.html
4. Bronkiewicz, M., Nalewajko, Ł.: Ocena Funkcjonowania Ronda Turbinowego na Przykładzie Ronda na Skrzyżowaniu Drogi Krajowej nr 46 z Łącznicą Węzła Autostradowego „Prądy". Zeszyty Naukowo-Techniczne SITK RP oddział w Krakowie **92**(151), 77–92 (2010)
5. Majer, S.: Analiza Zdarzeń Drogowych na Wybranych Rondach w Gorzowie Wielkopolskim. Zeszyty Naukowo-Techniczne SITK RP oddział w Krakowie **92**(151), 161–173 (2010)
6. Macioszek, E., Czerniakowski, M.: Road traffic safety-related changes introduced on T. Kościuszki and Królowej Jadwigi streets in Dąbrowa Górnicza between 2006 and 2015. Sci. J. Silesian Univ. Technol. Ser. Transp. **96**, 95–104 (2017)
7. Macioszek, E.: The Application of HCM 2010 in the determination of capacity of traffic lanes at turbo roundabout entries. Transp. Prob. **11**(3), 77–89 (2016)
8. Macioszek, E.: The comparison of models for follow-up headway at roundabouts. In: Macioszek, E., Sierpiński, G. (eds.) Recent Advances in Traffic Engineering for Transport Networks and Systems. LNNS, vol. 21, pp. 16–26. Springer, Switzerland (2018)
9. Macioszek, E., Lach, D.: Analysis of Traffic Conditions at the Brzezinska and Nowochrzanowska Intersection in Myslowice (Silesian Province, Poland). Sci. J. Silesian Univ. Technol. Ser. Transp. **98**, 81–88 (2018)
10. Pratelli, A., Souleyrette, R.R.: Visibility, perception and roundabout safety. WIT Trans. Built Environ. **107**, 557–588 (2009)
11. Shadpour, E.: Safety Effect of Roundabouts. Laurier Centre for Economic Research & Policy Analysis. Commentary No. 2014-2 (2014)
12. Harper, N.J., Dunn, R.C.M.: Accident Prediction at Urban Roundabouts in New Zealand—-Some Initial Results. 26th Australian Transport Research Forum, Wellington New Zealand, 1–3 Oct 2003. http://atrf.info/papers/2003/2003_Harper_Dunn.pdf (2003)
13. U.S. Department of Transportation: Federal Highway Administration: Intersection Safety Roundabouts—Safety. Technical Report FHWA-SA-10-006 (2010)
14. Statistical Data Received from the Provincial Police Headquarters in Katowice. Katowice (2017)
15. Podstawski, J.: The analysis of the roundabouts perception by drivers, pedestrians and cyclists with an assessment of their functioning in Silesian Voivodeship. Graduate thesis, Katowice (2017)

Evaluation of the Distribution of Sound Levels in a Public Transport Bus During the Ride and at Standstill

Malgorzata Orczyk

Abstract The article will present the results of noise measurements inside MAN city buses. Ten measuring microphones were placed in the passenger compartment of the tested vehicles, which, through the PULS measuring system, enabled simultaneous recording of acoustic signals during driving and at a standstill. The aim of the analyses was spatial and scoring analysis of the distribution of sound levels in the passenger compartment of the bus at a standstill and during the ride. In addition, for the acceleration phase, the locations with the highest and lowest sound levels have been identified, and graphic zones for distribution of sound levels were determined.

Keywords Travel comfort · Noise in the bus · Means of urban transport

1 Introduction

Communication noise has now become the most common and dominant source of noise. It is especially noticeable mainly in large- and medium-sized cities, where in recent years, a dynamically developing individual car communication could be observed. In Poland, in 2000, there were 250 passenger cars per 1000 inhabitants and in 2016 this number exceeded 550 passenger cars per 1000 inhabitants [1]. In urban agglomerations, where people move a lot, it is estimated that pollution of automotive origin (air pollution and noise) constitutes about 75% of all pollutants. The distribution of these pollutants varies in time and space and to a large extent depends on the intensity and nature of road traffic and the technical efficiency of vehicles [2]. Due to the air pollution increasing year by year and high levels of sound recorded, the authorities of many cities strive to minimize the presence of cars (passenger cars and delivery vans) in cities. Therefore, the actions undertaken are aimed at reducing the demand for journeys made by individual transport (passenger cars) and increasing the share of journeys made with public transport. This

M. Orczyk (✉)
Faculty of Machines and Transport, Poznan University of Technology, Poznan, Poland
e-mail: Malgorzata.Orczyk@put.poznan.pl

M. Suchanek (ed.), *Challenges of Urban Mobility, Transport Companies and Systems*, Springer Proceedings in Business and Economics,
https://doi.org/10.1007/978-3-030-17743-0_7

issue is related to the concept of sustainable urban mobility, which is understood as combining, complementing and improving the movement: on foot, by bike, tram, bus and trolley bus, as well as suburban train, in the city and its immediate surroundings (functional area). Mobility understood in this way gives an opportunity to reduce individual car traffic, contributes to the reduction in pollutants harmful to the environment and the consumption of fossil fuels. Public transport is an important element of the efficient functioning of each city, having a very large impact on its socioeconomic attractiveness. Its development would be impossible without undertaking research [3–7] related to various aspects of the functioning of public transport in cities. The most visible changes in this area are related to the replacement of existing fleet with low-floor fleet in both tram and bus transport. In Poland, over 85% of city buses are low-floor vehicles, and in many cities, there are no high-floor city buses at all. Regarding the rolling stock, the number of low-floor vehicles has also increased from 7% in 2009 to 28% in 2015 [8]. In addition, more and more attention of public transport operators in cities is directed to low-emission buses powered by: CNG, LPG, hybrid and electric engines. In the field of tram rolling stock, many new and so far not used solutions are being implemented, such as pneumatic suspension of running gear, liquid-cooled traction motors or independently rotating wheels, which improve passenger comfort. An important role is also played by issues related to the impact of noise and vibrations on passengers while driving. Managing the right comfort inside the vehicle plays a significant role in the attractiveness of a particular means of transport for passengers. This is indicated, among others, by the survey that was carried out in 2013 among passengers of local public transport operator (Miejskie Przedsiębiorstwo Komunikacyjne) in Poznan, Poland. Respondents—1054 people—were asked to indicate factors that they believe have a major impact on the comfort of the vehicle. Among the factors mentioned in the survey form, the respondents had to indicate three which in their opinion are: very important, important and least important. Answers to this question are presented in Fig. 1.

The respondents indicated that very important and important factors that affect the comfort of travel by means of public mass transport are: congestion of the vehicle (38% answered "very important" and 23% "important"), temperature in the vehicle (20% answered "very important" and the other 20% answered "important") and noise generated during the ride (16% answered "very important" and 10% answered "important"). For the other factors mentioned in the questionnaire, the distribution of responses did not exceed 10%.

Based on the conducted surveys, a research problem was formulated, which concerns the assessment of the noise level in the interior of buses and its distribution along the passage for passengers. The solution to this research problem requires the development of a research methodology, the implementation of tests and the analysis of the sound distribution, its level and changes during driving. The aim of the conducted research was the evaluation and spatial analysis of the distribution of the sound level in the interior of buses at standstill and during driving. The aim of the paper formulated in this way allows the following research hypotheses to be adopted:

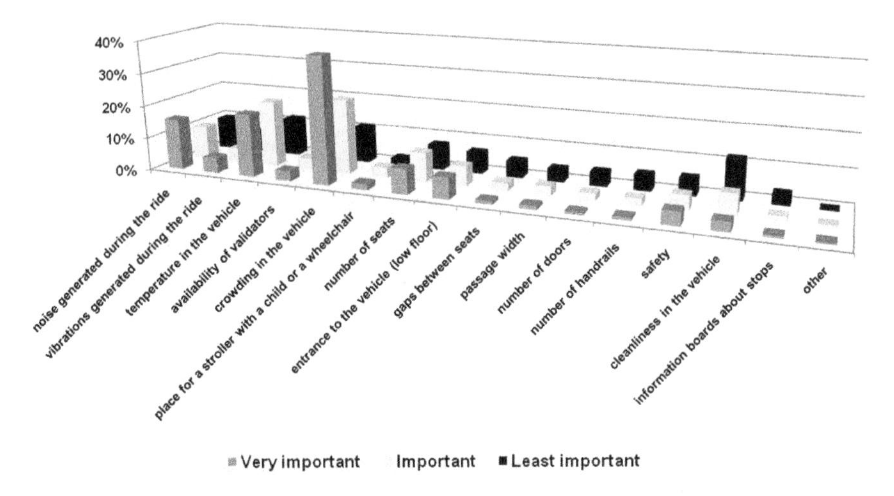

Fig. 1 Distribution of answers to the question "Which factors do you think have a major impact on the comfort of the vehicle?"

- the distribution of the sound level inside the bus is not even on its length,
- the noise level inside the bus depends on the driving speed and the individual driving phases: acceleration, constant drive and braking,
- the sound level distribution on the bus changes and depends on random events and road unevenness, which are the source of additional noise.

In order to achieve the defined objective of the paper and to verify the adopted research hypotheses, the internal noise in buses was measured. Ten measuring microphones were placed inside. Through the PULS measuring system, parallel recording of acoustic signals was carried out during stopping and passing of a selected measuring section.

The aim of the measurements was a spatial and point analysis of the distribution of sound levels at a standstill and while driving and the identification of places with the highest and lowest sound levels in the vehicle. During the ride, the analysis was carried out for three phases of the bus movement: acceleration phase, driving with constant speed and braking phase. In addition, graphical time sequences of the sound level distribution in the passenger area are presented for accelerating the bus and entering the road hole.

2 The Methodology of Noise Measurements Inside of the Bus

The objects of research were three MAN buses operated since 2009 by one of the municipal transport companies. The choice of these vehicles was related to the fact

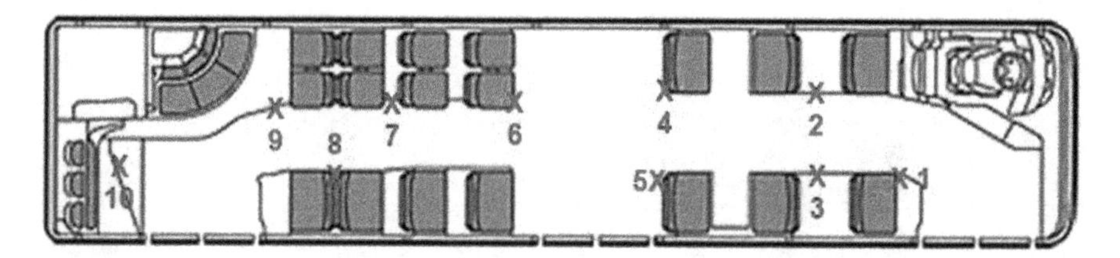

Fig. 2 Distribution of measuring points in the MAN bus 1, 2, 3, 4, 5, 6, 7, 8, 9, 10—locations of measuring microphones (Own elaboration based on [9])

that all of them came from the last delivery (2009) of MAN buses for the municipal transport company (it was the year when this company bought only three buses of this brand) and had a similar mileage of around 400,000 km. MAN buses are low-floor city buses manufactured by MAN SE since 2005. The vehicles supplied for the study were equipped with a vertical engine with a capacity of 10,500 m^3, meeting the EURO IV standard, and 4-speed automatic gearbox from VOITH. This 12-m bus is designed to carry 83 people: 28 sitting and 55 standing. Basic vehicle data are: length: 11,980 mm; width: 2550 mm; height: 2850 mm; total weight: 18,000 kg; door arrangement: 2-2-2 [9]. For the needs of this work, the company developed its own methodology for measuring noise inside the bus, based on the PN-90/S-04052 standard *Measurement of noise inside motor car. Limit of sound level* [10]. In the passenger compartment of the bus, ten measuring microphones were placed, located 1.6 m above floor level. Microphone system was preserved both for measurements at a standstill (when the bus engine was operating at idle) and during the ride. Figure 2 shows the distribution of measuring points in the MAN bus.

Measurement rides were done in empty vehicles under normal urban traffic conditions. During the tests, all windows in the buses were closed and the air conditioning in the passenger compartment was turned off. The buses supplied for the study were in various operational conditions; the mileages of the tested vehicles were similar and did not exceed 400,000 km. The measuring route was about 2 km, and it was driven through in such a way that it was possible to distinguish three driving phases of the vehicle: acceleration, driving with constant speed and braking. Measurements of noise at a standstill included situations when engine was operating at idle. The study was carried out in the spring and summer period, which provided favorable and comparable weather for all rides, mainly due to the lack of precipitation. The measurements were taken on a dry roadway, and the wind speed did not exceed 5 m/s. The sound levels recorded while driving were compared with the permissible level contained in PN-90/S-04052 standard. Measurements in the tested vehicles were made with a set of ten free-field B&K microphones 1/2″, type 4189. The recording of acoustic signals was carried out using the Brüel&Kjær PULSE® system. The measuring route featured a manual GPS Garmin eTrex 20 speed measurement device that allows precise measurement of the speed of moving vehicles.

3 Analysis of Measurement Results

3.1 Noise Assessment in the Bus at a Standstill

Measurements at a standstill in all buses delivered for study—three were carried out at the bus depot. The measurement was taken while the engine was idling. The duration of a single measurement session for each vehicle was 30 s. Before performing the basic tests, the acoustic background was measured in the bus area (with engine turned off), and it did not exceed 40 dB. Figure 3 presents distributions of sound levels in individual microphone locations in the tested vehicles at a standstill.

Basing on the measurements carried out in the passenger compartments of the studied buses, the interior of the entire vehicle can be divided into three zones. The first zone of the bus includes places where the lowest sound levels have been registered. There are measuring microphones Nos. 1, 2, 3 and 4. The second zone is the middle of the bus, with measuring microphones Nos. 5, 6 and 7. The third zone is the place where the highest sound levels were recorded. This zone is located next to the combustion engine. There are measuring microphones Nos. 8, 9 and 10. In MAN buses at a standstill, the equivalent sound levels measured were on average around

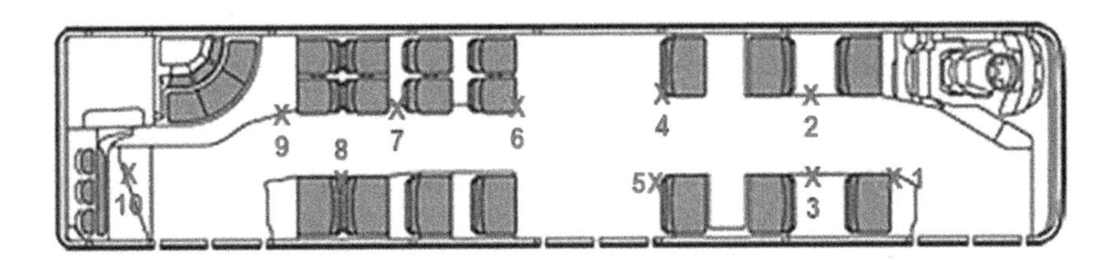

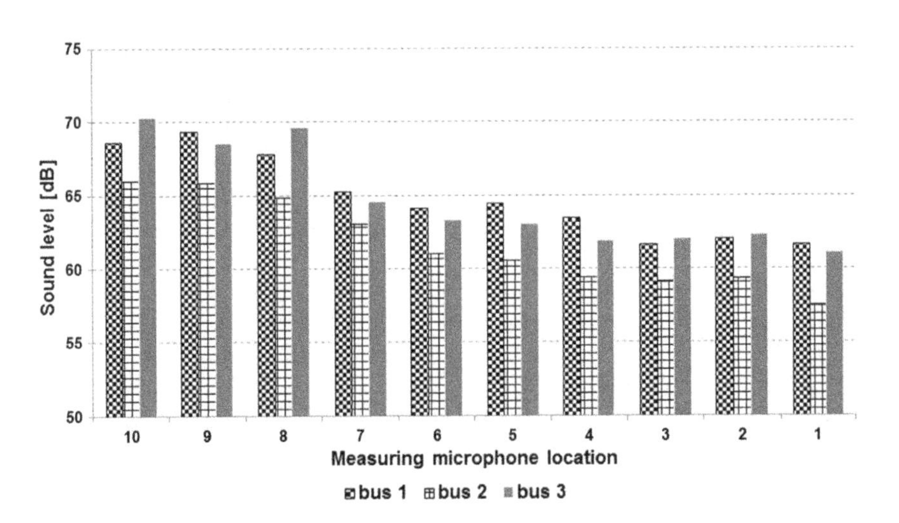

Fig. 3 Distribution of sound levels inside the MAN bus at a standstill

65 dB in the entire vehicle. The lowest sound levels at the stop were recorded in bus number 2; 62 dB was registered in this vehicle. In the other studied vehicles, 65 dB was registered. The difference in sound levels between the back and the front of the vehicle does not exceed 10 dB. At the front of the vehicle, at the location of the first four measuring microphones, the recorded sound levels ranged from 58 to 62 dB. At the back of the vehicle (near to the combustion engine), sound levels ranged from 66 to 70 dB. The quietest vehicle turned out to be the bus number 2. The recorded sound levels were 58 dB in the front of the vehicle and 66 dB in the back. The sound levels in buses 1 and 3 were at a similar level: 61 dB at the location of the first measuring microphone and 70 dB at the location of the tenth measuring microphone.

3.2 Noise Assessment in the Bus During the Ride

Studies during the ride were conducted on a selected measuring section of about 2 km. Measurements took place in normal traffic, so each of the tested buses was driven through the measurement section at a different speed. The driver assigned for the study drove the vehicle in such a way to distinguish three driving phases: acceleration, driving with constant speed and braking. The range of speeds for each phase is as follows:

- acceleration phase: 0–42 km/h—the acceleration phase included starting the movement of the vehicle (0 km/h) and reaching a constant speed;
- driving with constant speed: 43–50 km/h;
- braking phase: 42–0 km/h—this phase included braking from constant speed to 0 km/h.

This speed range was chosen due to the fact that the maximum speed with which the public transport bus can move in the city is about 50 km/h. Firstly, the distribution of sound levels in the studied buses during the whole ride was made. The obtained study results were compared with the permissible level indicated in the PN-90/S-04052 standard *Measurement of noise inside motor car. Limit of sound level.* The standard states that in city buses, the permissible level should not exceed 85 dB during the ride [10]. Figure 4 presents the results of the measurements performed for individual microphone locations and compares them to the permissible level of 85 dB.

The measurements indicate that during the ride, the distribution of sound levels inside the examined buses are not the same. The highest levels were registered in bus No. 2; they were in the range of 78–82 dB. The results obtained in this bus are higher than the results obtained in the other two buses by an average of about 8 dB. During the ride, the highest sound levels of 82 dB were recorded in the location of the measuring microphone No. 8. The same trend was observed in all tested buses. The highest sound levels in buses 1, 2 and 3 were in the location of the measuring microphone No. 8. In addition, in the bus Nos. 1 and 3, two zones of sound impact can be indicated during the ride. The first zone includes measurement microphones

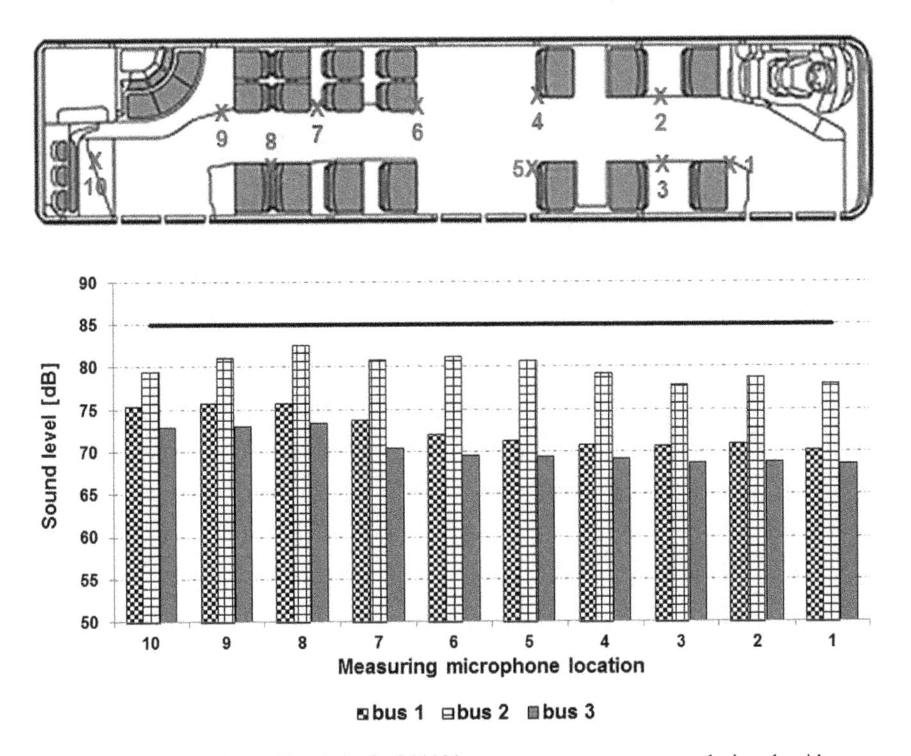

Fig. 4 Distribution of sound levels in the MAN bus passenger compartment during the ride

Nos. 1, 2, 3, 4, 5 and 6. The lowest sound levels were recorded in this zone. 70–72 dB was recorded in the bus No. 1, and 68–69 dB was recorded in the bus No. 3. The second zone of noise impact includes locations of microphones Nos. 8, 9 and 10. These are the places where the highest sound levels were recorded: 75–76 dB in bus number 1 and 73 dB in bus number 3. The location of the measuring microphone No. 7 was a transition zone. In this place, the recorded sound levels increased. Bus No. 3 turned out to be the bus with the lowest sound levels during the ride. In the bus No. 2, the sound impact zones cannot be indicated. The sound levels recorded in this bus during the ride at individual microphone locations were accidental. Comparing the recorded sound levels in individual microphone locations with an acceptable level of 85 dB, it can be concluded that in none of the examined buses, this level was exceeded. In addition, the difference in recorded sound levels between the standstill (when the engine is running at idle) and during the ride is around 11 dB in MAN buses. The second stage of work on the acoustic evaluation of the interiors of MAN buses was the evaluation of sound levels during the three driving phases: acceleration phase, driving at constant speed and braking phase. Figure 5 shows the distribution of sound levels inside the bus during the three driving phases.

In the three driving phases, two zones of sound impact can be also indicated. The first zone of sound impact includes microphones Nos. 1, 2, 3, 4, 5 and 6. These

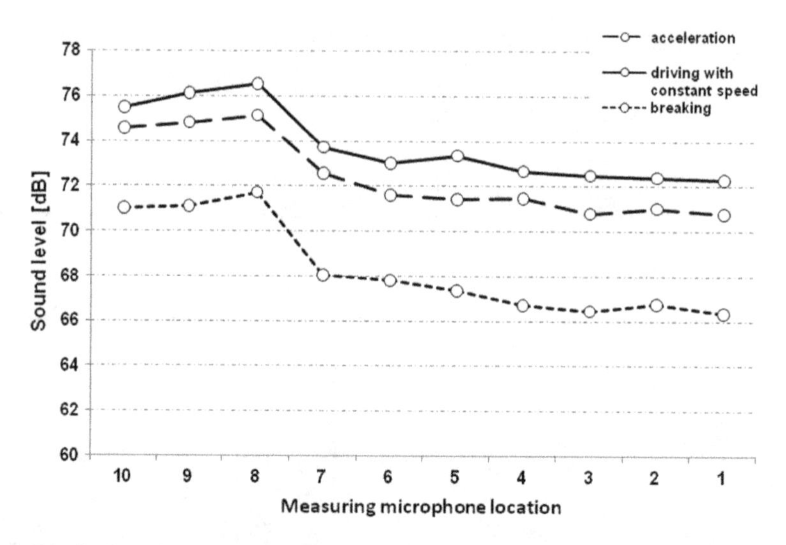

Fig. 5 Distribution of sound levels inside the MAN bus during the three driving phases

are zones with the lowest sound levels. Starting from the seventh location of the measuring microphone, the recorded sound levels increase. The second zone of sound impact includes locations of measuring microphones Nos. 8, 9 and 10. The highest sound levels were recorded in this zone. The driving phase with the highest recorded sound levels is the phase of driving with constant speed. In the constant speed phase, the buses were driven at about 45 km/h. At the same time, for this driving phase, a correlation was determined between the measured sound levels and the speed at which the examined buses moved. The correlation was at a level of around 0.2, which means a weak or low correlation. The second driving phase, which also featured high sound levels, was the bus acceleration phase. This phase was by 1–2 dB lower than the constant speed phase. The correlation between the individual microphones and the speed at which buses moved was high. Calculated correlation coefficient was 0.6. The last specified driving phase was the braking phase. In this phase, the lowest sound levels occurred. Differences in the recorded sound levels between this phase and the acceleration and constant speed phases were about 5 dB. The correlation coefficient for this driving phase varied a lot: for bus No. 1, it was 0.32 (low correlation), for bus No. 2—0.54 (moderate correlation) and for bus No. 3—0.83 (very high correlation). In the driving phases of the studied buses, the recorded sound levels were: 68–80 dB for the acceleration phase, 72–83 dB for the constant speed phase and 66–76 dB for the braking phase. Based on the recorded sound levels, maps of the sound level distribution in the interior of the bus during the acceleration phase and entry into the road hole were also made. The graphic sequences were made with a resolution of 0.2 s. Due to the large number of images, only characteristic time sequences are shown below. Sound distribution maps do not include the driver's cab and the interior

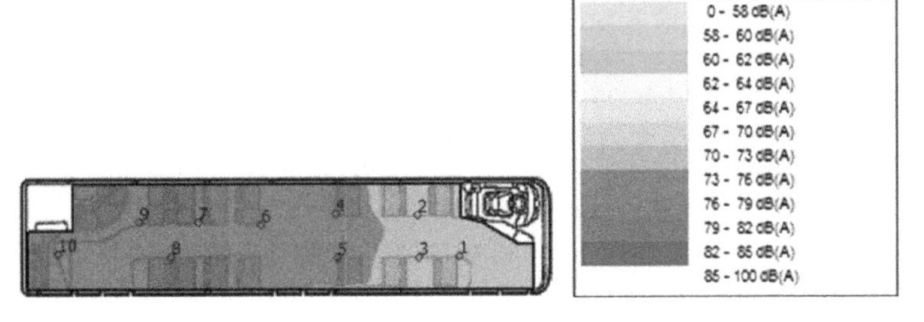

Fig. 6 Distribution of sound levels inside the MAN bus at a standstill 1, 2, 3, 4, 5, 6, 7, 8, 9, 10—locations of measuring microphones

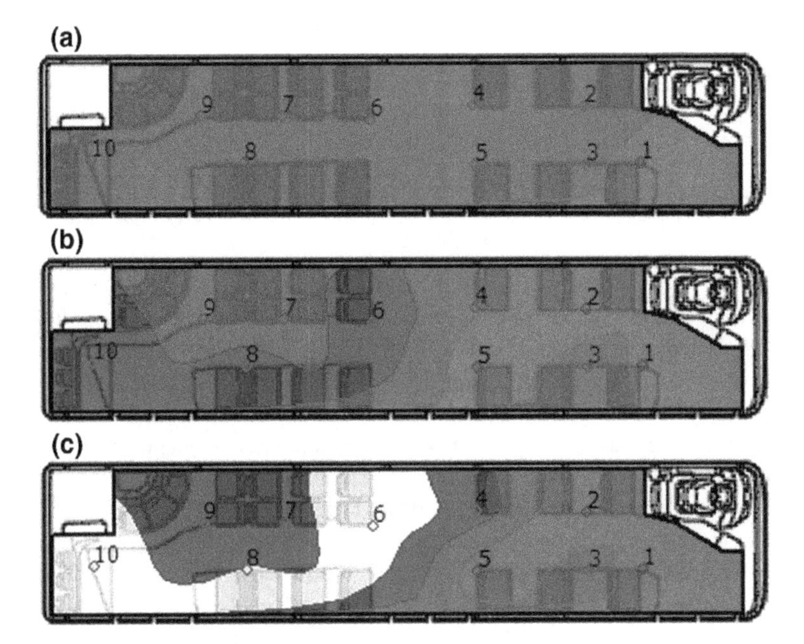

Fig. 7 Distribution of sound levels inside the MAN bus during acceleration 1, 2, 3, 4, 5, 6, 7, 8, 9, 10—locations of measuring microphones

of the engine compartment—the back of the bus. Figure 6 presents the distribution of sound levels in the bus at a standstill.

The maps allow to indicate three zones of sound impact in the bus. The first zone, marked orange, includes locations of measuring microphones Nos. 1, 2 and 3. In this zone, recorded sound levels were in the range of 70–73 dB. In the second zone of sound impact (bright red color), there are microphones Nos. 4, 5 and 6. There, the recorded sound levels were in the range of 73–76 dB. The last, third zone in which the highest sound levels were recorded was the location of the measuring microphones Nos. 7, 8, 9 and 10 (dark red color). 76–79 dB was recorded in these places. Figure 7 shows selected time sequences for the acceleration phase of the bus and entry into the road hole.

Figure 7a shows the beginning of the acceleration phase. In this phase, in the whole passenger space in the vehicle, apart from the location of the first measuring microphone, microphones Nos. 2, 3, 4, 5, 6, 7, 8, 9 and 10 recorded the same sound levels: 79–82 dB. The first measuring microphone recorded 73–76 dB. Figure 7b, c shows uneven sound distribution in the bus space. There are places in the bus where the recorded levels are lower, and places where they are much higher. The measuring microphones Nos. 1, 2 and 5 recorded 73–76 dB. The measuring microphone No. 3 also recorded constant sound levels: 79–82 dB. Microphones Nos. 6, 8 and 10 recorded the highest levels: 82–85 dB (rys. 6 b) and 85–100 dB (6 c). Microphones Nos. 7, 8 and 9 recorded 79–82 dB (rys 6 b) and 82–85 dB (rys 6 c). In addition, it is worth noting that the first measuring microphone recorded the lowest sound levels of approximately 70–76 dB.

4 Conclusion

The measurements of noise at standstill and while driving in selected MAN buses allow to draw the following conclusions: At standstill, three zones of sound impact can be identified in the studied buses. The first zone, where the lowest sound levels were recorded, includes measurement microphones Nos. 1, 2, 3 and 4. The second zone included measuring microphones Nos. 5, 6 and 7, and the third, loudest zone, located by the combustion engine, included microphones 8, 9 and 10. The sound levels recorded in the studied buses do not exceed 65 dB. The difference between the first and the tenth measuring microphone does not exceed 10 dB. During driving, the sound levels recorded in the buses selected for tests increased to the level of 71–83 dB and were more diverse. Two zones of sound impact could be distinguished in the studied vehicles. The first zone, with the lowest sound levels, includes measurement microphones Nos. 1, 2, 3, 4, 5 and 6. The second zone included 8, 9 and 10 measurement microphones and those were the places with the highest levels of sound. The prepared maps of the sound level distribution during the acceleration of the bus and the entry into the road hole showed that in the passenger area, the distribution of sound levels may increase to about 100 dB.

Acknowledgements The presented research and the paper is partly funded by Statutory Activities fund of the Institute of Combustion Engines and Transport, PUT (PL)5/52/DSPB/0259.

References

1. Central Statistical Office: Transport Activity Results in 2016. Zakład Wydawnictw Statystycznych, Warsaw (2017)
2. Drąg, Ł.: Modeling of car exhaust pollutant emission and dispersion. J. Arch. Automot. Eng. **1**, 21–41 (2007). (in Polish)
3. Firlik, B., Tabaszewski, M.: Assessment of the track condition using the Gray relational analysis method. J. Maintenance Reliab. **20**(1), 147–152 (2018)
4. Stroia, M.D.: Solutions to mitigate vibrations and noise produced by tramways (state of art). J. Analele Universitătii "EFTIMIE MURGU" REŞITA. **XVI**(1), 226–231 (2009)
5. Merkisz, J., Fuć, P., Lijewski, P., Pielecha, J.: Actual emissions from urban buses powered with diesel and gas engines. J. Transp. Res. Procedia **14**, 3070–3078 (2016)
6. Lopez-Martinez, J.M., Jimenez, F., Paez-Ayuso, F.J., Flores-Holgado, M.N., Arenas, A.N., Arenas-Ramirez, B., Aparicio-Izquierdo, F.: Modelling the fuel consumption and pollutant emissions of the urban bus fleet of the city of Madrid. J. Transp. Res. Part D-Transp. Environ. **52**(A), 112–127 (2017)
7. Sawicki, P., Kiciński, M., Fierek, S.: Selection of the most adequate trip-modelling tool for integrated transport planning system. J. Arch. Transp. 37(1), 55–66 (2016)
8. Wolański, M., Karolak, A., Pieróg, M., Mazur, B., Mikiel, P.: Report on the State of Public Transport in Poland in 2009–2015. Economic Chamber of Urban Transport. Izba Gospodarcza Komunikacji MIejskiej, Warsaw (2016). (in Polish)
9. Catalog MAN: The online statement. www.bus.man.eu (2018)
10. PN-90/S-04052: Measurement of noise inside motor car. Limit of sound level (1990)

Car-Sharing in Urban Transport Systems—Overview of Europe and Asia

Katarzyna Turoń and Grzegorz Sierpiński

Abstract The paper describes car-sharing, a model of short-term rental of vehicles used in urban transport systems. The article focuses on the analysis of specific factors pertaining to car-sharing systems, such as their availability, car fleet operated and fees. For in-depth identification of the problem, selected systems operating in Europe and Asia have been examined. The review also covered both large operators, whose systems operate in several cities of a given country, as well as small operators, who have just entered the market. The aim of the paper is to review existing car-sharing systems in European and Asian cities. The analysis highlighted the differences between functioning systems. The experience in operating those systems can be a source of knowledge for operators and people managing urban transport systems when implementing new car-sharing services, as well as improving already existing systems.

Keywords Car-sharing · Shared mobility · Car-sharing in Europe · Car-sharing in Asia · Urban transport systems · Sustainable transport

1 Introduction

Sustainable development is one of the main trends in contemporary urban transport systems [1–3]. It involves the development that combines people needs and due care of the environment. As regards traveling, proposed approaches include, among others, travel optimization, changes of modal split, emission and noise reduction

K. Turoń (✉)
Department of Automotive Vehicle Construction, Faculty of Transport, Silesian University of Technology, Katowice, Poland
e-mail: katarzyna.turon@polsl.pl

G. Sierpiński
Department of Transport Systems and Traffic Engineering, Faculty of Transport, Silesian University of Technology, Katowice, Poland
e-mail: grzegorz.sierpinski@posl.pl

M. Suchanek (ed.), *Challenges of Urban Mobility, Transport Companies and Systems*, Springer Proceedings in Business and Economics,
https://doi.org/10.1007/978-3-030-17743-0_8

[4–7]. Shared mobility, as an innovative transport strategy, has become increasingly popular for the past six years [8]. Depending on actual needs, shared mobility can take various forms in urbanized areas, e.g., bike-sharing, car-sharing, ride sharing or on-demand ride services. In fact, all types of shared mobility are good for sustainable transport, and some people claim that it is the 'key to sustainable mobility.' This is true especially in the case of car-sharing [9]. Thus, in the urban areas, regardless their geographic locations, an increase in the use of short-term rental vehicles, i.e., car-sharing systems can be noticed. These systems offer a number of advantages that have been so much appreciated by citizens. The most important advantages include [10–18]:

- reduced number of private motor vehicles in the urban modal split,
- more urban space available,
- improved environmental performance (less pollution, vibrations and noise),
- improved transport accessibility through additional transport service,
- support for promotion cars with unconventional drives (i.e., electric mobility) or powered by alternative fuels
- reduced cost related to use and maintenance of vehicle,
- promotion of sharing economy,
- fewer families with more than one car.

The aim of the paper is to present a comparison of selected systems operating in European and Asian countries. The authors have distinguished several key features, chiefly transport and economic ones, which provided a framework for the analysis of existing systems.

2 Car-Sharing

The concept of car-sharing involves the possibility of short-term rental of a vehicle in a city. It is one of many solutions under the umbrella concept of the sharing economy and it is similar to bike-sharing conception [19]. According to sources available, the idea of car-sharing in Europe dates back to 1948 [20]. It was, however, a single initiative implemented by a Swiss housing cooperative. More advanced car-sharing projects started in France in 1971, in the Netherlands in 1973 and in Sweden in 1983 [20]. In the United States, car-sharing systems emerged in 1983 [19]. Over the years, with the technological progress, car-sharing systems have become increasingly complex and mobile. Moreover, operators have evolved taking various organizational forms.

The car-sharing system does not follow a single operating standard. First of all, we can distinguish various business models, structures and operation schemes. As regards its form of operation, car-sharing can be provided by for-profit and nonprofit organizations and cooperatives [20]. In terms of its business model, we can distinguish three main types of enterprises, such as cooperatives, business-to-business

Fig. 1 Examples of car-sharing vehicles: **a** Fully-electric drive Hyiundai Ioniq; **b** Hybrid drive Toyota Yaris. *Source* Author's own collaboration with Warsaw-based car-sharing systems

(B2B) and peer-to-peer (P2P) [21]. Currently, private enterprises prevail in the car-sharing business [20, 21]. Yet another criterion used for the classification of car-sharing services is the spatial factor. It is possible to distinguish the following types of car-sharing [22]:

- stationary/classic (round-trip, RT CS)—the vehicle is rented and returned in the same location,
- unidirectional (one-way, OW CS)—the vehicle is rented at one station and returned at another, but the system is limited to rental stations established by the same operator,
- free (floating, FF CS)—when the car is rented and returned at any public parking in the city.

The fleet of vehicles within a car-sharing system may vary. Although initially, car-sharing systems operated vehicles with traditional combustion engines, hybrid and electric engine vehicles have become increasingly popular with the development of the "going green" trend. Examples of car-sharing vehicles with electric and hybrid engines are presented in Fig. 1a, b.

Car-sharing systems also vary depending on their type, application, requirements set by regulations or the service user liability, e.g., liability for the damage to the vehicle. Additionally, the type of car-sharing may vary from country to country.

3 Car-Sharing in Europe and Asia—General Overview and Research Methodology

The research on existing car-sharing systems helped to define the time framework for the development of this form of mobility in European and Asian cities. Regardless of the continent, advantages of car-sharing systems have been broadly recognized around the world. Currently, in Europe, Germany is the pioneer in car-sharing. In

2016, car-sharing systems operated in 600 German cities, whereas at the beginning of 2018, already 677 cities have such systems established [23, 24]. The establishing of the free-floating system [25] marked a breakthrough in the development. Car2Go, a prevailing car-sharing system in Germany, had 790,000 registered users in 2017 [26]. According to the statistics, in 2017, the total number of registered users in all German car-sharing systems was 1715 million [24]. At the beginning of 2018, the number of users exceeded 2 million to reach in the middle of the year 2,110,000 users registered with 165 German car-sharing service providers [24].

Car-sharing has also gain popularity in Asian countries. However, due to the size of Asian cities, the system has developed in certain cities rather than countries. The development of car-sharing systems in the Asian market took place after 2010 [27–29]. For example, in Shanghai, China, a local operator EVCard had over 1.2 million users registered in 2017 [30]. Due to a large diversity of car-sharing systems in Europe and Asia, the authors decided to compare car-sharing systems in selected European and Asian countries only.

According to the research methodology, the study consists of several stages. Firstly, the study defined organizational, infrastructural and economic characteristics of car-sharing systems. Secondly, several existing systems were selected and matched with the list of their key features on each continent. Finally, the analysis covered specific time of vehicle rental and distance traveled. Thus, it was possible to compare the cost of individual systems.

4 Comparative Analysis of Car-Sharing Systems

In order to analyze car-sharing systems, the authors selected ten operators, five in Europe and five in Asia. Operators selected represented such countries as Poland, Germany, Latvia, England, China, Malaysia and Kazakhstan.

Criteria compared included the range of the operator (local or global), the type of fleet, the type of fees, including fees charged per distance and travel time, as well as parking charges. Additionally, the authors also compared characteristic features of those systems. Analysis findings for selected five European and five Asian systems are presented in Tables 1 and 2, respectively.

The analysis shows that car-sharing systems, regardless of the country of origin, do not share any operational standard. Depending on the requirements of a given operator, the fee charged for using the service depends on the following:

- distance expressed in kilometers or miles, and
- time of using the vehicle per minute or hour.

To facilitate the comparison, the authors adopted a standard car-sharing trip. The assumption was that the user rents the vehicle and travels directly from point A to point B in 20 min over a distance of 5 km.

The user does not make any stops along the way and uses the most basic service provided by the operator and the cheapest car, the trip is taken at a daytime tariff, there

Table 1 Comparison of selected car-sharing systems in Europe

Country	City	Name	Range	System type	Fleet type	Fee per travel time	Fee per distance	Parking fee	Characteristic feature
Poland	Wroclaw	Vozilla	Local	FF CS	Electric (segment C,VAN)	0.21 €/min (car) 0.27 €/min (VAN)	–	0.02 €/min	– 1st e-car-sharing in Poland
Poland	Cracow, Warsaw, Tricity, Poznan	Traficar	Local	FF CS	Classic (segment B, kombivan)	0.19 €/min	0.12 €/km	0.02 €/min	–
Germany	Berlin, Frankfurt, Hamburg, München Rhineland, Stuttgart	Car2Go	Global	FF CS or OW CS	Classic (segment A, B, C) and electric (segment A)	0.26–0.34 €/min	200 km—free then 0.29 €/km	–	– possibility of mileage package tariff, – possibility of long distance loan
England	London, Bristol, Cambridge	Zipcar	Global	FF CS	Classic (segment B, VAN) and electric (segment C)	3.37 €/h (segment B) or 8.98–13.46 €/h	0.29 €/mi or (60 mi–free then 0.28–0.33 €/mi	–	– hourly rate, – business days/weekends rates – monthly or multi-day rates,
Latvia	Riga	CARGURU	Local	FF CS	Hybrid (segment B-C)	0.21–0.25 €/min	100 km–free, then 0.15 €/km	0.01–0.02 €/min, night/day	– 24/house rental – liability for damage

Source author's own collaboration

Table 2 Comparison of selected car-sharing systems in Asia

Country	City	Name	Range	System type	Fleet type	Fee per travel time	Fee per distance	Parking fee	Characteristic feature
Malaysia	Kuala Lumpur, cities around	SoCar	Local	FF CS	Classic (segment A, C, crossover)	1.66–5.19 €/h	15km/h—free then 0.05 €/km	–	Door-to-door concept
China	Chongqing	CAR2GO	Global	FF CS or OW CS	Electric (segment A)	0.04 €/min	0.23 €/km	0.01 €/min	–
China	Shanghai	EvCARD	Local	FF CS	Electric (segment A,B)	1.92 €/30min then 0.06 €/min	–	–	–
Malaysia	Kuala Lumpur, cities around	GOCAR	Local	FF CS	Classic (segment A,C, VAN)	3.09–3.92 €/h	15 km/h—free then 0.10 €/km	–	–
Kazakhstan	Almaty	DOSCAR CLUB	Local	FF CS	Classic (segment C, crossover)	4.59–9.96 €/h	200 km—free after 0.23 €/km	0.51 €/h	Different day/night rates

Source Author's own collaboration

Table 3 Fees per 20-min car rental at a distance of 5 km, basic car-sharing service, while using the least expensive car in selected systems in Europe

Country	System	Fees (€)
Poland	Vozilla	4.29
Poland	Traficar	3.33
Germany	Car2Go	5.20
England	Zipcar	4.67
Latvia	CARGURU	4.20

Source Author's own collaboration

Table 4 Fees per 20-min car rental at a distance of 5 km, basic car-sharing service, while using the least expensive car in selected systems in Asia

Country	System	Fees (€)
Malaysia	SoCar	1.66
Malaysia	GOCAR	3.09
China	Car2Go	1.95
China	EvCard	1.92
Kazakhstan	DOSCAR CLUB	4.59

Source Author's own collaboration

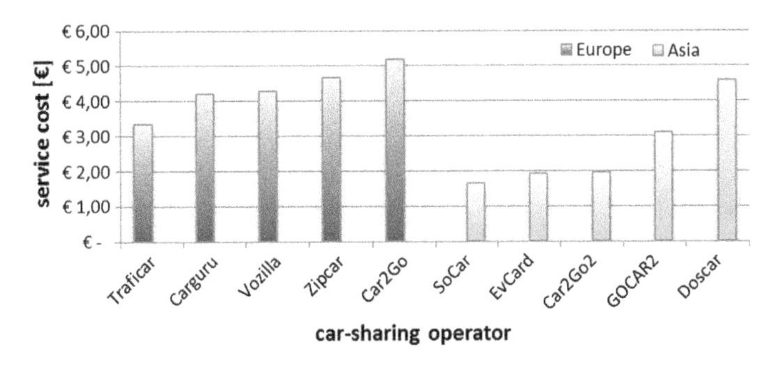

Fig. 2 Comparison of car rental cost per 20 min and distance of 5 km in European and Asian systems. *Source* Author's own collaboration

are no additional specific costs, and no validation fee and the parking fee is included in the total cost. Cost items related to the renting of the vehicle are presented in Table 3 (for selected European car-sharing systems) and Table 4 (for selected Asian car-sharing systems). The comparison is graphically presented in Fig. 2.

The authors interpreted findings of the case study based on statistical parameters that provide a methodical description of the statistical structure. It helped to determine summary statistics, such as minimum and maximum car-sharing prices in Europe and Asia, arithmetical average and the first, second (median) and third quartiles. Additionally, the statistical dispersion was established as a standard deviation and the coefficient of variation. The latter was as follows (1):

Table 5 Selected statistical parameters determined for the case study (€)

Country	Average value	1st quartile	2nd quartile	3rd quartile
Europe	4.34	4.20	4.29	4.67
Asia	2.64	1.92	1.95	3.09
Country	Minimum	Maximum	Standard deviation	Coefficient of variation
Europe	3.33	5.20	0.69	0.16
Asia	1.66	4.59	1.22	0.46

Source Author's own collaboration

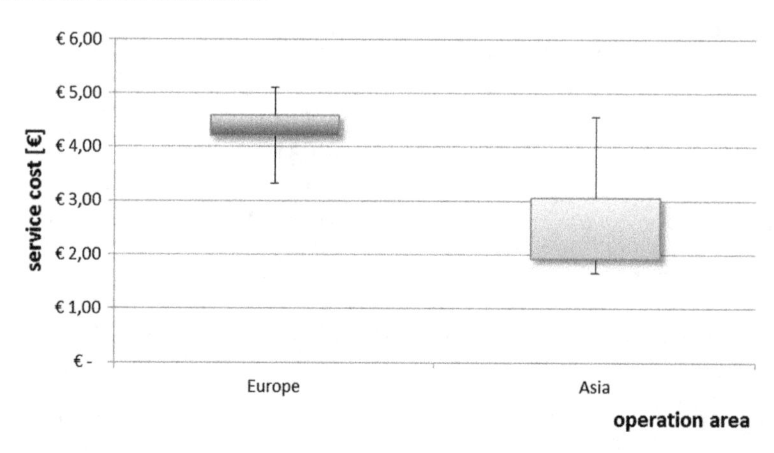

Fig. 3 Box plot comparing the cost of renting a vehicle for 20 min and distance of 5 km in European and Asian systems. *Source* Author's own collaboration

$$W = \frac{S}{\bar{x}} = \frac{\sqrt{\frac{1}{n-1}\sum_{i=1}^{n}(x_i - \bar{x})^2}}{\frac{1}{n}\sum_{i=1}^{n} x_i} \quad (1)$$

where

S—standard deviation
$\bar{x}$—sample mean
x_i—feature value
n—sample size

Values determined are presented in Table 5.

The case study clearly shows that the use of car-sharing systems in Asia is cheaper than in Europe (Fig. 3). The arithmetic average cost of renting a vehicle in Europe in systems concerned is 4.34 €, while in Asia it is 2.64 €. The largest price difference between rental options is 3.54 € per 20-min trip at a distance of 5 km.

Prices are similar in the case of the most expensive offers in Europe and Asia. The difference is only 0.61 €. European operators have more uniform prices, which is marked by a nearly twice lower standard deviation and a very low coefficient of variation. Additionally, taking into account, the possibility of parking the vehicle before exhausting the rental time and continuing the trip afterward (parking fees) has not always been provided by all operators. However, if it is offered, the cost remains at a similar level in both Europe and Asia.

When analyzing various arrangement of the fleet, combustion engine vehicles prevail in both markets, while electric vehicles represent an additional complementary part of the fleet, restricted mainly to one segment of vehicles. Moreover, vehicle segments B and C dominate in Europe, whereas in Asia, it is chiefly segment A.

A common feature of car-sharing systems in Europe and Asia is the system type. Free floating has been found the most popular among all car-sharing systems analyzed. Additionally, some systems also offered a one-way car-sharing option.

Special features of some of the systems are particularly interesting. The access to vehicles is charged according to 24-h tariffs or against a monthly subscription, which significantly reduces the cost of using a vehicle. Moreover, in some systems, the cost of the service varies depending on the day or night tariff or depending on the day of the week or weekend or, alternatively, the user can choose a package tariff.

The possibility of selecting a service fee depending on the time or distance traveled is also interesting. It is worth mentioning that the user may also chose a package that reduces cost incurred in the event of a damage to the vehicle or a door-to-door service, i.e., delivering the vehicle directly to the user, which is a unique feature of the Malaysian system [31].

5 Summary

Summarizing, the analysis of car-sharing systems operating in Europe and Asia has shown that there is no standardized method of providing car rental services. The feature that connects all systems is the type of car-sharing, namely free floating. As regards vehicles used, the systems varied from each other with both a fleet of cars, electric or conventional and a segment of vehicles used.

From the fees point of view, systems in Europe have proved to be more expensive than Asian ones. Additionally, depending on the system, various forms of charging are applied based on the rental time, distance covered or both. Moreover, in European systems, it is possible to select a tariff or a multi-hour package. The latter reduces the rental cost. Moreover, not every system provides the possibility of returning a car after it has been parked (and parking fee paid). However, if this option is offered, the cost is virtually the same in Europe and Asia.

It is worth noting that there are certain additional options, such as the possibility of delivering the vehicle directly to the user (Asia), the possibility of buying a special tariff, which reduces the cost of possible fines and liquidated damages in the event of an accident caused by the user (Europe) or various options to make the payment—either per distance or time (Europe). The above can be useful for operators and people managing urban transport systems when implementing new car-sharing services or improving already existing systems.

Further research in this field has been planned. The research will focus on the determination of the best solution ensuring balance between cost, time and distance for car-sharing users.

References

1. White Paper: Roadmap to a Single European Transport Area—Towards a Competitive and Resource Efficient Transport System. COM, 144 (2011)
2. Clean Power for Transport: A European alternative fuels strategy. COM, 17 (2013)
3. White Paper on the Future of Europe, Reflections and scenarios for the EU27 by 2025. COM, 2025 (2017)
4. Borkowski, P.: Towards an optimal multimodal travel planner—lessons from the European experience. In: Sierpiński, G. (ed.) Intelligent transport systems and travel behavior, advances in intelligent systems and computing, vol. 505, pp. 163–174. Springer, Cham (2017)
5. Sierpiński, G.: Revision of the modal split of traffic model. In: Mikulski, J. (ed.) Transport Systems Telematics 2013. Activities of Transport Telematics, Communications in Computer and Information Science, vol. 395, pp. 338–345. Springer, Heidelberg (2013)
6. Szarata, A., Nosal, K., Duda-Wiertel, U., Franek, Ł.: The impact of the car restrictions implemented in the city centre on the public space quality. Transp. Res. Procedia **27**, 752–759 (2017)
7. Pawłowska, B.: Zrównoważony rozwój transportu na tle współczesnych procesów społeczno-gospodarczych. Wydawnictwo Uniwersytetu Gdańskiego, Gdańsk (2013). (in Polish)
8. Fulton, L., Mason, J., Meroux, D.: Three revolutions in urban transportation. How to achieve the full potential of vehicle electrification, automation and shared mobility in urban transportation systems around the world by 2050. Climate Works Foundation, William and Flora Hewlett Foundation, Barr Foundation, Institute for Transportation and Development Policy, New York (2017)
9. Muheim, P., Reinhardt, E.: Car-sharing: the key to combined mobility. Energy, 58–71 (2000). BFE Swiss Federal Office of Energy, Bern, Switzerland (1998)
10. Cervero, R., Aaron G., Brendan N.: San Francisco City CarShare: Longer-Term Travel-Demand and Car Ownership Impacts. Prepared for Department of Transportation & Parking City of San Francisco, University of California pp. 1–52. Berkeley Institute of Urban and Regional Development (2006)
11. Schmöller, S., Weikl, S., Müller, J., Bogenberger, K.: Empirical analysis of free-floating car sharing usage: the Munich and Berlin case. Transp. Res. Part C **56**, 34–51 (2015)
12. Schwietermana, J.P., Bieszczat, A.: The cost to carshare: a review of the changing prices and taxation levels for carsharing in the United States 2011–2016. Transp. Policy **57**, 1–9 (2017)
13. Turoń, K., Czech, P., Juzek, M.: The concept of a walkable city as an alternative form of urban mobility. Sci. J. Silesian Univ. Technol. Ser. Transp. **95**, 223–230 (2017)
14. Le Vine, S., Polak, J.: The impact of free-floating carsharing on car ownership: early-stage findings from London. Transport Policy, Elsevier (in press)
15. Ziwen Ling, Z., Cherry, C.R., Yang, H., Jones, L.R.: From e-bike to car: a study on factors influencing motorization of e-bike users across China. Transp. Res. Part D **41**, 50–63 (2015)
16. Chen, F., Turoń, K., Kłos, M., Czech, P., Pamuła, W., Sierpiński, G.: Fifth-generation bike-sharing systems: examples from Poland and China. Sci. J. Silesian Univ. Technol. Ser. Transp. **99**, 5–13 (2018)
17. Samociuk, W., Krzysiak, Z., Bartnik, G., Skic, A., Kocira, S., Rachwał, B., Bąkowski, H., Wierzbicki, S., Krzywonos, L.: Analysis of explosion hazard on propane-butane liquid gas distribution stations during self tankage of vehicles. Przem. Chem. **96**(4), 874–879 (2017)

18. Okraszewska, R., Nosal, K., Sierpiński, G.: The role of the polish universities in shaping a new mobility culture—assumptions, conditions, experience. In: Proceedings of ICERI2014 Conference Case Study Of Gdansk University Of Technology, Cracow University Of Technology And Silesian University Of Technology, pp. 2971–2979. ICERI Press, Seville (2014)
19. Harms, S., Truffer, B.: The emergence of a nationwide carsharing co-operative in Switzerland, Research Report, EAWAG, Dübendorf (1998)
20. Millard-Ball, A.: Transit cooperative research program (TCRP) Report 108. In: Car-sharing: Where and how it Succeeds. Transportation Research Board (2005)
21. Münzel, K., Boon, W., Frenken, K., Vaskelainen, T.: Carsharing business models in Germany: characteristics, success and future prospects. Inf. Syst. e-Business Manag. **2**, 1–21 (2017)
22. Shaheen, S.A., Chan, N.D., Micheaux, H.: One-way carsharing's evolution and operator perspectives from the Americas. Transportation **42**(3), 519–536 (2015)
23. Fleet Europe: Germany enacts car-sharing law. Available at: https://www.fleeteurope.com/en/news/germany-enacts-car-sharing-law. Access 18.04.2017
24. Current data and data on the CarSharing service in Germany. Available at: https://carsharing.de/alles-ueber-carsharing/carsharing-zahlen/aktuelle-zahlen-daten-zum-carsharing-deutschland. Access: 20.02.2018
25. Giesel F., Nobis C.: The impact of carsharing on car ownership in German cities. Transp. Res. Procedia **19**, 215–224 (2016)
26. Car2Go: pioneer and market leader in free-floating carsharing. Available at: https://www.car2go.com/media/data/germany/microsite-press/files/factsheet-car2go_november-2017_en.pdf. Access: 20.02.2018
27. Anne Yu, A., Pettersson, S., Wedlin, J., Jin, Y., Yu, J.: A user study on station-based EV car sharing in Shanghai. In: EVS29 International Battery, Hybrid and Fuel Cell Electric Vehicle Symposium Montréal, Québec, Canada, 19–22 June 2016, pp. 1–10 (2016)
28. Qian, C., Li, W., Ding, M., Hui, Y., Xu, Q., Yang, D.: Mining carsharing use patterns from rental data: A case study of Chefenxiang in Hangzhou, China. In: World Conference on Transport Research—WCTR 2016 Shanghai. 10–15 July 2016. Transp. Res. Procedia **25**, 2583–2602 (2017)
29. Weithmann, S., Klug, S.: Integrative Sustainable Concepts for Individual Mobility in Asia—A Qualitative Analysis of Carsharing and Taxi Services in Singapore, pp. 1–19. Available at:https://opus.bibliothek.uni-wuerzburg.de/opus4-wuerzburg/frontdoor/deliver/index/docId/14598/file/Weithmann_Klug_Individual_Mobility_Asia_Carsharing_Taxi_Preprint.pdf. Access: 20.02.2018
30. China Daily: EVCARD announces subscriber numbers. Available at: http://usa.chinadaily.com.cn/china/2017-09/01/content_31433805.htm. Access: 20.02.2018
31. SoCar: Door to door car-sharing. Available at: https://www.socar.my/faq.php. Access 20.02.2018

Public Transport Fares as an Instrument of Impact on the Travel Behaviour: An Empirical Analysis of the Price Elasticity of Demand

Anna Urbanek

Abstract The studies on the elasticity of demand for urban public transport services have a very high theoretical and application value, proven by the fact that despite the knowledge already possessed new studies in this field are permanently undertaken worldwide. The paper presents the hitherto results of studies in the field of price elasticity of demand for public transport services. Also results of studies on price elasticity of demand have been presented, based on empirical data related to the sales of tickets for public transport services in the Upper Silesian conurbation in Poland. Moreover, the paper identifies the non-price factors that determine changes in demand for urban public transport services. The obtained results became the basis to discuss estimation limitations of demand elasticity in the urban public transport and also allowed to determine directions of further studies.

Keywords Urban public transport · Public transport fares · Elasticities · Fare elasticity · Demand

1 Introduction

The world literature of the subject pretty often considers the issues related to studies on the price elasticity of demand. Results of studies from various countries and for various market conditions can be shown here, e.g. [1–5]. In Poland, the studies on elasticity of demand for public transport services, including price elasticity, are not that extensive in the transport economics theory as in the world literature. Papers of Ciesielski [6], Grzelec [7], and Hebel et al. [8] deserve special attention. Polish literature of the subject does not comprise many result of studies based on empirical data, which would be the basis to formulate conclusions on the specific nature of Polish cities in this field.

A. Urbanek (✉)
Department of Transport, University of Economics in Katowice, Katowice, Poland
e-mail: anna.urbanek@ue.katowice.pl

M. Suchanek (ed.), *Challenges of Urban Mobility, Transport Companies and Systems*, Springer Proceedings in Business and Economics,
https://doi.org/10.1007/978-3-030-17743-0_9

New studies on elasticity of demand for the urban public transport are permanently undertaken worldwide. It happens like that primarily because the issue of demand elasticity is very important not only from the theoretical research point of view, but first of all from the application point of view for entities managing the mobility in cities.

The paper is aimed at the discussion of hitherto results of studies on price elasticity of demand as well as at an attempt to study the price elasticity of demand based on empirical data related to the sales of tickets for public transport services in the Upper Silesian conurbation in Poland. Moreover, the paper attempts at identifying the non-price factors determining changes in demand for urban public transport services.

2 Travel Demand and Transport Elasticities

The demand for transport services is a demand for a specific amount and type of transport services notified in a defined time. The mobility-related needs are the original source of travel demand. In accordance with the law of demand, the demand for transport is obviously affected by the price, which is one of the main factors directly influencing the demand size. However, it is not the only factor affecting the demand. In the group of factors that influence the travel demand and travel behaviour in a given area in specific transport system, it is possible to distinguish primarily the demographic factors as well as preferences and trends (including, as shown also by the studies, the awareness and ecological sensitivity), the level of household incomes, economic activity, geography and land use patterns, transport options and their quality, information about options, and also factors related to the demand management strategies in the given area [9]. Litman distinguishes five main groups of factors affecting the travel demand in a given area, which should be considered in transport planning and modelling and which can be used in the process of demand management (Table 1).

The demand elasticity is a measure of demand response to changes in factors that affect it. The demand elasticity is a unitless economic ratio, which can be defined as a ratio of a relative (percentage) change in the demand size to a relative (percentage) change in the demand-affecting factor [9, 10]. Most generally, it can be presented by means of a formula, where Q means demand and x is a demand-affecting variable [10]:

$$E = \frac{\Delta Q/Q}{\Delta x_i/x_i} \quad \text{or} \quad E = \frac{\Delta Q x_i}{\Delta x_i Q} \tag{(1)}$$

In the theory of economics, the following are the most important and most frequently used types of demand elasticity [11]:

- price elasticity of demand, which is a measure of demand change intensity under the influence of price changes,

- and income elasticity of demand, which is a measure of demand response to a specific good or service under the influence of customer income changes.

In the transport sector, also other types of elasticity are frequently studied, which are specific to studies on the economics of transport, e.g. [10, 12]:

- public transport service elasticity, which is the degree of increase in demand for public transport services due to changes in the transport offer,
- or travel speed elasticity or travel time elasticity, as measures of response intensity of the demand for transport services under the influence of changes in the travel speed or the travel time in the latter case.

A few methods of demand elasticity calculations can be found in the literature. In the case of scientific research on the elasticity in the transport sector, the following are used most frequently:

- point elasticity E^{Point} illustrating the elasticity in a given point of demand curve, which is used in particular in the case, when the demand changes result from a relatively small change in factor x [10]:

$$E^{\text{Point}} = \frac{\lim}{\Delta x_i \to 0}\left(\frac{\Delta Q/Q}{\Delta x_i/x_i}\right) = \frac{\partial Q x_i}{\partial x_i Q} \tag{2}$$

- or arc elasticity E^{Arc}, which illustrates demand changes in a given section of demand curve [9]:

$$E^{\text{Arc}} = \frac{\Delta \log Q}{\Delta \log X} = \frac{\log Q_2 - \log Q_1}{\log X_2 - \log X_1} \tag{3}$$

The demand elasticity can have a negative, zero or a positive value. A negative value means changes in the opposite direction, for example, an increase in specific

Table 1 Factors affecting travel demand

Demographics	Commercial activity	Transport options	Land use	Demand management	Prices
Number of people (residents, employees and visitors), Employment rate, Wealth/incomes, Age/lifecycle, Lifestyles, Preferences,	Number of jobs, Business activity, Freight transport, Tourist activity,	Walking, Cycling, Public transport, Car-sharing, Private cars, Taxi services, Telework, Delivery services,	Density, Walkability, Connectivity Public transport service proximity, Roadway design,	Road use prioritisation, Pricing reforms, Parking management, user information, Promotion campaigns	Fuel prices and taxes, Vehicle taxes and fees, Road tolls, Parking fees, Vehicle insurance, Public transport fares,

Source Litman ([9], p. 2)

factor implies a decrease in demand. Instead, a positive value means changes in the same direction. The demand elasticity can have the following values [9]:

- $|E| > 1$ demand is elastic, i.e. responding more than proportionally to changes in specific factor,
- $|E| = 1$ demand is unit elasticity, which is the factor changes resulting in a proportional demand change,
- $|E| < 1$ demand is inelastic, responding to factor changes less than proportionally.

The demand elasticity is most frequently studied and considered within a defined period of time—short- and long-run elasticities. In the long run, the demand usually responds stronger and is more elastic than in a short run, because consumers have more time to make decisions and to change their consumption patterns [13].

Because of a large scope of substitution, in particular in the urban public transport, very often also cross-elasticity is studied, which measures the response of demand for one transport type to changes in substitutive transport prices [9, 14].

3 Price Elasticity of Demand for Urban Public Transport Services

3.1 The Price Elasticity of Demand: A Literature Review

The studies carried out so far do not give unequivocal results in relation to the price elasticity value of the demand for urban transport in cities. The price elasticity of transport services, and in particular of public transport in cities, is usually specified as rather low. Meta-analysis of public transport demand shows that the coefficient of price elasticity of demand for the urban public transport can range from −0.009 to −1.32, with an average value of around −0.38 [10].

The longer is the period of time, the higher is the price elasticity of demand. In the long run, the price elasticity of demand for urban transport services can achieve a value smaller than −1, which means that these are goods and services with elastic demand and that demand is relatively strongly diminishing under the influence of price increase. Table 2 presents results of various studies on the price elasticity of the bus transport and metro in long and short run.

The price elasticity factor varies also because of geographical character of the area, in which public transport is used for travels. The study carried out in the UK shows that in urbanised (metropolitan) areas the demand for public transport responds weaker to price changes than in the case of lower urbanisation areas (Table 3).

A relatively high price elasticity of demand in highly urbanised metropolitan areas is confirmed also by studies carried out in 1981 in London, New York, and Paris (Table 4).

In the studied big cities, the demand for public transport responded to price changes not elastically, which is proven by a low value of demand price elasticity coefficient, on average around −0.30. However, it is worth noting that passengers of the rapid rail service were much less sensitive to price than bus passengers, which probably resulted from a better transport offer, primarily from the travel time point of view.

Hence, the value of demand price elasticity coefficient depends on very many factors, among which the following can be primarily mentioned [4, 9, 17]:

a. Type of passenger, in particular income, age, professional activity, ownership of a car—persons having a possibility to choose between public and individual transport are more sensitive to price than persons somehow dependent on the public transport.
b. The accessibility of alternative methods of travelling and costs of using those alternatives (fuel costs, parking costs), i.e. the degree of freedom to choose the method of travel.
c. Travel type, its obligatory or optional nature and related travel length and time, and also whether it is carried out during off-peak or peak hours—off-peak public transport price elasticity is usually higher than in peak hours.
d. Geographical features of the city, the level of urbanisation, the population density—in big cities, and metropolises travellers are less sensitive to fare changes. The percentage of persons dependant on the urban transport is higher in big, densely populated cities than in suburbs and areas of low urbanisation level. This dependence results primarily from the fact that in big cities and metropolises travels by car are much worse due to the congestion and parking problems, and vehicle parking costs.
e. Analysed time period (short run, medium run or long run).
f. Mode of transport and distance.

Table 2 Fare elasticities in UK

Time period	Bus				Metro			
	Range of value reported		Mean	Number of studies	Range of value reported		Mean	Number of studies
	From	To			From	To		
Short run	−0.07	−0.86	−0.42	33	−0.15	−0.55	−0.30	15
Long run	−0.85	−1.32	−1.01	3	−0.61	−0.69	−0.65	2

Source Paulley et al ([4], p. 297)

Table 3 Bus fare elasticities in UK by type of area and time period (UK, 1999)

Time period	Metropolitan areas	Shire counties
Short run	−0.21	−0.51
Long run	−0.43	−0.70
	London	Outside London
Short run	−0.42	−0.43

Source Dargay and Hanly ([15], p. 88)

Table 4 Public transport fare elasticities by mode in New York, London, and Paris

City	Bus service	Rapid rail service
New York	−0.32 ∓ 0.11	−0.16 ∓ 0.04
London	−0.33	−0.16
Paris	−0.20	−0.12
Mean and standard deviation	−0.30 ∓ 0.10	−0.15 ∓ 0.13

Source Lago et al. ([16], p. 44)

g. The current level of urban public transport service prices as well as the type of ticket, which price changes—users of single-travel tickets are more sensitive to price changes than season ticket users.
h. The transport offer and its quality (travel times, comfort, frequencies, coverage, etc.)
i. The method of paying for travels. Persons buying tickets/paying cash for travel are more sensitive to price changes than persons using prepaid public transport fare instruments, or electronic tickets or paying for the travel by means of payment cards [18].

Diversification of factors affecting the demand price elasticity causes difficulties in comparing the obtained indicators of demand price elasticity between cities and regions.

4 Results of Empirical Studies in the Upper Silesian Conurbation

Municipal Transport Union of the Upper Silesian Industrial Region (KZK GOP) is the urban transport organiser in the area of 29 municipalities in the central part of Silesian Voivodeship in Poland. This is the biggest, in terms of operations area, public transport organiser in Poland and one of biggest in Europe. KZK GOP performs functions typical of an urban public transport organiser, which include the competence to determine the ticket prices as well as their distribution and sales. Basic sources of urban transport services financing consist of tickets sale revenue and subsidies of municipalities, KZK GOP members. In the years 2009–2013, KZK GOP was raising the tickets prices every year. The KZK GOP policy carried out in the field of service prices to a large extent was determined by the increasing transport expenditures, which resulted to a lower extent from the operational work and to a larger extent from the need to perform indexation of transport rates based on the inflation, changes in fuel and electricity prices, and also expenses related to the carried out investment projects (Table 5).

Table 5 Basic data on KZK GOP expenditures and revenue on tickets sale in the years 2009–2014

Years	Expenditure on transport services provision (PLN million)	The number of vehicles-km in the KZK GOP area (million)	Revenue on ticket sales (PLN million)	Average tickets price change	All-items HICP (%) previous year = 100	Price of 1 L of diesel oil (PLN)
2009	438.1	84.0	225.4	0.0%	103.5	3.80
2010	464.6	84.4	224.9	5.5%	102.6	4.64
2011	498.5	84.4	233.3	11.8%	104.3	5.61
2012	541.4	84.3	242.2	6.9%	103.7	5.64
2013	555.7	83.8	247.4	7.7%	100.9	5.46
2014	579.8	85.7	243.5	0.0%	100	4.81
Change (%) 2013/2009	26.84	−0.19	9.77	35.80	11.97	43.7
Average yearly rate of changes (%) in the period 2009–2013	6.12	−0.05	2.36	7.95	2.87	9.48

Source Own study based on GUS [19] and KZK GOP data [20]

The sales of monthly tickets in the years 2009–2013 were analysed to study the demand price elasticity in KZK GOP. Such an analysis period was selected because during that time regular ticket price rises were performed and at the same time no significant changes were made in the field of available ticket types, their entitlements as well as the distribution technology (paper tickets). In the studied period, the ticket prices were raised in various months of the year (March, April or August), hence the need to study elasticity in the long run; the sales figures for October every year were analysed. In the studies on the demand for urban public transport services in Poland, October is considered a representative month, i.e. due to a standard number of holidays as well as to the fact that this is a month of school and academic youth education.

Two most popular in KZK GOP types of season tickets were analysed to study the demand price elasticity: a monthly ticket for one selected city and a monthly ticket for the entire available network of KZK GOP connections. Both tickets exist in two price versions: normal tickets with a standard price and concessionary tickets. The analysis of price elasticity of demand for season tickets provides the basis to formulate conclusions on the price impact on the behaviour of permanent passengers of the urban public transport. Table 6 specifies the amounts of analysed tickets sales in October in the years 2009–2013 together with prices effective in a given year

Table 6 Number of tickets sold in October in the years 2009–2013, with nominal and real price

Ticket type	October 2009	October 2010	October 2011	October 2012	October 2013	Change 2013/2009 (%)
Monthly ticket for one city, regular SM/AT						
Price (PLN)	88	94	108	116	126	43.2
Real price (PLN)	88	90.43	102.31	111.83	121.56	38.1
Number of tickets sold	13,281	10,651	9942	9598	8500	−36.0
Monthly network ticket, regular SC/AT						
Price	104	112	128	138	150	44.2
Real price	104	107.75	121.25	133.04	144.72	39.1
Number of tickets sold	29,935	23,828	20,958	19,498	15,067	−49.7
Monthly ticket for one city, reduced SM/AT						
Price	44	47	54	58	63	43.2
Real price	44	45.22	51.15	55.91	60.78	38.1
Number of tickets sold	18,764	15,473	13,910	12,124	11,068	−41.0
Monthly network ticket, reduced SC/AT						
Price	52	56	64	69	75	44.2
Real price	52	53.87	60.63	66.52	72.36	39.1
Number of tickets sold	44,241	35,441	30,081	25,463	20,250	−54.2

Source Own study based on KZK GOP internal data

in nominal and real terms. The study on price elasticity, especially in the long run, should be based on real prices; therefore, nominal ticket prices were adjusted by the index of average monthly gross pay rise. The increase in the average monthly gross pay in Poland in the years 2009–2013 was more dynamic than inflation changes; therefore, the adjustment by the index of average monthly gross pay changes better reflects the real tickets price in that period (Tables 5 and 7). Based on the collected data, also approximate curves of demand for the analysed ticket types in the years 2009–2013 were drawn (Fig. 1).

A decline of analysed ticket types sales is visible in the years 2009–2013. The performed analysis became the basis for calculations of demand price elasticity for individual ticket types (Table 8).

The performed analysis shows that the demand for the analysed ticket types elastically responded to price changes in the studied period. In the case of an exponential demand function, the exponent determines a point elasticity of the demand function.

Table 7 Basic data for the Silesian Voivodeship in the years 2009–2013

Years	Population of Silesian Voivodeship		Number of cars in Silesian Voivodeship per 1000 residents	Price of 1 L of unleaded petrol (PLN)	Average gross salary (PLN)	Minimum gross salary (PLN)
	Million	Share of persons up to 24 years of age				
2009	4.641	27.5%	420.2	4.29	3,101.74	1276
2010	4.635	26.9%	440.5	4.81	3,224.13	1317
2011	4.626	26.4%	463.6	5.45	3,403.51	1386
2012	4.616	25.9%	478.1	5.53	3,530.47	1500
2013	4.599	25.4%	494.4	5.38	3,659.40	1600
Change 2013/2009	−0.9%	−2.1 p.p.	17.7%	25.4%	18.0%	25.4%
Average yearly rate of changes [%] in the period 2009–2013	−0.2	–	4.2	5.8	4.2	5.8

Source Own study based on GUS [19]

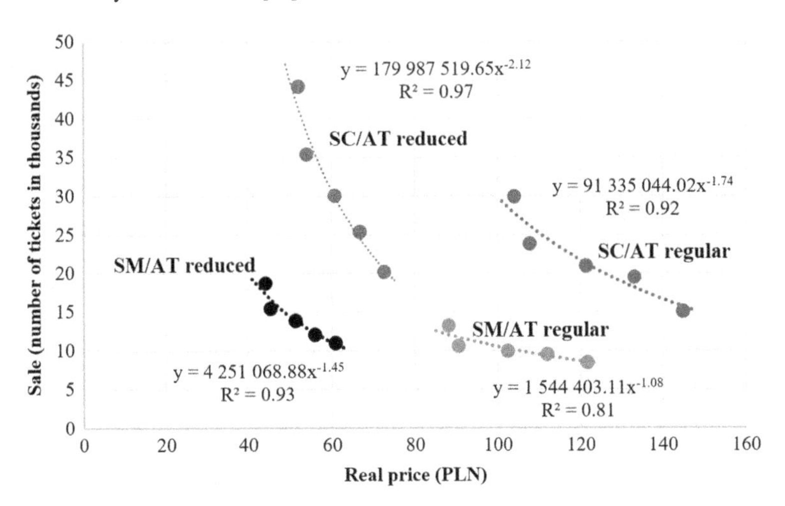

Fig. 1 Curves of demand for studied ticket types *Source* Own study based on KZK GOP internal data

Table 8 Coefficients of demand price elasticity for studied ticket types

Ticket type	Elasticity	October 2009	October 2010	October 2011	October 2012	October 2013
SM/AT monthly pass, one selected city, regular fare	Arc	–	−8.096	−0.558	−0.396	−1.456
	Point	−1.08				
SC/AT monthly pass, entire network (all cities), regular fare	Arc	–	−6.444	−1.087	−0.779	−3.064
	Point	−1.74				
SM/AT monthly pass, one selected city, reduced fare	Arc	–	−7.075	−0.863	−1.544	−1.092
	Point	−1.45				
SC/AT monthly pass, entire network (all cities), reduced fare	Arc	–	−6.264	−1.389	−1.797	−2.723
	Point	−2.12				

Source Own study

In the analysed cases, the exponential trend was the best matching function, which is proven by a high value of determination coefficient R^2. The analysis of arc elasticities shows that the demand changed most elastically in the upper part of the demand curve (the lowest price level) and in the lower part (the highest price change in the studied period).

It is also visible that the demand for tickets for the entire network (comprising all KZK GOP municipalities) featured a higher demand elasticity than the demand for tickets for one selected city. It is possible to presume that in longer-distance travels, going beyond one city, passengers more frequently decided to choose a means of transport competitive to the public transport, i.e. a car. What is interesting, in the studied tickets group the concessionary tickets featured a definitely higher elasticity, which in a definite majority of cases were bought by pupils or students. Because of obligatory nature of transport needs of this passenger group, one can presume that the reason for a decline of demand for such tickets can consist of—in the case of students—the choice of a car in daily commuting to the place of study or choosing other ticket types, e.g. single-travel of medium-term season tickets, which in many cases could be related to travelling without a valid ticket. In this case, also the issue of society ageing and a diminishing share of persons up to 24 years of age in the population structure is important (Table 7).

Nevertheless, it is necessary to consider that the presented analysis is only a kind of model approach. The analysis of price elasticity factors, neglecting the qualitative and quantitative changes in the transport offer, changes in the demand structure, and also the influence of other factors on the demand, can result in erroneous conclusions. After all, the demand for urban public transport is shaped by other, non-price factors, which will shift the demand curve. In 2014, KZK GOP did not decide to raise the prices, but the tickets sale, including the analysed monthly tickets, went down (Table 4). In the urban public transport, not only within the KZK GOP area, but also on the scale of entire Poland, we have been observing a systematic decrease in the passengers number for many years. In the KZK GOP area and in the entire Silesian

Voivodeship, this is approximately 20% a year. This is affected by such factors, such as the process of the society ageing, diminishing population of Silesian Voivodeship, the process of suburbanisation related to the depopulation of city centres, as well as a very strong competition of individual motorisation, which to a large extent is conditioned by very good conditions to move through Silesian cities by car [21]. Moreover, it is necessary to consider that a car is a normal good, for which the demand grows with the increasing income of buyers (a positive income flexibility of the demand). The urban public transport is a good of lower usability, for which the demand decreases together with the increasing income (negative income flexibility of the demand) [11].

The determination, what had a bigger impact on the demand change—ticket prices or another non-price factor, is undoubtedly a very important and necessary direction for further studies and analyses.

5 Discussion and Conclusions

Studies on the demand elasticity in the public transport feature various limitations and are burdened with an error resulting primarily from demand studying methods. For the needs of this analysis, the demand size was determined based on the number of sold paper tickets. However, it is necessary to consider that in paper ticket systems the sales do not represent a fully real demand for such tickets. In paper ticket-based systems, the organisers of public transport know only the number of tickets sold from their own warehouses to individual distributors, and not the actual sales. Only an analysis in a longer period, e.g. a year and from historical sales perspective, allows to take into account some correction related to the ticket distributors adjustment to the real demand on the market.

In studies on the elasticity of demand for public transport services, also the data on travels are frequently used to estimate the demand. So far, especially in Poland, the source of such information consisted in studies on vehicles occupancy carried out by means of manual counting during a selected period of time, considered a representative period. For example, KZK GOP carried out such measurements every year only on approximately a half of lines and they were performed on one selected working day and on one Saturday and on one Sunday. Results obtained on those individual days were then averaged onto the whole year, which was burdened with a big error, for obvious reasons. It is also worth adding that the cost of such studies in KZK GOP ranged between PLN 400,000 and 450,000 a year. Moreover, from the studies on the demand price elasticity point of view the data about demand obtained in this way have additional drawbacks in the form of missing information on the demand structure, i.e. about individual customer segments, but also of not considering in the analysis the passengers who do not pay for tickets (e.g. due to possessed entitlements), who are not sensitive to price changes.

Electronic fare collection systems, which are more and more widespread in the public transport in cities worldwide, including Poland, provide an opportunity for a new quality of demand elasticity studies. The information provided by such systems creates a possibility to carry out studies on the hitherto unavailable scale of observations number, time, and precision.

The knowledge of demand price elasticity is very important in the context of pricing policy carried out by the public transport organisers. Price elasticity indicators in individual ticket segments should be considered during the process of tariff development and potential price changes. Moreover, the value of demand price elasticity decides about the importance of price management when creating the demand for public transport.

The value of demand price elasticity coefficient depends on very many factors. Also the factors affecting the urban public transport are very diversified and complex. The multitude and diversity of factors affecting the demand price elasticity cause that the elasticity should be studied always in the context of other non-price factors, which frequently result from historical and geographical conditions of city development. The carried out analysis is a starting point for further detailed studies, primarily on the force of individual factors influence on the demand, and also on their hierarchisation, i.e. identification of factors with the greatest impact and those of secondary importance. Such studies are very significant not only from the transport economics, but also from the application point of view. Results of such studies are important not only for transport policy entities of various tiers, but also for those entities, which in their financial model incur the risk of obtaining revenue on tickets sale on an appropriate level.

References

1. Webster, F.V., Bly, P.H. (eds.): The Demand for Public Transport. Report of an International Collaborative Study. Transport and Road Research Laboratory, Crowthorne, Berkshire (1980)
2. Oum, T.H., Waters II, W.G., Yong, J.-S.: Concepts of price elasticities of transport demand and recent empirical estimates. An interpretative survey. J. Transp. Econ. Policy **26**, 139–154 (1992)
3. Goodwin, P.B.: A review of new demand elasticities with special reference to short and long run effects of price changes. J. Transp. Econ. Policy **26**, 155–170 (1992)
4. Paulley, N., Balcombe, R., Mackett, R., Titheridge, H., Preston, J., Wardman, M., Shires, J., White, P.: The demand for public transport: the effects of fares, quality of service, income and car ownership. Transp. Policy **13**, 295–306 (2006)
5. De Rus, G.: Public transport demand elasticities in Spain. J. Transp. Econ. Policy **24**(2), 189–201 (1990)
6. Ciesielski, M.: Elastyczność popytu w transporcie miejskim, Zeszyty Naukowe. Seria 1. Akademia Ekonomiczna w Poznaniu nr 204, 118–129 (1992)
7. Grzelec, K.: Zmiany elastyczności cenowej popytu w warunkach marketingowego kształtowania ofert przewozowej w komunikacji miejskiej. Biuletyn Komunikacji Miejskiej nr 67, IGKM (2002)
8. Hebel, K., Wołek, M., Wyszomirski, O.: Elastyczność cenowa popytu jako determinanta kształtowania taryf w transporcie miejskim. Zeszyty Naukowe Uniwersytetu Gdańskiego. Ekonomika Transportu i Logistyka 75, 115–125 (2017)

9. Litman, T.: Understanding transport demands and elasticities. How prices and other factors affect travel behavior. Victoria Transport Policy Institute. http://www.vtpi.org/elasticities.pdf (2017)
10. Holmgren, J.: Meta-analysis of public transport demand. Transp. Res. Part A Policy Pract. **41**(10), 1021–1035 (2017)
11. Mankiw, N.G., Taylor, M.P.: Microeconomics. Thomson Learning, London (2006)
12. Lago, A.M., Mayworm, P., McEnroe, J.M.: Transit service elasticities: evidence from demonstrations and demand models. J.Transp. Econ. Policy **15**(2), 99–119 (1981)
13. Cowie, J.: The Economics of Transport: A theoretical and Applied Perspective. Routledge, London (2010)
14. Hensher, D.A.: Establishing a fare elasticity regime for urban passenger transport. J.Transp. Econ. Policy **32**(2), 221–246 (1998)
15. Dargay, J.M., Hanly, M.: The demand for local bus services in England. J. Transp. Econ. Policy **36**(1), 73–91 (2002)
16. Lago, A.M., Mayworm, P., McEnroe, J.M.: Further evidence on aggregate and disaggregate transit fare elasticities. Transp. Res. Rec. **799**, 42–47 (1981)
17. De Grange, L., González, F., Muñoz, J.C., Troncoso, R.: Aggregate estimation of the price elasticity of demand for public transport in integrated fare systems: the case of Transantiago. Transp. Policy **29**, 178–185 (2013)
18. Litman, T.: Transport Elasticities: Impacts on Travel Behaviour. Sustainable Urban Transport Technical Document# 11. GIZ, Bonn. www.sutp.org/GIZ_SUTP_TD11_Transport-Elasticities_EN.pdf (2013)
19. GUS Local data bank. https://bdl.stat.gov.pl/BDL/start
20. KZK GOP. http://bip.kzkgop.pl/
21. Urbanek A.: Micro- and Macroeconomic factors of fares changes in urban public transport. In: Suchanek M. (ed.) Sustainable Transport Development, Innovation and Technology. TranSopot 2016. Springer Proceedings in Business and Economics, pp. 165–175. Springer, Cham (2017)

Intelligent Transport Systems (ITS) in Smart City

Barbara Kos

Abstract Contemporary cities are areas with diverse functions on which economic and social activities are concentrated. The concept of a Smart City as a progressive city of the future assumes sustainable urban development based on innovative technologies, the application of which is aimed at increasing the functionality of cities by economical, effective and ecological management. The purpose of intelligent technologies used in the city is to support the residents and provide them with comfortable, economical and safe lives. This concept covers economic, infrastructural, organizational and social issues. One of the elements of the Smart City concept is smart mobility in the sense of safe, effective and efficiently integrated management systems of transport, logistics, public transportation, bicycle traffic and parking—due to the use of intelligent transport systems, which constitute a wide collection of various tools based on information and telecommunications technology. The aim of the paper is to present selected ITS solutions that increase the efficiency and integration of the urban transport system.

Keywords Smart City · Smart mobility · Intelligent transport systems

1 Introduction

Contemporary cities are areas with diverse functions on which economic and social activities are concentrated. Decision-making, administrative, economic, political, educational, cultural, health care and others functions are concentrated in multimillion cities, agglomerations and metropolises. The issue of urban development is a very important and complex one, especially from the point of view of sustainable development because [37]:

- half of the world's population—3.5 billion people—live in cities today,
- by 2030, almost 60% of the world's population will live within urban areas,

B. Kos (✉)
University of Economics, 40-287 Katowice, Poland
e-mail: barbara.kos@ue.katowice.pl

M. Suchanek (ed.), *Challenges of Urban Mobility, Transport Companies and Systems*, Springer Proceedings in Business and Economics,
https://doi.org/10.1007/978-3-030-17743-0_10

- cities in the world occupy only 3% of the Earth's surface and account for 60–80% of energy consumption and 75% of carbon emissions,
- rapidly progressing urbanization processes exert excessive pressure on freshwater resources, sewage systems, living conditions and public health care.

Eradication of unfavourable phenomena related to the expansion of urban centres is one of the priorities of global development, recognized in a number of international documents, including the Agenda 21 (i.e. a document showing an approach to implementing sustainable development in local life), adopted at the First Earth Summit in Rio de Janeiro (1992); the Territorial Agenda of the European Union (the so-called Leipzig Charter), created at an informal meeting of ministers of EU Member States on Territorial Cohesion and Urban Development (2007); the Treaty of Lisbon (2007), as well as the renewed EU sustainable development strategy (2006); Declaration from Marseille (2008); or the Declaration of Toledo (2010). These documents clearly state that the management of urban areas should be regarded as a priority and that it should be based on the strategy of coherent and sustainable development conducive to achieving greater economic competitiveness, environmental effectiveness, social cohesion and strengthening civic attitudes [29]. The most common issues covered by urban strategies for sustainable development are [29]:

- economic development, including job creation,
- creation of conditions for new business investments,
- enhancement of the intellectual capital of the city,
- increase in the quality of life in the city (condition of the natural environment, level and quality of medical services and social assistance, level of public safety),
- modernization and extension of the city's technical infrastructure system,
- revitalization of the area,
- increase in the recreation and leisure offer as well as the tourist attractiveness of the city.

Smart City as a modern city of the future optimizes sustainable urban development based on innovative technologies which are aimed at increasing the functionality of cities by economical, effective and ecological management. Intelligent technologies used in the city enhance the quality of urban services and improve the quality of life. The concept of a Smart City covers a wide range of economic, infrastructural, organizational and social issues. One of the most important elements of the Smart City concept is smart mobility understood as safe, effective and efficiently integrated systems monitoring and managing traffic and transportation systems, using intelligent transport systems, consisting of many various tools based on information and telecommunications technology.

The aim of the article is to review and analyse theoretical issues related to the definition and evolution of the Smart City concept as a direction of urban development. Hypothesis that without the implementation of innovative technological solutions modern, especially multi-million cities will not be able to solve their growing problems (among others in the area of economic and social urban mobility) can be put forward. For the purposes of the article, methods of descriptive and comparative analysis were used to present and evaluate selected ITS solutions that increase the efficiency and integration of the city's transport system.

2 The Smart City Concept

The concept of a Smart City appeared in year 1992 to signify the turn in urban developments towards technology, innovation and globalization [6]. The concept of Smart City as an urban area of sustainable development is the subject of serious consideration in the literature. There are many definitions that differ from each other. The Smart City concept is commonly used in different nomenclatures and contexts and with different meanings. Apart from the Smart City concept, it is possible to indicate other similar concepts relating to modern urban areas such as [5]:

- Intelligent City,
- Knowledge City,
- Sustainable City,
- Talented City,
- Wired City,
- Digital City,
- Eco-City.

Many publications review the definition of smart cities [2, 14, 19, 22, 23], pointing to changes that have occurred in them as a result of the development of the concept.

Among the definitions of the Smart City, one can find those that emphasize technical, social, institutional or environmental aspects, e.g.:

- "The Smart City is a new way of leaving and considering the cities. The optimization of available and new resources as well as of possible investments is required. The achievement of Smart City objective can be reached through the support of various information and communications technologies. These can be integrated in a solution considering the electricity, the water and the gas consumptions, as well as heating and cooling systems, public safety, wastes management and mobility" [18].
- "The Smart Cities concept is connected to notions of global competiveness, sustainability, empowerment and quality of life, enabled by broadband networks and modern ICTs. Its implementation requires the development of migration paths regarding Internet infrastructures, test bed facilities, networked applications and stakeholder partnerships" [13].
- "Key conceptual components of Smart City are three core factors: technology (infrastructures of hardware and software), people (creativity, diversity and education) and institution (governance and policy). Given the connection between the factors, a city is smart when investments in human/social capital and IT infrastructure fuel sustainable growth and enhance a quality of life, through participatory governance" [23].

The more complex concept of the Smart City factors is presented by Chourabi et al. [2]. They list eight clusters of factors including (1) management and organization, (2) technology, (3) governance, (4) policy, (5) people and communities, (6) the economy, (7) built infrastructure and (8) the natural environment.

Mosannenzadeh and Vettorato [22] analysed existing Smart City definitions and concluded that the concept of Smart City was mostly developed in three areas: academic, industrial and governmental. The review of the literature also allowed to identify that the Smart City concept itself is not fixed but usually assigns ICTs a significant role in Smart urban development. In addition, one should pay attention to the differences occurring in the above-mentioned three areas (academic, industrial and governmental) on the semantic interpretation of the word "Smart". In the academic area, it usually refers to the features of technological solutions, in the industrial area to the characteristics of intelligent products and services [23], and in the governmental documents to the concept of "Smart Growth" that is an urban planning theory originating from USA and dealing with prevention of urban sprawl [8].

It is generally believed that a city can be treated as "intelligent" when it invests in human and social capital and communication infrastructure in order to actively promote sustainable economic development and high quality of life, including wise management of natural resources, through civic participation.

For the purpose of the article, it was assumed that Smart City is a concept focused on sustainable urban development based on innovative technologies, primarily IT and communication. It aims to improve the functionality of cities by economical, effective and ecological management. It is a concept that covers economic, infrastructural, organizational and social issues. It is possible to identify six basic components that make up a Smart City [3, 7, 12, 27]:

- smart governance as a transparent exchange of information between residents, a city, central units and municipal services units, police, fire brigades, emergency medical services,
- smart economy (e-business and e-commerce) enabling efficient flow of goods, services and knowledge at the city level and between cities,
- smart mobility as safe, effective and efficiently interconnected systems for management of transport, logistics, public transport, bicycle traffic, parking,
- smart environment as intelligent management of environmental resources (smart meters, devices supporting energy storage, devices supporting the reduction in energy consumption, rational management of electricity, intelligent lighting systems, implementation of renewable energy sources, waste management),
- smart people—smart education/residents information (access to training, education through modern communication and information technologies; support of resources, creativity and human potential, enabling active participation of residents in the life of the urban community),
- smart living—smart lifestyle enabling to raise the quality of lifestyle based on ICT: a wider, more diverse cultural and service offer, better insight into the offer of healthcare facilities, housing offer.

Currently, the attractiveness and quality of life in cities are assessed using a series of complex indicators and indices—the so-called Urban Indicators, which are tools for monitoring sustainable urban development. Among them, there are the following: City Development Index (CDI), the Sustainable Cities Index, the Green City Index, the Smart City Index, the Quality of Life Index, the IESE Cities in Motion Index, etc. They differ in the choice of indicators to calculate subindices, the number of subindices and the method of selecting cities for comparison. Therefore, the rankings of cities based on individual indicators are often incomparable. In addition, even within one index, there are problems in comparing the position of the city in subsequent years, because the number of indicators to be assessed for particular subindices, the number of cities assessed or the introduction of new indicators change. Only formulation of relatively stable models for the assessment in a given timeframe will allow for a full comparative analysis.

To indicate the position of Polish cities, IESE Cities in Motion Index (CIMI) [9–11] were chosen. In the 2016 edition of the CIMI, 181 cities (including 72 capitals) were assessed from 80 countries in 10 areas: economy, human capital, technology, the environment, international outreach, social cohesion, mobility and transportation, governance, urban planning and public management (overall 77 indicators). In the 2017 edition of the CIMI, 180 cities (including 73 capitals) from 80 countries (overall 79 indicators) were assessed. In 2018 edition of the CIMI, there are 165 cities (including 74 capitals) from 80 countries. In 2018, changes were introduced expanding the number of indicators (altogether 83 indicators), including levels of compliance with ISO 37120 (known as a Smart City standard) and combining two areas of governance and public management into one called governance.

In the 2016 edition, the following cities were at the top of the ranking (first 5) New York, London, Paris, San Francisco and Boston. In the 2017 edition, the first 5 cities were almost the same, only San Francisco and Boston changed places: New York, London, Paris, Boston and San Francisco. In 2018 edition, the first 5 cities were: New York, London, Paris, Tokyo and Reykjavik (in the previous editions Reykjavik was not evaluated).

Table 1 presents data from CIMI in 2016–2018 for two Polish cities: Warsaw and Wroclaw. However, Wroclaw was not included in the CIMI 2018 edition. Due to changes in the calculations of individual subindices, it is difficult to unequivocally assess changes in particular areas, which is most evident for Warsaw for the environmental subindex. Nevertheless, it can be noticed that the worst values for these two cities are found in economy, public management and technology subindices. Wroclaw is also low in human capital and international outreach areas, while both cities have the best values in urban planning (in 2017 both cities were positioned in the top ten).

The aforementioned ISO 37120: 2014 standard, Sustainable development of communities—Indicators for city services and quality of life measures the development of cities from a social, economic or environmental perspective, the quality of urban services and the standard of living of residents through 100 indicators, of which 46 are considered basic and 54 additional ones. All indicators were divided into 17 main thematic groups describing the concept of sustainable development and qual-

Table 1 Position of Polish cities according to CIMI in 2016–2018 [9–11]

Index	Warsaw			Wroclaw		
	2016	2017	2018	2016	2017	2018
Cities in motion	74	54	64	94	95	–
Economy	113	95	108	121	127	–
Human capital	121	59	65	146	123	–
Social cohesion	34	29	58	77	68	–
Environment	66	16	83	71	19	–
Public management	140	142	–	170	169	–
Governance	78	93	50	94	108	–
Urban planning	12	6	15	14	10	–
International outreach	87	83	30	116	105	–
Technology	68	70	114	53	68	–
Mobility and transportation	61	41	58	126	155	–

ity of life. These are: economy, education, energy, environment, recreation, safety, shelter, solid waste, telecommunications and innovation, finance, fire and emergency response, governance, health, transportation, urban planning, wastewater, water and sanitation. Certification is carried out on five levels: aspiring, brown, silver, gold and platinum, and the level of certification depends on the number of indicators reported by the city. In Poland, the certificate of compliance with the ISO 37120 standard has been awarded to three cities: Gdynia and Gdansk in 2017 and Kielce in 2018.

The implementation of the ISO 37120 standard aims at more efficient city management, including provision of higher-quality urban services and support for informed decision-making based on data and verified information. It is a tool that allows for objective assessment of the compliance with the standards of a modern management in the city.

In 2016, the next standard ISO 37101:2016 Sustainable development in communities—Management system for sustainable development—Requirement with guidance for use was published.[1]

3 Intelligent Transport Systems (ITS)

Smart mobility as one of the elements of the Smart City concept is essential for the functioning and development of the city and the quality of life of the residents. The efficient transport system of the city based on the use of intelligent transport systems, which depend on information technology and telecommunications, is a

[1]ISO 37101:2016 Sustainable development in communities—Management system for sustainable development—Requirements with guidance for use, https://www.iso.org/standard/61885.html.

factor influencing decisions regarded as decisive for the competitiveness of cities and regions, such as investment location, both from perspective of investors and people seeking a place to live.

Directive 2010/40/EU of the European Parliament and of the Council of 7 July 2010 on the framework for the deployment of Intelligent Transport Systems in the field of road transport and for interfaces with other modes of transport [4] has defined Intelligent Transport Systems (ITS) as information and communication systems to provide services related to different modes of transport and traffic management, allowing better information for different users and ensuring safer, more coordinated and "smarter" use of transport networks. They are used in traffic management and mobility management, and they can cooperate with similar systems used in other modes of transport. In addition, the Directive provides a common framework for the coordinated implementation of ITS in the EU, with particular emphasis on the cross-border cooperation between Member States. The application of ITS in road transport and its interactions with other modes of transport is the main concern of this document. The Directive sets the necessary conditions and specifies the directions of activities in the priority areas with the development of appropriate standards when needed. This applies in particular to areas such as:

- optimization of the road, traffic and travel data usage,
- continuity of ITS services related to traffic management and freight transport,
- ITS applications related to road safety and security,
- connection of the vehicle with the used transport infrastructure.

Intelligent Transport Systems (ITS) constitute a wide range of various tools based on information technology, telecommunications and vehicle electronics used to increase the efficiency and integration of the entire transport system of the city. ITS implementation aims to improve traffic efficiency by reducing travel time, reducing nuisances such as pollution and noise, and increasing safety in transport [28].

The main areas supporting Intelligent Transport Systems are as follows [1, 16, 20, 21, 25, 28, 31, 35, 38, 39]:

- traffic management,
- management of public transport,
- management of cargo transport and fleet of vehicles,
- traffic safety management and monitoring system for regulations' violation,
- management of road incidents and emergency services,
- information services for travellers,
- electronic payment services [15] and electronic systems for collecting toll for road use.

ITS operation is based on the processes of collecting a lot of different information about the status of individual subsystems and then their processing and analysis in order to take the best decision in a given situation. Various entities may be the recipients of the information, e.g. traffic management authorities, public transport managers, local government units, infrastructure users, emergency services, passengers, etc.

From the point of view of urban mobility management, the subsystems related to transport management play an important role in [27]:

- central gathering of information on traffic flows; control of traffic lights and flows depending on the current traffic situation, road events in real time, improving the capacity of the city's road transport system,
- collective transport management, enabling rolling stock monitoring, effective passenger information, building dynamic schedules allowing more comfortable use of public transport, better resource planning and savings associated with more efficient use of rolling stock,
- managing parking spaces together with a guidance system for road users, enabling efficient guidance on free parking spaces, detection of exceeding the set parking time, reservation of seats.

Thanks to the ITS application, the benefits can be achieved by both residents and entrepreneurs conducting business activities in the city. The basic benefits of using Intelligent Transport Systems in the city include increasing street capacity, reducing time losses in the street network and improving traffic safety [16].

Benefits of using Intelligent Transport Systems are as follows [25]:

- increase in the capacity of the street network by 20–25%,
- improvement in road traffic safety (reduction in accidents by 40–80%),
- reduction in travel times and energy consumption (by 45–70%),
- improvement in the quality of the natural environment (reduction in exhaust emissions by 30–50%),
- improvement in the comfort of travel and traffic conditions for drivers, travellers and collective transport,
- reduction in the costs of road fleet management,
- reduction in expenses related to maintenance and renovation of the surface,
- increase in the economic benefits in the region.

Today, Polish agglomerations, following the example of European cities, implement modern traffic management systems, including Intelligent Transport Systems that bring many benefits related to improving the local quality of life. Intelligent Transport Systems contribute to the optimization of transport mobility of the society and the quality of transport services, which in turn gives positive economic effects, reduction in pressure on the environment and improvement in road safety [36].

The ITS projects currently implemented in Poland are usually large and complex solutions, consisting of many subsystems. The most common domain subsystems are as follows [30]:

- area motion control by means of traffic lights,
- information on traffic conditions, transmitted via various media—electronic text tables and variable message signs (VMS), Web portals and mobile applications,
- electronic boards, guiding parking lots and informing about the number of vacancies,

- supervision of compliance with road traffic regulations—measurement of instantaneous and part speed, entry at red traffic light, exceeding the permissible total weight of the vehicle,
- video traffic monitoring and event detection through automatic image analysis,
- dynamic stop information on electronic boards about the estimated time of departure of public transport vehicles,
- supervision of transport traffic: location of vehicles while driving, deviations in running in relation to the timetable, dedicated dispatch communication systems,
- charging for road tolls, using public transport, parking.

The most activities that could be considered as implementation of intelligent solutions were undertaken by the following cities in Poland: Bialystok, Bydgoszcz, Czestochowa, Gdynia, Gdansk, Gliwice, Lodz, Poznan, Rzeszow, Sopot, Szczecin, Kraków, Warsaw, Wroclaw and Zielona Góra. However, these cities mainly carry out activities related to the construction of infrastructure, in particular intelligent traffic management systems at intersections (assigning priority to the public transport vehicles), parking spaces occupancy monitoring systems, systems "smart transition" that increase the visibility of pedestrians, intelligent stops, as well as dynamic passenger information systems, etc.

Implemented electronic city cards enabling payments for various types of public services in the cities (many Polish cities, both large and small, already have such cards) or Open Payment Systems which operate in Jaworzno, Lodz and Wroclaw are examples of modern solutions. In addition, various types of mobile applications allowing access to timetables, dynamic passenger information or purchasing tickets for travel can be indicated as further examples of modern technological solutions employed.

Although many cities undertake various initiatives, they cannot be classified as systematic and orderly. As far as the implementation of extensive programs is concerned, local authorities are most often limited by the lack of financial resources.

4 Conclusion

In recent years, the concept of "Smart City" has become very popular. Many cities in the world and in Poland implement innovative, modern solutions to improve the quality of life of their residents.

The concept of Smart City as a modern city of the future assumes sustainable urban development based on innovative technologies, the application of which is aimed at increasing the functionality of cities in the aspect of economical, effective and ecological management.

Currently, the attractiveness and quality of life in cities are assessed using a series of complex indicators and indices—the so-called Urban Indicators, which are tools for monitoring sustainable urban development. Among them, City Development Index (CDI), the Sustainable Cities Index, the Green City Index, the Smart City Index, the Quality of Life Index, the IESE Cities in Motion Index, etc., and ISO 37120:2014 and ISO 37101:2016 norms can be included. When analysing the position of Polish cities, it should be stated that despite many initiatives, they still have much to do to be recognized as modern cities in accordance with the models used as a base for the various assessments. Nevertheless, the fact that three Polish cities were able to obtain a certificate of compliance with the ISO 37120:2014 standard gives hope for further development of Polish smart cities.

The use of ICT systems in transport in urban areas is part of the concept of smart mobility, which is one of the Smart City components. From the point of view of creating smart mobility, available resources, financial, personal and information resources should be involved in order to properly shape attitudes and communication behaviours of city dwellers towards more frequent use of public transport, walking and cycling, and joint use of individual vehicles [17, 24, 26, 33, 34].

In Polish conditions, the mobility management process should aim at providing city residents with high-quality public transport services [32], transport interchange, an adequate network of safe bicycle paths, extensive, well-marked and safe pedestrian routes that will retain current public transport users and encourage other residents to undertake more pro-ecological journeys, in particular the resignation from the use of passenger cars for urban travel. But despite many initiatives undertaken by cities, it is still hard to classify these activities as systematic and orderly actions.

References

1. Barwiński, S., Kotas, P.: Inteligentne Systemy Transportowe w wybranych miastach Polski. Autobusy: technika, eksploatacja, systemy transportowe **10**, 26–29 (2015)
2. Chourabi, H., Nam, T., Walker, S., Ramon Gil-Garcia, J., Mellouli, S., Nahon, K., Pardo, T.A., Jochen Scholl, H.: Understanding smart cities: an integrative framework. In: 45th Hawaii International Conference on System Sciences, pp. 2289–2297. CPS, Los Alamito (2012)
3. Czupich, M., Kola-Bezka, M., Ignasiak-Szulc, A.: Czynniki i bariery wdrażania koncepcji Smart City w Polsce. Studia Ekonomiczne **276**, 223–235 (2016)
4. Directive 2010/40/EU of the European parliament and of the council of 7 July 2010 on the framework for the deployment of intelligent transport systems in the field of road transport and for interfaces with other modes of transport, 14 Mar 2018. http://eur-lex.europa.eu/legal-content/EN/TXT/?uri=CELEX:32010L0040 (2010)
5. Directorate general for internal policies policy department a economic and scientific policy: mapping smart cities in the EU, 14 Mar 2018. European Union Internet: http://www.europarl.europa.eu/RegData/etudes/etudes/join/2014/507480/IPOL-ITRE_ET(2014)507480_EN.pdf (2014)
6. Gibson, D.V., Kozmetsky, G., Smilor, R.W.: The Technopolis Phenomenon, Smart City, Fast Systems, Global Networks. Rowman & Littlefield Publishers, Washington D.C. (1992)
7. Giffinger, R., Fertner, C., Kramar, H., Kalasek, R., Milanoviü, N., Meijers, E.: Smart Cities Ranking of European Medium-Sized Cities. Centre of Regional Science (SRF), Vienna UT (2007)
8. Herrschel, T.: Competitiveness and Sustainability: Can 'Smart City Regionalism' Square the Circle? Urban Studies **50**, 2332–2348 (2013)
9. IESE Business School: IESE Cities in Motion Index 2016, 9 Aug 2018. https://www.iese.edu/research/pdfs/ST-0396-E.pdf (2016)

10. IESE Business School: IESE Cities in Motion Index 2017, 9 Aug 2018. https://www.iese.edu/research/pdfs/ST-0442-E.pdf (2017)
11. IESE Business School: IESE Cities in Motion Index 2018, 9 Aug 2018. https://www.iese.edu/research/pdfs/ST-0471-E.pdf (2018)
12. Jankowska M.: Smart City jako koncepcja zrównoważonego rozwoju miasta – przykład Wiednia. Studia i Prace Wydziału Nauk Ekonomicznych i Zarządzania Uniwersytetu Szczecińskiego **42**(t. 2), 174–175 (2015)
13. Komninos, N., Schaffers, H., Pallot, M.: Developing a policy road map for smart cities and the future internet. In: eChallenges e-2011 Conference Proceedings, IIMC International Information Management Corporation (2011)
14. Komninos, N.: The Age of Intelligent Cities. Smart Environments and Innovation-for-All Strategies. Routledge, London and New York (2014)
15. Kos, B.: Development of electronic payments in Poland using the example of local and regional collective transport. In: TST 2015 Tools of Transport Telematics, CCIS 531, pp. 352–361. Springer, Berlin Heidelberg (2015)
16. Koźlak, A.: Inteligentne systemy transportowe jako instrument poprawy efektywności transportu. Logistyka **2**, CD (2008)
17. Koźlak, A.: Kierunki zmian w planowaniu rozwoju transportu w miastach jako efekt dążenia do zrównoważonego rozwoju. Transport Miejski i Regionalny **7/8**, 39–43 (2009)
18. Lazaroiu, G.C., Roscia, M.: Definition methodology for the smart cities model. Energy **47**, 326–332 (2012)
19. Marolla, C.: Climate Health Risks in Megacities, Sustainable Management and Strategic Planning. CRC Press, Boca Raton (2016)
20. Mikulski, J.: Wizja rozwoju Inteligentnych Systemów Transportowych w Polsce. Przegląd ITS **17**, s. 8–9 (2009)
21. Modelewski, K.: Czym jest ITS?, 14 Mar 2018 http://www.itspolska.pl/?page=11
22. Mosannenzadeh, F., Vettorato, D.: Defining Smart City. A conceptual framework based on keyword analysis. TeMA, Journal of Land Use, Mobility and Environment, Special Issue, 683–694 (2014)
23. Nam, T., Pardo, T.A.: Conceptualizing Smart City with dimensions of technology, people, and institutions. In: Proceedings of the 12th Annual International Digital Government Research Conference: Digital Government Innovation in Challenging Times, ACM, pp. 282–291 (2011)
24. Nosal, K., Starowicz, W.: Wybrane zagadnienia zarządzania mobilnością. Transport Miejski i Regionalny **3**, 26–31 (2010)
25. Oskarbski, J., Jamroz, K., Litwin, M.: Inteligentne Systemy Transportu - zaawansowane systemy zarządzania ruchem, 14 Mar 2018. http://docplayer.pl/409431O-Inteligentne-systemy-transportu-zaawansowane-systemy-zarzadzania-ruchem.html (2006)
26. Osyra, B.: Zarządzanie mobilnością miejską – Instrumenty i podstawowe etapy wdrażania zrównoważonych planów zarządzania (SUMP). Zeszyty Naukowe Politechniki Częstochowskiej Zarządzanie **22**, 218–229 (2016)
27. Polskie miasto przyszłości: Inteligentne rozwiązania w Twoim mieście, 14 Mar 2018. http://polskiemiastoprzyszlosci.pl/ (2014)
28. Proper, K., Allen, T.: Intelligent Transportation System Benefits: 2000 Update. U.S. Department of Transportation, Washington D.C. (2001)
29. PWC: Miasta na drodze do zrównoważonego rozwoju. Najlepsze praktyki polskich miast nagrodzonych w Konkursie "Miasto Szans-Miasto Zrównoważonego Rozwoju", 14 Mar 2018. https://www.pwc.pl/pl/biuro-prasowe/assets/miasto_szans_opisy_dobrych_praktyk.pdf (2013)
30. Qumak: Inteligentne systemy transportowe, 14 Mar 2018. http://www.qumak.pl/wp-content/uploads/2014/01/Qumak_ITS_2015.pdf (2015)
31. Siergiejczyk, M., Tatar, K.: Koncepcja wdrażania usług Inteligentnych Systemów Transportowych w obszarze miejskim. Prace Naukowe Politechniki Warszawskiej. Transport **113**, 443–453 (2016)

32. Starowicz, W.: Jakość usług w miejskim transporcie publicznym. Wydawnictwo Politechniki Krakowskiej, Kraków (2008)
33. Starowicz, W.: Zarządzanie mobilnością wyzwaniem polskich miast. Transport Miejski i Regionalny **1**, 42–47 (2011)
34. Szołtysek, J.: Kreowanie mobilności mieszkańców miast. Wolters Kluwer, Warszawa (2011)
35. Świderski, A., Kamiński, T., Zelkowski, J.: Aspekty Inteligentnych Systemów Transportowych w miastach. Gospodarka Materiałowa i Logistyka **5**, 697–707 (2016)
36. Tomaszewska, E.J.: Inteligentny system transportowy w mieście na przykładzie Białegostoku. Zeszyty Naukowe Uniwersytetu Szczecińskiego, Problemy Zarządzania, Finansów i Marketingu **41**(t. 2), 317–329 (2015)
37. UN GLOBAL COMPAKT: Zrównoważone miasta. Życie w zdrowej atmosferze. Network Poland (2016)
38. Wałek, T.: Inteligentne Systemy Transportowe jako instrument poprawy bezpieczeństwa. Security, Economy & Law **2**(XI), 67–73 (2016)
39. Wojewódzka-Król, K., Rolbiecki, R.: Inteligentne systemy transportowe w świetle europejskiej polityki transportowej. Zeszyty Naukowe Uniwersytetu Szczecińskiego, Ekonomiczne Problemy Usług **57**, 65–72 (2010)

Ergonomics of Designing Means of Public and Individual Transport Taking into Consideration the Ageing Process of Society

Anna Miarka

Abstract The aim of the article is to describe the basic guidelines for designing public and individual means of transport together with their environment and pointing to the necessity of a holistic approach to the system. In the light of the society ageing, it is crucial to work on implementing procedures which support the elderly, but first of all on appropriate construction of assumptions and design forms for the emerging means of transport and their environment. In the article, we present the basic guidelines for design determined by the needs of users, i.e. the elderly, as well as broader directives including the needs of extreme users. The analysed guidelines are constructed on the basis of adequate data from the literature and on the basis of research done by the author exploring the needs of the elderly within the scope of ergonomic design of products introduced between the years of 2010–2017.

Keywords Elderly · Designing for the elderly · Universal design · Anthropomorphic data of the elderly for designing

1 Introduction. Aim of the Article

The ageing of the society is quite a big challenge for the industrial design artists who also design means of public transport, firstly, because still there is no base of anthropomorphic data of the elderly complex enough and there is not enough knowledge about their needs, possibilities, as well as perceptive, mental and physical limitations. The hereby text aims at presenting and analysing the most important anthropomorphic data which can be used when designing means of transport and their environment, as well as gathering key assumptions, which should serve the process of design. The publications mentioned in the text supplement the knowledge in the analysed topic and broaden the perspective of looking at the problem with the

A. Miarka (✉)
Faculty of Industrial and Interior Design, Design Ergonomics Laboratory, Strzemiński Academy of Art in Łódź, Łódź, Poland
e-mail: amiarka@asp.lodz.pl

M. Suchanek (ed.), *Challenges of Urban Mobility, Transport Companies and Systems*, Springer Proceedings in Business and Economics,
https://doi.org/10.1007/978-3-030-17743-0_11

issues connected with universal design, i.e. one which includes a larger percentage of users with special needs.

In this text, the author wishes to emphasise the need for a very broad view of the ageing process of Polish society, especially in the context of planning common spaces with means of transport and their surroundings. To achieve the best results, it is necessary to cooperate in interdisciplinary teams from broadly understood design areas, including designers of industrial design, geriatricians and specialists in the field of anthropometry, which should be confirmed below.

In the text, we focus only on municipal means of public transport such as buses and trams and individual such as passenger cars. A deeper analysis of cases is not the aim of the hereby article; however, some guidelines depicted can be used universally also with reference to trains, train stations, planes and airports.

2 Old Age in Definitions

In the context of the analysed topic, it is essential to define old age, as over the course of the last 30 years the concept has evolved in connection with the development of modern technologies. Thanks to contemporary technology and medicine together with knowledge of healthy eating and lifestyle, people formally regarded as the elderly are not that dependent on the environment and not self-reliant as they used to be 20–30 years ago. Social view of age which means a person is considered a senior citizen, is also very different than the classification made by WHO, which says it is a person over 60 years old. For today's society, a 60-year-old is in his/her prime, most often still working and rather physically and intellectually capable. However, in medical terms old age starts when a person is 60–65 years old and is characterised by general biological changes in the human body, which are mainly connected with organs being worn out and cells not regenerating so well [1].

Nowadays, in the developed societies people live 70–90 years with women living statistically longer. The information itself can be the reason to think about creating new guidelines for designing means of transport. When we combine the data with the information from Statistics Poland (GUS) about the prognosis of the length of life and the number of representatives in each age group in the population, it easy to notice that in 30 years citizens aged 60 and above will be a considerable group using means of individual and public transport in the common area and will constitute as much as 1/4 of population, and most of them will be women [2].

3 Most Important Publications and Data Connected with Designing for the Elderly

For many years, the whole of Europe is struggling with the problem of societies growing older. In Poland we have also recognised the situation and more and more steps are taken to improve the everyday lives of senior citizens. Nevertheless, current knowledge makes us—designers—look significantly wider on the context of the target users of our work. In order to achieve better results and extend the group of users, designers might also use the rules of universal design. A model example of such use is Norway, where the scope of design works is not limited to solving senior citizens' problems only, but much broader procedures, described as universal designing, are implemented as well [3].

Ron Mace—the creator of the theory and the propagator of the idea—describes universal design in the following words: "it is the design of products and environments to be usable by all people, to the greatest extent possible, without the need for adaptation or specialized design".

In Poland, the most complex publication about universal design is "Things are for People. The Idea of Universal Design/Rzeczy są dla ludzi. Idea projektowania uniwersalnego" by Maciej Błaszczak and Łukasz Przybylski, where the authors describe the issues connected with disabilities and the process of ageing of societies, which leads to more and more people with some disabilities appearing in public space. They stem both from innate or acquired health problems and from the natural cycle of life. The authors also present seven rules of universal design. According to the publication, using those guidelines when designing all products and services should ensure adapting the environment for the whole population, including the needs of the so-called extreme users [4]. Nevertheless, in reality we very often have too many other assumptions and problems to solve in order to follow at least some of the seven guidelines. The research done by the author of the hereby text indicates that using rules based on anthropomorphic data itself while designing can have positive influence on the utility process and the comfort of users. However, in highly developed countries we should aim at constantly improving the quality of environment and public space common for all users. That is why, universal design should be promoted more and more. It should be assumed that the basic goal while designing a product is the possibility for it to be used by all people, obviously providing for personal technical aids such as wheelchairs and hearing aids. We should not, however, use solutions particularly dedicated to people with disabilities; the design form should not emphasise in any way that the product was designed especially for people with lower functionality. It is very important to try to resign from solutions which compensate for shortcomings, i.e. chair lifts or separate "adapted entrances" [3].

One of the most important European institutions which focuses on universal design is the European Institute for Design and Disability—EIDD, which defines its mission in the following way: "To improve the quality of life by implementing the rules of design for all (Design for All-DfA)". The motto of the institution is "good design enables, bad design disables" [5]. This sentence fits perfectly with the issues connected with designing means of transport. If they are badly designed (together with their environment) and do not take into account the needs and possibilities of senior citizens and other extreme users, it can lead to a number of accidents and injuries and sometimes even to isolating the elderly from the society. The consequences can be very serious in terms of finances as well—treatments, rehabilitation and care are expensive and might be essential for too many people—the health system might not be able to handle it. Thus, in compliance with the confirmed prognosis of the ageing society, we should already pay particular attention to complex and holistic design of means of transport and the surrounding infrastructure.

In accordance with the rules of ergonomic design, it is known that the best effects are achieved when we design on the basis of anthropomorphic data of a population for which we work. That is why when designing for the Polish market, it is well based to use the anthropomorphic data of the Polish people. For many years, such data were supplied by the research conducted by the Industrial Design Institute in Warsaw. The research requires a large sample, it is expensive and time-consuming, and hence, it is not updated very often—the most recent data that we have go back to 1998. According to the described demographic characteristics of people in the post-productive age, it is visible that most of them are women, that is why the research was conducted on the population of women. The fact that only 106 women between the age of 60 and 96 who lived in Warsaw were the focus of the research itself is a proof that there is no such data or there is not enough of it [6].

The results of the research were also published in the Atlas by Ewa Nowak issued in 2000. The advantage behind this publication was prognosticating data until the year 2010; however, that data are also a bit less up to date now [7].

A very popular anthropomorphic atlas which covers the data regarding senior citizens as well is: "The measure of man and woman: human factors in design". Although it is very complex, it is incorrectly used in Poland when designing for the Polish population. In the publication, we can find anthropomorphic data of the elderly (aged 65–79) in America; nevertheless, the data for 80- to 90-year-olds are inaccessible [8].

It should be noticed that we should not design for one population on the basis of anthropomorphic data of another, all the more, if the sizes and the scope might be significantly different. What is more, the difference in anatomy of various races can be crucial too. While the difference between the Polish, Czech or German population will not be that critical, the comparison with Portuguese, Spanish or Italian populations might reveal dissimilarities that could influence the guidelines and assumptions for designing—discrepancies between populations inhabiting different continents might be more vital.

4 Most Important Parameters Which Should Be Taken into Consideration When Designing

Using selected—even adequate—anthropomorphic data for constructing means of transport does not guarantee the right solutions. We should use the holistic approach for this purpose, taking into account the whole infrastructure accompanying means of public and individual transport. We ought to pay special attention to bus/tram stops, parking space, and parking metres by adapting them to physical, mental and perceptual abilities of senior citizens. The analysis of the utility process and the scripts of users' behaviour can be very helpful in this.

4.1 Bus/Tram Stops

The most popular form of a bus/tram stop for people who are waiting for public transport is a shelter with a place to sit on—a bench. The mistakes that often appear here include the wrong height of the seat and the lack of support needed when standing up. The elderly have a slightly bigger problem when standing up, that is why taking this into account, we could use a bit higher seats, which will make the movement easier. According to anthropomorphic data, the correct height of a seat for adult population in Poland should be around 42 cm, while for senior citizens it is 34 cm. However, this assumption is more useful when thinking about a seat for working, at home and under the condition that it has support which facilitates standing up. A seat in public place used for a couple minutes while waiting for a bus/tram the height might be a bit bigger to make it easier to get up from it without the help of other people. Thus, seats of 42–45 cm are appropriate; however, they should definitely be equipped with grips supporting elbows, which will aid standing up if necessary.

The information which includes timetables should be put in a place where it will be easy to come close without standing over somebody sitting or the seat itself. Currently, the timetables are very small and their legibility for senior citizens is not sufficient. To some extent, it is compensated for by the electronic displays installed at some bus/tram stops, which inform people how long they have to wait for a bus/tram. However, they only show information for the next couple or dozen minutes. What is more, for functional reasons the displays are put quite high, so they can be not very visible for the elderly who might have problems with their cervical spine, which make it hard to raise your head high.

Additional information which is located at the bus/tram stops and is visible from a greater distance is the numbers of buses and trams which halt at these particular stops. This makes it easier to have a clear picture of the system of public transport for people who are just starting to get to know the place and the transport itself, but it also helps in big cities, where people cannot know all the districts and lines that go there.

The surroundings of a stop and its location are also extremely important. What is crucial is the height of the pavement from which people get on the bus/tram. In many cities, where it is possible, while renovating the tracks elevated stops are built, which make it easier to enter the vehicles as the height difference between the steps and the pavement is eliminated. This is a very good solution, especially in connection with the low-floor buses and trams. It facilitates getting aboard not only for senior citizens but also for mothers with baby strollers and people on wheelchairs.

It is necessary to work on a legible system for displaying the numbers of buses and trams. Nowadays, the numbers are digitally displayed in the top part of the front windowpane and on the side windowpanes near the entrance. Also, the information about the changes in routes, finishing the course at the bus/tram terminal or about breakdowns is displayed in the same place—the only difference being a red frame around the number. It is necessary to create a better system, which will be clearly visible from afar and adapting the currently used led diodes for this seems to be absolutely feasible. In terms of how the information about the numbers and modifications is displayed, a research should be conducted among the elderly, in order to define the real problems they encounter.

4.2 Means of Public Transport—Buses and Trams—Planning Interior Space

More and more cities have low-floor trams and buses, which indeed make it easier for the elderly to move around. However, the interior solutions in some of these vehicles can be troublesome for senior citizens.

Starting with the systems for opening the doors, which are located in different places in various models of trams/buses—most often on the door or just beside it. Usually, it is a button which lights up in red or green. Unfortunately, the marking of its location is often not sufficient in the context of the colours used on the vehicle (city colours or an advertisement). We should create a protection—display area for such buttons, thanks to which they will be more visible.

The situation is similar with the buttons located inside the buses/trams where we can observe two types of buttons: those that open the door and stop buttons—dedicated to signalling the will to get off on the on-demand stops. Their location and colours are not permanent either, and they are often wrongly used.

Clear marking of the buttons which open the door, especially in situations when a person is running to the bus/tram at the last moment, is important not only for senior citizens, but for all users of public transport. It is a very uncomfortable situation when a person is in a hurry and is unable to get on the bus or tram because s/he cannot locate the button to open the door.

The handles which are used for support during the ride are quite well designed. Most often there are so many of them on various heights that each user can find a place to hold in accordance with his/her needs and abilities. The only difficulty noticed in buses is the big differences between various levels of the floor.

The construction of low-floor buses is often made in a way that some seats are accessible only when using a step. Depending on a model of the vehicle, the step can be as high as 20 cm and its depth can sometimes be insufficient to put the whole foot on it in a stable way. It is a big inconvenience for an older person who wants to sit, but also for the people who are standing and holding onto handles on these seats, as everything is elevated by the height of the step. It is inadvisable to hold your arm above the line of the heart for a longer period of time, as it can cause ischaemia and tingling not only in senior citizens, but also in younger people. The steps which are located inside trams or buses, apart from those which are there for getting on and off the vehicle, basically should not be used at all. However, if it is necessary to use steps which lead to a bus or tram, their height should not be more than 15 cm. Sharp corners of the steps, very often protruding towards circulation areas, can cause injuries and discomfort while travelling.

Thus, we should aim at such construction of the floor that the possible differences in the levels could be eliminated with declines rather than steps. It would also be useful to have wider seats, which take into account the size of senior citizens and for other users a place for putting one's bag or backpack. The height of the seats in the means of public transport is most often correct; nevertheless, it is a bit unfortunate that they are located near circulation areas and people who are standing. It also often happens that the distance between the rows of seats is too small in order for a sitting person's legs to fit in freely, and it is particularly difficult to leave the seat situated under the window.

4.3 *Means of Individual Transport—Passenger Cars—Including Taxis*

A completely different area in the scope of adapting means of transport for the needs of senior citizens are passenger cars used for private use or functioning as taxis. We should make a definite division between them because there is an assumption that the user has influence on the model of a private car when s/he decides to buy it. However, we do not have much influence on the model of the taxi we order—the only limitation is the number of passengers.

"The most optimal" vehicle to function as a taxi is the model of a car which we know from the streets of London. These cars are an example of perfectly applied rules of universal and non-excluding design. Each vehicle is adapted for entering a wheelchair inside, equipped with handles and a mobile ramp. The space inside provides freedom of movement and comfort of use for the passengers.

In private vehicles bought by or for the elderly, we should pay attention to the following parameters: the height and position of the seat—the systems used in SUV, VAN or MPV cars are definitely better solutions, as the position of the driver and passengers is a bit higher and it is easier to get in or get out of such a car. The system for regulating the height of the driver's seat and the position of the driver's as well as the passengers' seats are also crucial. The regulation should be smooth and the button for controlling it should be easily accessible from the seating position.

The construction of such vehicles makes it possible to provide better visibility in front of the car bonnet and in the rear window. Nowadays, more and more car models have various systems for supporting the driver, which can be really useful for senior drivers, e.g. a parking assistant or a 360° camera. The systems for steering with one's voice are better and can also be helpful for the elderly. Such solutions can limit the necessity to turn one's head around so much or move one's hands a lot while parking the car correctly. Moving the buttons for controlling the radio or your telephone (connected to the car with the use of bluetooth) onto the steering wheel turned out to be an extremely good solution. In case of any buttons used for controlling their size is also important—they cannot be too small or too close to one another, as it forces precision and requires focusing of attention, which can have negative influence on the safety of driving. Touch screens, which are more and more popular, can require too much precision and certainly not all senior drivers will learn to use them quickly.

It is essential to put the handle for opening the door or the system for regulating side mirrors on the inside part of the door. The handle should not be placed in a way that would require the user to bend his/her hand a lot when opening the door. The more accessible the steering elements are, without the need to reach out far away, the bigger the comfort of driving and adjusting it for your own needs and abilities.

Another useful and more popular system is the keyless access to the car and the start stop system, once installed in the cars of the highest class, nowadays more often present in city models.

4.4 Parking Space, Parking Meters

Parking space constitutes a big problem in most of the cities, and an attempt to solve it would take a separate text if not a PhD dissertation. However, the form of paying for parking spots in cities can and should be changing. The commonly used parking meters are badly constructed and very complicated. The information is not very clear and using the parking meter is not very intuitive. In many models, the buttons and the place where you get the ticket or put the payment card are placed incorrectly. The topic of a design of a user-friendly parking meter is very broad and requires further research and design works.

5 Conclusions

All the systems for supporting drivers used in private vehicles become more and more common and financially accessible. Many of the solutions start to become standard equipment, which improves safety and comfort of using the vehicles by senior citizens.

In terms of public transport and the surrounding infrastructure, we can notice that the situation in cities is definitely better than in the country, where the bus stop shelters are not adapted, the buses are old and not appropriate for senior passengers—who, in reality, use them most often. Another problem is the low frequence of the buses combined with the lack of seats at the bus stops.

As the current data about populations are harder to access, we design on the basis of old data or using the sizes of products already tried and used. Nevertheless, in connection with big demographic changes and the changes in eating habits which influence the emergence of a larger number of overweight people, which is also a dominant feature in senior citizens, such approach can lead to incorrect solutions. Thus, it seems necessary to conduct anthropomorphic research of the Polish population; however, any population can make use of such data, as it will lead to designing products and services effectively adjusted to its needs in terms of size.

Extended behavioural and perceptive research can also offer a lot of knowledge about how senior citizens see the world and about their expectations, which will make it easier to design for them. We should emphasise the fact of significant improvement in self-esteem of the elderly in a situation when they can be independent, self-sufficient and do not need the help of others.

In order to make an impact and improve the quality of design of means of public transport and the infrastructure, we ought to propagate using the simulator of old age among students of design. It will help them empathise with a senior citizen and function as one (with some conditions) for a moment.

The study was prepared on the basis of observation and analysis of Łódź means of public transport and their surroundings. This is preliminary research conducted for the need of the hereby text in order to check whether it is a topic which should be looked into further. The conclusions are: we should do extended research, even though the current state is not as bad as we suspected.

References

1. Kiljan, M.: Growing old of societies as a challenge for the contemporary world. Praca Socjalna **1**(styczeń-luty), 21–34 (2010)
2. http://swaid.stat.gov.pl/Demografia_dashboards/Raporty_predefiniowane/
3. Thematic Report. Universal Design. Explanation of the Concept. Norwegian Ministry of Environmental Protection, Norway, Nov (2007). http://www.universell-utforming.miljo.no/file_upload/uniutf%20a4%20polsk-eng_v8.pdf
4. Błaszczak, M., Przybylski, Ł.: Rzeczy są dla ludzi. Idea projektowania uniwersalnego, Warszawa (2010)
5. http://dfaeurope.eu/wp-content/uploads/2014/05/stockholm-declaration_english.pdf
6. Jarosz, E.: Dane antropometryczne osób starszych dla potrzeb projektowania, Instytut Wzornictwa Przemysłowego, Prace i Materiały Zeszyt 153 (1998)
7. Nowak, E.: The Anthropomorphic Atlas of the Polish Population. Data for Design Institute of Industrial Design, Warsaw (2000)
8. Tilley, A.R., Dreyfuss, H., Associates: The Measure of Man and Woman: Human Factors in Design. Wiley (2001)

Railway Stations in Creating the Competitive Advantage of Agglomeration Public Transport

Grazyna Chaberek-Karwacka and Marek Chrzanowski

Abstract The share of individual motorization in everyday commuting still increases. This situation is dictated by the fact that alternative ways of urban transportation are still not competitive, especially due to the time and convenience of moving in a "door-to-door" relation. Railways are the fastest means of agglomeration transport; however, very often the time benefits achieved during the travel are lost at interchange points. Thus, one of the key factors is railway stations and stops, which very often are a barrier to pedestrian traffic and do not fully utilize their integration potential in the passenger transport system. The aim of the chapter is to present a new Railways Stations Strategy with a new categorization of railway stations managed by Polish State Railways (PKP SA) and to show the contribution of this strategy to creating the competitiveness of alternative solutions in urban transport in relation to individual motorization.

Keywords Public transport · Urban transport · Interchanges nodes · Railway stations · Interchanges design

1 Introduction

Despite many activities, cities constantly struggle with the problem of congestion, smog and pedestrian safety, caused by excessive individual motorization of residents. The European Union in 2007 pointed to the development of alternative means of urban transport as the way of reducing residents' dependence on their own car [1]. However, previous campaigns undertaken by cities in Poland have not contributed to reducing the number of cars on their streets. A good example illustrating this

G. Chaberek-Karwacka (✉)
Spatial Management Department, University of Gdansk, Gdansk, Poland
e-mail: grazyna.chaberek-karwacka@ug.edu.pl

M. Chrzanowski
Head of research development and innovations unit, Polish State Railways S.A., Warsaw, Poland
e-mail: Marek.Chrzanowski@pkp.pl

M. Suchanek (ed.), *Challenges of Urban Mobility, Transport Companies and Systems*, Springer Proceedings in Business and Economics,
https://doi.org/10.1007/978-3-030-17743-0_12

problem is city of Gdansk, which in recent years has dynamically invested in the development of the tram network. Additionally, the city is strongly committed to the construction of bicycle infrastructure and campaigns promoting cycling commuting. Investments in the development of tram lines in recent years have connected distant districts with the city center but have not influenced on increase usage of public transport in daily commuting. According to the latest Traffic Survey in Gdansk in 2016, the share of individual motorization in daily commuting increased again, with a slight increase in the share of bicycles and the falling number of commuting by public transport [2]. Therefore, information about factors influencing on choosing the means of commuting is important.

The report European Future of Transport [3] is the result of a study on factors discouraging car users to use public transport. Respondents who said they used a car as their main mode of transport were presented with a list of potential reasons that could stop them from using public transport and were asked to rank the importance of each one. A large majority (71%) of car users felt that public transport was not as convenient as a car. A similar proportion (72%) of car users said that a lack of connections stopped them using public transport. A low frequency of services was considered important by 64% of car users, and 54% said the same about a lack of reliability. About half (49%) of car users stressed a lack of information about schedules, and a similar proportion (50%) said that public transport was too expensive. Finally, security concerns were considered as an important reason not to use public transport by 40% of car users. Also, the latest research [4] indicates that the main components of buses and subway service quality are represented by the following aspects: security, reliability, comfort, travel time and waiting conditions. Similar results were obtained in local studies. For example, in long-year research on communication behavior of inhabitants in Gdynia, car users always indicated time and convenience as the most important determinant of car selection in everyday commuting [5]. In the already mentioned Gdansk Traffic Survey from 2016 [2], respondents admitted that the main reasons for choosing a car include: greater convenience (40.40% of respondents), travel time (25.90% of respondents) and the need to transport purchases (20.2% of respondents). Travel time was also indicated as the main factor in the commuting decisions in the study carried out in June 2017 in Gdansk [6]. Unfortunately, along with the urban sprawl, the time which residents spend for commuting constantly increases and more and more burdens their daily time balance. This is a cumbersome, unfavorable and socially unacceptable phenomenon, as various studies have shown that travel time should not exceed half an hour [7, 8].

Travel time and convenience should be considered in the "door-to-door" relation, which means all the times of movement stages should be considered together: time to reach the bus stop, waiting for the vehicle, transit, transfer to the destination. The number of necessary transfers and transfer conditions affect directly the time and convenience of traveling [5, 9–11]. Therefore, in the process of everyday mobility not only development of new routes and travel time and comfort are important, but also the transfer points and passenger service centers. The places of transfer primarily affect the individual satisfaction from the transport process and then influence the future commuting decisions, and when the commuting experience was satisfying, passengers start encourage others to try a given commuting solution [12]. So far, the development of the network of individual means of public transport in Gdansk took place independently. At the same time were developing tram lines and rapid urban rail routes. Linear infrastructure and rolling stock were modernized but without the interchange points infrastructure integration of these two modes of transport.

A survey carried out in October 2017 in Gdansk [13] showed that the fastest connecting journeys with using public transport in this city are journeys using the rapid urban railway. Rapid urban railway is the fastest mean of public transport in the scale of the whole Gdansk agglomeration. Nevertheless, the total travel time is not spectacularly shorter than time of traveling using other modes of transport. Very often, time benefits gained during the travel by urban rail are lost at interchange points, caused by the need to walk long distances, to reach from one to another means of transport and the need to wait for next mode of transport. Such situation is caused by lack of integration between bus, tram and railway stations. Passengers wishing to get to the rapid urban railway platform must travel long distances from bus or tram stops and always have additional barriers such as high stairs.

Elimination of these factors which increase the total travel time requires integration and coordination of various forms of individual and collective transport within the whole agglomeration. First, there are no intermodal nodes, which would enable efficient transfer between various means of transport, particularly lacking nodes that would allow quick and barrier-free transfer to the urban railway platform. Thus, one of the key factors here is railway stations and stops, which very often are barriers to pedestrian traffic and do not fully exploit their integration potential in the passenger transport system. A new railway station strategy adopted by the Polish State Railways (PKP SA) seems a chance for changing the situation.

2 Methodology

The aim of the chapter is to present a new PKP SA Railway Stations Strategy with new categorization of railway stations managed by PKP SA and to show the contribution of this strategy to creating the competitiveness of alternative solutions in agglomeration's transport in relation to individual motorization.

The study is a case study of Polish State Railways SA company, due to the presence of only one entity responsible for the creation and implementation of rail transport policy in Poland. The study uses a comparative analysis of the prepared PKP SA Railway Station Strategy with a model guideline for planning effective intermodal nodes in passenger transport. The study puts forward a thesis that the new Railway Station Strategy meets all postulates currently proposed in the field of proper planning and construction the intermodal passenger nodes in urban areas. Thus, the Railway Station Strategy is part of the sustainable urban mobility strategy in the European Union, including Poland.

Therefore, the research process included two steps:

- Study of the literature and policy documents to collect together current political, urbanistic and social postulates regarding the organization of passenger transport in urban areas—Standards guide for effective interchange nodes
- Indication of entries that are present in the PKP SA Railway Station Strategy, which correspond to the model postulates.

3 Standards Guide for Effective Interchange Nodes

Several studies are aimed at identifying the factors that are crucial for passengers in the interchange nodes and that directly affect the assessment of the entire process of public transport commuting. Studies carried out as part of the European NICHES+ projects: Mainstreaming Urban Transport Innovation [14, 15], are a kind of guide collecting the most important demands that interchange nodes should meet to support the increase in the use of public transport in passenger transport. And according to them, passenger friendly interchanges should:

- provide a great opportunity to use and be familiar with public transport (PT) modes for daily commuters and for tourists as well as for first-time users;
- be safe, well-lit and clean;
- offer accessible, up-to-date information (e.g., timetable, smart guidance) where and when required;
- be basically designed to provide an accessible ("easy to reach, easy to use") environment.

Among these four factors, the most important in the assessment of interchanges or in general the whole travel process is indicated in the first place: information [16], especially in railway transportation [17], and safety and security [12]. The adequate travel information before and during the journey is the main factor to positively influence not only travelers experience but also lead to increasing efficiency of transport services and transport network as a whole [18]. Commission Staff Working Document: "Towards a roadmap for delivering EU-wide multimodal travel information, planning and ticketing services," underlines that "multimodal information allows travelers to make better informed choices by making them aware of all possible travel options, allowing them to make the best choices for their needs (e.g., means of transport, routes, cost, travel time and even environmental impact) and helping them to complete their journey successfully by providing reliable information before and during the trip" [19].

Based on Polish research, including Broniarska and Zakowska [10] conclusion "well-planned and carefully structured interchanges facilitate integration between different modes of transport, allow passengers to shorten traveling time and reduce the effort required to change the means of transport." Hernandez and Monzon [20] also emphasize that travelers make decisions based on perceived walking and waiting times. The changeover time depends on: distance between stops, waiting time for the vehicle, time of transition between stops and transition conditions, such as: the difference in levels between stops, collision with other traffic participants, traffic lights [21]. Additionally, any environmental and design aspects of interchanges nodes should contribute to improving attitudes toward security and information [12]. Moreover, according to Olszewski et al. [22], the process of designing interchange nodes should consider such factors as:

- node compactness, measured by average length and average transition time between platforms in the node,
- the clarity of the node, which considers the assumption that orientation in the node is facilitated by mutual visibility of stops,
- additional equipment, such as roofing of platforms and pedestrian crossings, benches, litter bins, ticket machines, shops, toilets, taxi stands, bicycle stands, car parks,
- the quality of the basic infrastructure, which will be determined by the degree of compliance with the rules and guidelines for the design of public transport stops,
- accessibility for the disabled, expressed by the degree of meeting the accessibility requirements such as inclines or elevators in places of stairs, handrails along the incline, warning pavement tiles with ridges and contrast marking along the edges of platforms and stairs, lowered curbs and warning plates when crossing roadways, and sound signal at passages with traffic lights,
- personal security which goes with sufficient lighting and video monitoring of node elements,
- traffic safety applies to pedestrian crossings by roadways,
- passenger information, including timetables and tariff information at each stop, plan of the node and its surroundings, diagram of network, directional marking on platforms and branches of pedestrian crossings.

Most studies indicate that the interchange must meet all the demands of a very diverse group of travelers. The needs for interchanges are different depending on the purpose of the journey, age, sex and frequency of use of a given transfer node by each passenger [23]. The European vision for passengers: communication on passenger rights in all transport modes [24] underlines also the importance of linking the individual transport with public solutions in multimodal interchanges. The interchanges must contribute to the needs of [14]:

- daily commuters: they want to travel smoothly, reliably and fast, which can be guaranteed by providing smart guidance, and short distances/transfer times between transport modes;
- tourists and first-time users: they require safety, cleanliness, service staff, support for orientation and complementary services. To this end, understandable, accessible multi-language information, high-quality infrastructure and guidance, restaurant, shopping and leisure facilities should be provided;

- elderly and children: they want to travel easily and safe. Therefore "easy-to-reach, easy-to-use" design is crucial as well as avoiding level difference and providing sufficient lighting. Service staff should be available.

From the point of view of daily commuters, a short transfer path is the most important. For tourists, the availability of information (positioning, up to date) is most relevant, while for families, children and older people safety and easy access are crucial. Finally, interchange nodes should also be friendly for mobility impaired people, or those who want to spend their waiting time usefully, before or after traveling [15].

4 Assumptions of the New PKP SA Railway Station Strategy

The first contact of passengers with railways is usually railway stations, which are often referred to the "window to the world" of rail transport. For this reason, PKP SA as the manager of the clear majority of railway stations in Poland undertook the preparation of the Railway Stations Strategy. The main aim of the strategy is to define the tools for managing every category of railway stations allowing to implement the company's strategic goals.

The initial data used to build the Railway Stations Strategy and categorization came from "Qualitative exploration study of station users," which allowed to answer key doubts of passengers' perception of stations and defined desirable services or factors which build the satisfaction of stations' users. The detailed research objectives were aimed in getting to know the stereotypes and experiences of passengers, associated with the railway stations, learning about the behavior of stations' users and, most importantly, identifying expectations of the stations with an in-depth analysis of the differences and similarities between different categories of stations. The key conclusions of the study were:

- for passengers, information security is the most important with the emphasis on passenger information,
- passengers perceive stations not only the building itself, but also its surroundings, platforms, access paths and parking lots, regardless of the entity that manages it (Fig. 1),
- the key areas of passengers' perception at the station are safety, cleanliness and comfort,
- the respondents pointed an appropriate minimal standard of service at railway stations, indicating cash registers, timetables and toilets,
- the commercial offer should be tailored to the type of station, but with the provision of "must have" services such as drinks, food or newspapers.

The study showed that size of the station was defined by respondents in the context of agglomeration size. In addition, passengers recognize type of traffic (e.g., tourist or agglomeration traffic) as well as the location of the station in relation to the

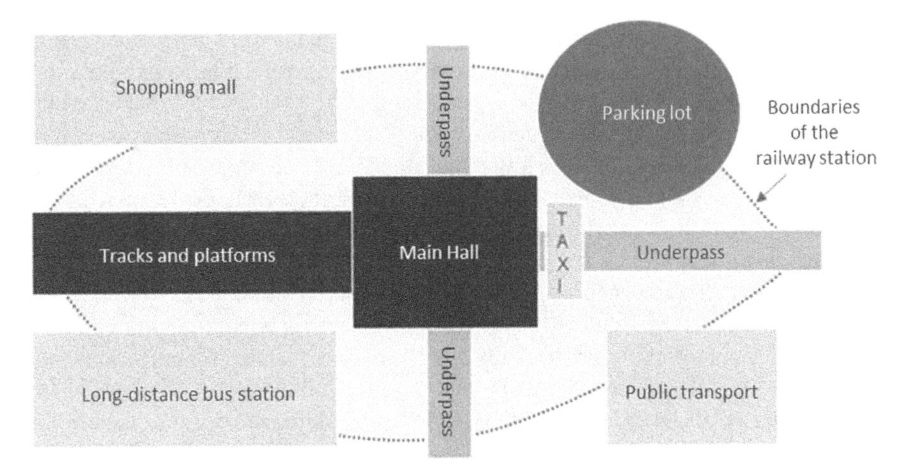

Fig. 1 Perception of the railway stations by users. *Source* Millward Brown [25], s. 36

city and surrounding infrastructure or its technical condition. Expectations toward railway stations among its users can be limited to the needs converging with the Maslow's Pyramid. As it was mentioned above, users defined as important access to passenger information and ensuring cleanliness and security. The basic services include cash desks, toilets, kiosks and catering outlets. For stations located in larger cities, a positive aspect may be wider commercial offer, which, however, has no negative impact on basic passenger services. It should also be emphasized that the facility must be accessible for people with limited mobility, which is in line with the implementation of the TSI-PRM standards [26].

An important element of the Railway Station Strategy is to organize the internal processes in the PKP SA, which will be clearly aimed at ensuring a high standard of services for the passenger. The most important factor which determines a standard of station and differentiates stations among themselves is their categorization. The project is the main element of the strategy implementation, to which other initiatives have been depended. Initiatives within the categorization of stations have been undertaken in the following order.

- Categorization—defining categorization criteria and data collecting
- Standard minimum—defining current standard and defining target standard per category
- Service Level Agreement—assigning a specific SLA to the standard and implementation.

Categorization of railway stations in the management of PKP SA is based on a developed algorithm which, based on objective data, assigns the station to the specific category. The input data of the algorithm are the type of traffic, the number of train stops, the number of carriers and the location of the station on the macroscale (category of territorial unit, population and region) and microscale related to the station location within the territorial unit. Territorial units have been categorized into powiat cities (county town), urban communes, urban–rural communes and rural communes. The location within the agglomeration or region was differentiated by the distance of unit to large urban centers, where the cities with populations of over 250,000 and 100,000 were the starting point. In defining station, location within the city took into consideration whether the station is in the center of the unit, in the residential or industrial areas or outside the center. For such criteria, weights were assigned which were significant determinants in the algorithm.

As the result of the categorization, six categories of railway stations were created. Each category received a defined description of the characteristics and the type of user of stations. Table 1 summarizes the most important features for each category.

A service basket has been built for each category of the station, including the target standard and catalog of services provided at the station. Offered services have to response to specific needs according to the type of the station and the character of passenger. According to Millward Brown report [27], over 80% of people arrive at the station within 30 min before the departure of the train. However, the difference should be noted between persons traveling long distance and those traveling by regional trains. Passengers going on a long trip arrive earlier to the station and thus have different expectations toward services located there. The set range of services at the stations corresponds as well to the conclusions of the qualitative research, which allowed segmentation of services according to priorities:

- "Must have"—services considered critical, which constitute the minimum of services provided at the station. The basket includes essential services that provide basic passenger services. The clear majority of these are services in the field of information security (timetables or voice information) access to the toilet, the possibility of purchasing tickets. Additionally, all the basic maintenance services such as security, cleaning and waiting rooms in a certain standard.
- "Should have"—essential services, which, however, were not considered critical. Set of complementary services, including strictly commercial services such as Wi-Fi, power sockets, newspapers, gastronomy and ATMs.
- "Could have"—services which enrich the station offer, however, are not required for implementation. Often referred to "good to be." The catalog contains strictly commercial services such as a currency exchange, a pharmacy or a florist.
- "Want have"—services, the implementation of which does not determine the success of the categorization, but only addresses additional, one-off needs of passengers. As part of this group, among others are laundry services or the possibility of shopping for jewelry.

The assignment of individual services within the baskets led to the construction of a specific package for each category in accordance with the actual needs of every group of railway station users. As a result, each category has received its unique service basket that adapts to the passenger profile, train traffic or the needs of the local community. As part of specific services, the initial step was to determine the appropriate "Service Level Agreement," which specifies in detail the standard of the

Table 1 Categories of railway stations and their most important features

	Category	Main features
1.	Premium stations	A station in a big city, serving international, inter-voivodeship and regional traffic, an important communication node at the national level, perceived as a showcase of a country or a city, facility offering travelers a wide range of commercial services (gastronomy, press, ATM, available space for children to play, car rental), wide range of passengers/users in terms of: reasons for travel, family situation (travel with children), time pressure
2.	Voivodeship level stations	A station in a big city, serving international, inter-voivodeship and regional traffic, an important communication node at the inter-voivodeship level, perceived as a "visit card" of a city or region, an object offering travelers basic services (gastronomy, press, ATM), a wide range of passengers/users in terms of: reasons for travel, family situation (travel with children), time pressure
3.	Agglomeration stations	A station serving daily local traffic, away from the center of a large city by no more than 50 km, important communication node at the agglomeration level (possible integration with inter-voivodeship transport), favorable conditions for micro location (large enterprises, housing estates), a permanent group of people who travel regularly on the same route—mostly with season tickets, significant importance of the time of stay at the station (minimum, mainly on the platform)
4.	Regional stations	Station in a medium or small town, serving mainly regional and local traffic, an important communication node at the level of a commune or voivodeship, residents using specific connections, the need to adjust the level of services to local transport conditions
5.	Local stations	Railway station with a small potential for the development of rail traffic, which is used by people traveling systematically to larger cities (to work, school), lack of favorable conditions for micro location (large workplaces, schools), people traveling cyclically to larger cities (work, school), minimum standard of services
6.	Tourist stations	Station in a small town or out of town, serving traffic of all types, an important transport hub in the tourist season, a station that can be a showcase of the city, tourist passengers including functions for families with children and the elderly

service provided. As a result, a specific level of service was specified, depending on the category.

The second important aspect of the new Railway Station Strategy is investment task, aimed at increasing the availability of facilities and implementing the adopted categorization. To increase accessibility, it is necessary to act in the field of innovation. In the design process of station buildings, it is important to search for innovation in the field of materials, information and communication technology (ICT), energy and architecture. An example of such solutions is the project of Innovative Stations System (ISS), which offer a high spatial standard with the simultaneous use of ecological and innovative material solutions. ISS meets the requirements set in the categorization of the stations with ensuring the accessibility for people with reduced mobility in accordance with the TSI-PRM standards.

The project has currently been expanded with the concept of a "micro-design," which in smaller locations will allow to provide a minimum standard. The assumption of the investment process in the PKP SA is not only the reconstruction or construction of stations themselves, but also the adjustment of the surroundings to the identified needs. Trends related to the mobility of society have been included in the design tasks of PKP SA. In the case of railway stations, assumed investments are related to building the availability of services for urban mobility. A visible need is the construction of integrated interchange nodes, including the construction of accessibility for Park and Ride solutions. One of the planned directions of redevelopment and construction of railway stations is to use the universal design, which opens to connect the stations with the urban tissue. The main purpose of the reconstruction is to connect stations with the urban environment, which will constitute a platform conducive to every kind of travel. As part of the investment strategy, PKP SA aims to build interchanges centers with other branches of transport, primarily with public transport and regional bus transport. At the same time, the company undertook the idea of building the connecting points with individual forms of travel, such as mentioned already Park and Ride parking lots, parking spaces, easy access for taxis and bicycle racks. The company has also started projects related to sharing economy services. The model aims to provide car-sharing and bike/scooter-sharing services at railway stations, which will be primarily dedicated to rail passengers.

5 Conclusion

Presented assumptions of the new PKP SA Railway Stations Strategy match not only the conclusions of the conducted scientific research on the assessment of interchanges, but also fit into the postulates of European policy represented by Polis. Polis is a network of European cities and regions working together to develop innovative technologies and policies for local transport. The aim of the network is to improve local transport through integrated strategies that address the economic, social and environmental dimensions of transport [28]. Analyzing all the assumptions of the Railway Station Strategy, it can be stated that as a result of new investment activities undertaken by PKP SA railway stations have the potential to become fully functional, efficient and passengers' friendly interchanges. The assumptions of the Railway Station Strategy are in line with the postulates resulting from several scientific studies on factors that influence the assessment and transport decisions of passengers. The strategy assumes meeting all the conditions necessary to improve the assessment of the transfer process and thus to increase the competitiveness of public transport and multimodal travels connected with use of individual transport, based on private cars of agglomeration users as well.

References

1. European Commission: Green paper—towards a new culture for urban mobility (2007)
2. Gdanskie Badanie Ruchu: Report of City of Gdansk. Available from: http://www.gdansk.pl/urzad-miejski/Wyniki-Gdanskiego-Badania-Ruchu-2016,a,65107 (2016)
3. The Gallup Organization for European Commission: Future of transport. Analytical report. Flash Eurobarometer 312 (2011)
4. Guglielmetti Mugion, R., Toni, M., Raharjo, H., Di Pietro, L., Petros Sebathu, S.: Does the service quality of urban public transport enhance sustainable mobility? J. Clean. Prod. **174**, 1566–1587 (2018)
5. Wyszomirski, O. (ed.): Zarządzanie komunikacją miejską. Gdanska Fundacja Kształcenia Menadżerów, Gdansk (1999)
6. Chaberek-Karwacka, G.: Determinants of bicycle-commuting behavior diffusion model—Gdansk case study. Presentation at the 6th EUGEO Congress on the Geography of Europe, Brussels, 4–6 Sept (2017)
7. He, M., Zhao, S., He, M.: Tolerance threshold of commuting time: evidence from Kunming, China. J. Transp. Geogr. **57**, 1–7 (2016)
8. Vincent-Geslina, S., Ravalet, E.: Determinants of extreme commuting. Evidence from Brussels, Geneva and Lyon. J. Transp. Geogr. **54**, 240–247 (2016)
9. Szołtysek, J.: Kreowanie mobilności mieszkańców miast. Wolters Kluwer Polska, Warszawa (2011)
10. Broniarska, Z., Zakowska, L.: Multi-criteria evaluation of public transport interchanges. Transp. Res. Procedia **24**, 25–32 (2017)
11. Kujala, R., Weckström, C., Mladenović, M.N., Saramäki, J.: Travel times and transfers in public transport: comprehensive accessibility analysis based on Pareto-optimal journeys. Comput. Environ. Urban Syst. **67**, 41–54 (2018)
12. Loisa, D., Monzón, A., Hernández, S.: Analysis of satisfaction factors at urban transport interchanges: measuring travellers' attitudes to information, security and waiting. Transp. Policy (in press) (2017)
13. Chaberek-Karwacka, G.: Organizational and infrastructural solutions in creating competitive advantage of sustainable urban transport—Gdansk case study. Presentation at the Conference "Organizowanie Współczesnej Przestrzeni Miejskiej", Warszawa, 24–25 Nov (2017)
14. NICHES: Efficient Planning and Use of Infrastructure and Interchanges. Guidelines for Implementers of Passenger Friendly Interchanges. https://www.polisnetwork.eu/publicdocuments/download/1727/document/21582_policynoteswg2_1_low.pdf
15. NICHES: Innovative Urban Transport Concepts Moving from Theory to Practice. http://www.rupprechtconsult.eu/uploads/tx_rupprecht/NICHES_overview_concepts_EN.pdf
16. Ge, Y., Jabbari, P., Mackenzie, D., Tao, J.: Effects of a public real-time multi-modal transportation information display on travel behavior and attitudes. J. Public Transp. **20**(2), 40–65 (2017)

17. Matsumoto, T., Hidaka, K.: Evaluation the effect of mobile information services for public transportation through the empirical research on commuter trains. Technol. Soc. **43**, 144–158 (2015)
18. Tovey, M., Woodcock, A., Osmond, J. (eds.): Designing Mobility and Transport Services. Routledge, New York (2017)
19. Commission Staff Working Document: Towards a roadmap for delivering EU-wide multimodal travel information, planning and ticketing services, Brussels (2014)
20. Hernández, S., Monzón, A.: Key factors for defining an efficient urban transport interchange: users' perceptions. Cities **50**, 158–167 (2016)
21. Dźwigoń, W.: Analysis of transition times of pedestrians and passengers in an interchange node. Sci. J. Silesian Univ. Technol. Ser. Transp. **92**, 31–40 (2016)
22. Olszewski, P., Krukowska, H., Krukowski, P.: Metodyka oceny wskaźnikowej węzłów przesiadkowych transportu publicznego. Transport miejski i regionalny **6**, 4–9 (2014)
23. Dell'Olio, L., Ibeas, A., Cecín, P., dell'Olio, F.: Willingness to pay for improving service quality in a multimodal area. Transp. Res. Part C Emerg. Technol. **19**(6), 1060–1070 (2011)
24. European Commission: European vision for passengers: communication on passenger rights in all transport modes (2011)
25. Millward Brown: Jakościowe badanie eksploracyjne użytkowników dworców (2014)
26. European Commission: Commission regulation (EU) No. 1300/2014 of 18 November 2014 on the technical specifications for interoperability relating to accessibility of the Union's rail system for persons with disabilities and persons with reduced mobility (2014)
27. Millward Brown: Raport z IV fali Badania satysfakcji z podróży w pociągach PKP Intercity (2015)
28. POLIS: https://www.polisnetwork.eu/about/about-polis

Transport Behaviour in the Context of Shared Mobility

Michał Suchanek, Aleksander Jagiełło and Marcin Wołek

Abstract The article presents the results of the research on the consumer profile as a key factor affecting the willingness to use the shared mobility services such as carpooling or bikesharing. A research based on 265 respondents shows that the customers who are participating in shared mobility have a different customer profile than those who are using the traditional transport services. The Scale of Customer Motivations (SMK) was used to analyse the customer profile, and then, the Kruskal–Wallis ANOVA was used to verify the differences of the profiles of people who use carpooling as a driver, carpooling as a passenger and bikesharing services.

Keywords Sharing economy · Shared mobility · Bikesharing · Transport behaviour

1 Transport Behaviour as a Type of Consumer Behaviour

The research on transport behaviour, a specific type of consumer behaviour, is focused on the determinants of how the passengers evaluate their needs and choose goods (e.g. passenger cars) and services (e.g. railway transportation) which allow to satisfy those needs [1]. The specifics of transport behaviour are a result of the specifics of transport needs. Transport needs result from the desire to satisfy primary needs of lower and higher tiers. Transport behaviour is the observable effect of transport decisions, thus being a vital part of two decision-making processes. The first process focuses on the wish to satisfy or abstain from satisfying the primal need for relocation. The second process focuses on the choice of the relocation mode [2].

M. Suchanek (✉) · A. Jagiełło · M. Wołek
Faculty of Economics, University of Gdansk, Sopot, Poland
e-mail: m.suchanek@ug.edu.pl

A. Jagiełło
e-mail: aleksander_jagiello@wp.pl

M. Wołek
e-mail: mwol@wp.pl

M. Suchanek (ed.), *Challenges of Urban Mobility, Transport Companies and Systems*, Springer Proceedings in Business and Economics,
https://doi.org/10.1007/978-3-030-17743-0_13

There is such a variety of factors affecting transport behaviour that the researchers divide them into groups. Therefore, transport behaviour is analysed as a consequence of changes in the following groups of factors [3]:

- urban,
- socio-demographic,
- psychological,
- economic.

Thus, the transport behaviour is affected, among others, by [4]:

- number of journeys,
- purpose of journeys,
- number of cars in the household,
- lifestyle,
- place of residence,
- price of public transport,
- private car expenses,
- quality and accessibility of transport infrastructure,
- duration of travel by different modes of transport.

The research indicates that the transport behaviour is characterized by low variability in the short run. In the long run (a time horizon of 5–10 years), along with the changes in primary needs and in the behaviour of the whole society, the transport behaviour changes as well.

The research on carpooling as a mode choice and the connected transport behaviour was initiated during the WWII when the shortages of fuel and rubber forced the US citizens to reduce their transport needs and to share their private vehicles [5]. Further research allowed to identify the factors which are correlated with the changes in carpooling popularity and thus the changes in the share of car journeys in which only the drivers are travelling in the car. The factors which negatively affect the carpooling popularity are [6]:

- an increase in the motorization ratio,
- a decrease in the number of people in one household (which decreases the possibility of using carpooling to satisfy the transport needs of one household—so-called fampooling),
- a decrease in fuel prices and an increased energy efficiency of passenger cars,
- suburbanization which leads to a more dispersed building structure and journey destinations (including the location of the companies).

The factors which affect the transport behaviour and result in an increased desire to use carpooling are among others:

- a desire to economize, by dividing the transport costs between more people who use the same mode of transport [7, 8],
- wanting to save time by decreasing congestion caused by a high share of journeys by passenger cars with only the driver in the car [9],

Table 1 Household expenditure by consumption purpose in EU-28, share of total (%)

	2007	2008	2009	2010	2011	2012	2013	2014	2015	2016
Housing	22.8	23.3	24.2	24.3	24.8	25.1	24.8	24.8	24.7	24.5
Transport	13.4	13.3	12.8	12.8	13.1	13.0	12.9	13.0	12.8	12.9
Food	12.2	12.1	11.3	11.3	11.3	11.2	11.2	11.5	11.5	11.5

Source Eurostat [12]

- a care for the natural environment and the desire to reduce external costs of transport [7, 9],
- a dislike for solitary travel and a satisfaction from joint travel with other passengers [10].

Despite the changes in transport behaviour in the last years (mostly the shift from public transport onto the cars [11]), the share of household transport expenditure has remained relatively stable. According to the 2016 Eurostat data, the transport expenditures had a share of 12.9% of all household expenditures in the European Union (EU-28). Therefore, transport expenditures were the second largest category of the expenditures, right after the expenditures connected directly with the household needs (24.5%). Furthermore, the expenditures on transport are comparable with the expenditure on food and non-alcoholic beverages (12.2%). The share of transport expenditures in the years 2007–2016 was relatively stable—it assumed values between 12.8% (in 2015) and 13.4% (in 2007) (Table 1).

The changes in the transport behaviour in Europe can be seen as a change in the motorization ratio. Its value affects the modal split, i.e. the share of all the journeys done by car, public transport as well as the journeys with the use of shared mobility services. In the years 2000–2015, there has been a steady increase in the number of passenger cars per 1000 inhabitants in the EU-28. This trend led to an increase in the number of cars per 1000 inhabitants from 410 in 2000 to 499 in 2015 (Fig. 1). At the same time, the number of city buses and coaches per 1000 inhabitants remained at a similar level of 1.6 vehicles.

However, apart from the objective factors, the transport decisions which result in a given transport behaviour are also affected by the non-economic factors, such as the customer profile [14]. The goal of the article is to prove that the customers who are willing to use shared mobility in order to satisfy their transport needs have a different customer profile than the customers who choose traditional transport modes.

2 Materials and Methods

The goal of the article resulted in a research design which included the preparation and use of a questionnaire which consisted of 34 questions, 24 of which were adapted from the Customer Motivation Scale [15] by Poraj-Weder and Maison. The remaining

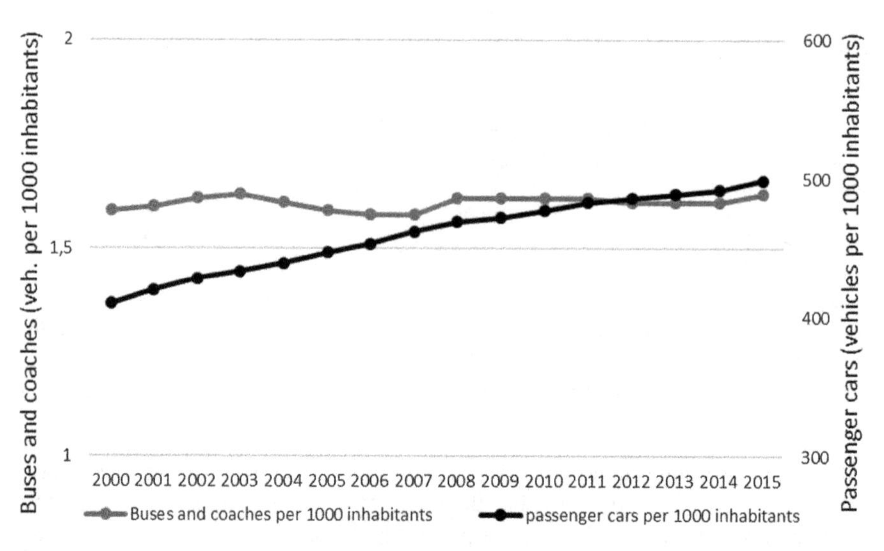

Fig. 1 Passenger cars, buses and coaches per 1000 inhabitants in EU-28. *Source* European Environment Agency [13]

Table 2 Eigenvalues above 1.0 resulting from the principal component analysis

Value	Eigenvalues (shared mobility)			
	Eigenvalue	% of total variance	Accumulated eigenvalue	Accumulated %
1	4.432576	18.46907	4.43258	18.46907
2	2.720606	11.33586	7.15318	29.80493
3	2.693500	11.22292	9.84668	41.02785
4	2.160806	9.00336	12.00749	50.03120
5	1.111790	4.63246	13.11928	54.66366

Source Own estimation

questions focused on whether the customers used shared mobility services and how often. The questionnaire was distributed using the snowball method [16] starting within the group of students of a University in the Tricity area in the north of Poland. A total of 265 observations were collected.

Initially, a principal component analysis was performed on the data regarding the customer profile and a scree test was performed to assess how many factors should be analysed (Fig. 2).

Based on the results of the test, five factors with eigenvalues above 1.0 were included in further research. In total, they determine 55% of all the variability of customer decisions (Table 2).

Factor loadings were analysed for the five factors which allowed to identify five characteristic latent attitudes which form the customer profile in this case. In order

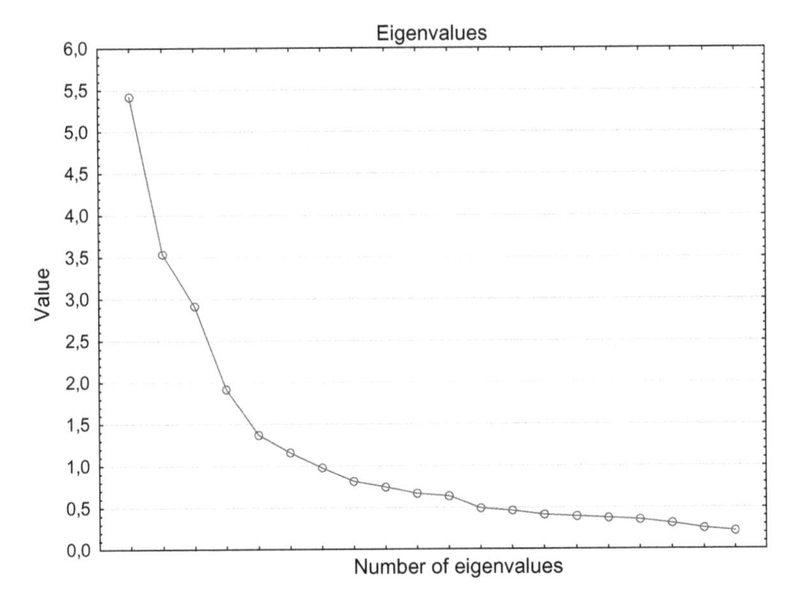

Fig. 2 Scree test for the customer profile questions. *Source* Own estimation

to achieve a result which is more easily analysed, a raw biquartimax rotation was applied (Table 3).

3 Discussion and Conclusions

The first factor which forms the customer profile represents the latent attitude of making customer decisions based on the expected reactions of the environment. The questions which indicate whether the customer wants to be noticed through his customer decisions had the highest loadings in the first factor. This factor represents the intensity of the egocentric attitude of a customer.

The second factor determines the latent attitude of making customer decisions in order to improve the well-being. Buying things in order to improve to comfort oneself, to lift the spirit or to take the mind off of current issues is characteristic for this attitude, and those are also the questions which had the highest factor loadings in the second factor, which represents the strength of recreational attitude in a customer.

The third factor is connected with the approach to the consequences of the customer decisions. In this case, the leading characteristic is the focus on brands which are already known and tested by the customer. It shows the strength of routine as a decision-making factor.

The fourth factor within the customer profile reflects the level of focus on the needs of others during the decision-making process. Questions reflecting the importance

Table 3 Factor loadings—customer profile—biquartimax rotation

Variable	Factor loadings—biquartimax rotation				
	Factor 1	Factor 2	Factor 3	Factor 4	Factor 5
I like when my things attract the attention of other people	0.896450	0.161210	0.031192	0.131259	0.143444
I like when my things grab attention	0.875801	0.144066	−0.036875	0.090111	0.218301
I like when my things set me apart	0.684863	−0.047257	−0.153467	0.079199	0.362317
I like when my things emphasize my position	0.455604	0.072955	−0.003892	0.015072	0.581095
I like the brands that I own to emphasize my position	0.272336	0.132618	0.059576	−0.025559	0.758093
I choose the brands which allow me to feel special	0.215773	0.057719	0.038368	0.016388	0.694724
I like to have things not many people can afford	0.239244	0.052897	0.112721	−0.051654	0.759146
I choose the brands which emphasize my individuality	0.337604	−0.015119	−0.015306	0.100106	0.420803
I choose reliable products	0.024616	0.028749	0.850171	0.087927	−0.014838
I don't take risks. I choose products which are tested	−0.075584	−0.032544	0.882472	0.106406	0.042950
I try to choose the same, reliable brands	−0.030675	0.047950	0.766062	0.087650	0.080212
I always choose the reliable things in life	0.005014	−0.056588	0.683496	0.063563	0.115240
I like to buy things which will make my close ones happy	0.155738	0.111641	0.160822	0.689890	0.047523
I like to buy things with my close ones in mind	0.090305	0.092006	0.061710	0.847718	0.045450
When choosing products and brands, I try to take into account the needs of my relatives	0.108010	0.135025	0.139536	0.789175	−0.014094

(continued)

Table 3 (continued)

Variable	Factor loadings—biquartimax rotation				
	Factor 1	Factor 2	Factor 3	Factor 4	Factor 5
I enjoy buying things for other people more than for myself	−0.006520	0.043262	0.056373	0.572502	−0.109998
I choose products and brands which I can share with other people	0.119300	0.084324	0.079584	0.616200	−0.099773
I sometimes buy things to feel better	0.177136	0.765218	0.044306	0.119388	0.076113
When I feel bad I buy things to feel better	0.104191	0.947803	−0.009786	0.130741	0.051718
I sometimes buy things which take my mind away from current issues	0.108043	0.648404	−0.047584	0.054364	0.057200
If something makes me happy, I don't care how much it costs	0.145971	0.176882	−0.033723	−0.087505	0.309687
I want most of my things to be of the highest quality possible	0.131240	−0.125314	0.090297	0.077711	0.464814
When I buy things which make me happy, I don't regret the money I've spent	0.143403	−0.009226	0.220970	−0.068778	0.154839
I appreciate the quality guaranteed by reliable brands	0.128270	−0.041244	0.191437	0.053537	0.374639
Share	0.875801	0.144066	−0.036875	0.090111	0.218301

of other people and their needs have the highest factor loadings in this factor. It represents the strength of empathic attitude in the customer.

The last factor which determines the customer profile shows the latent attitude of self-appreciation as a result of customer decisions. The questions regarding the ability to own luxurious, unique goods have the highest loadings in this factor. It represents the strength of the hedonic attitude.

The differences in the levels of these five attitudes represented by the factors were verified using the Kruskal–Wallis ANOVA. This nonparametric method was used due to a distribution of the data which was not normal. The results in the strength of the factors were analysed three times, for the groups of people who use or do not use the shared mobility services: carpooling as a driver, carpooling as a passenger, bikesharing. Significance level of 0.05 was assumed.

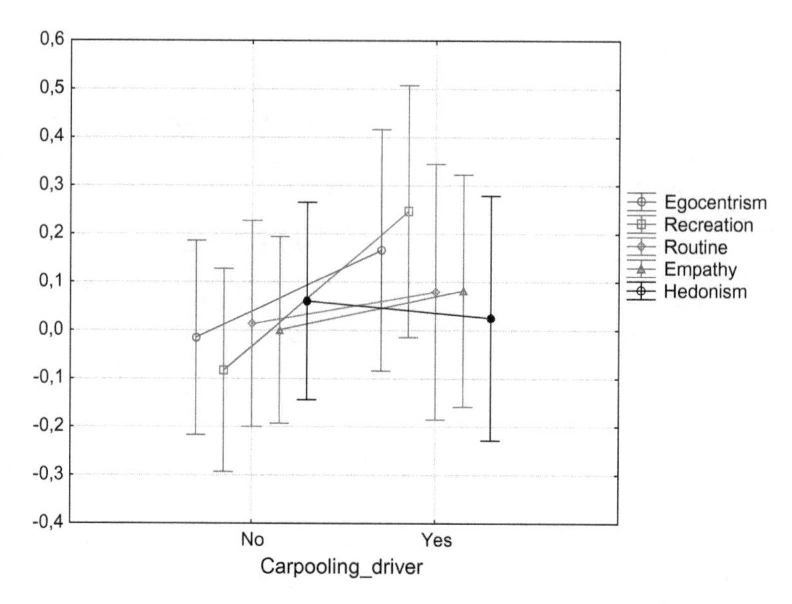

Fig. 3 Differences in attitudes of people who use carpooling as a driver. *Source* Own estimation

Customers who use carpooling as a driver are characterized by a stronger recreational and egocentrical attitude. The differences in other attitudes are not statistically significant (Fig. 3).

The customers who use carpooling as a passenger have a statistically weaker hedonist attitude. The differences for other attitudes are not statistically significant (Fig. 4).

The customers who use bikesharing services are characterized by a statistically stronger empathic attitude, while also being more egocentric and hedonist (Fig. 5).

There is an observable difference in the customer profile of people who use sharing mobility services and people who do not. It can be concluded that the people who use them tend to be more open to new ideas, which is reflected by their profile. However, shared mobility services can be a potential substitute for the increased car use in urban areas. Thus, they should be publicized in a way which is attractive to a wider array of customers, not only those with a very specific customer profile. Until they do, it is likely that they will remain a niche and their share in the modal split will remain relatively low.

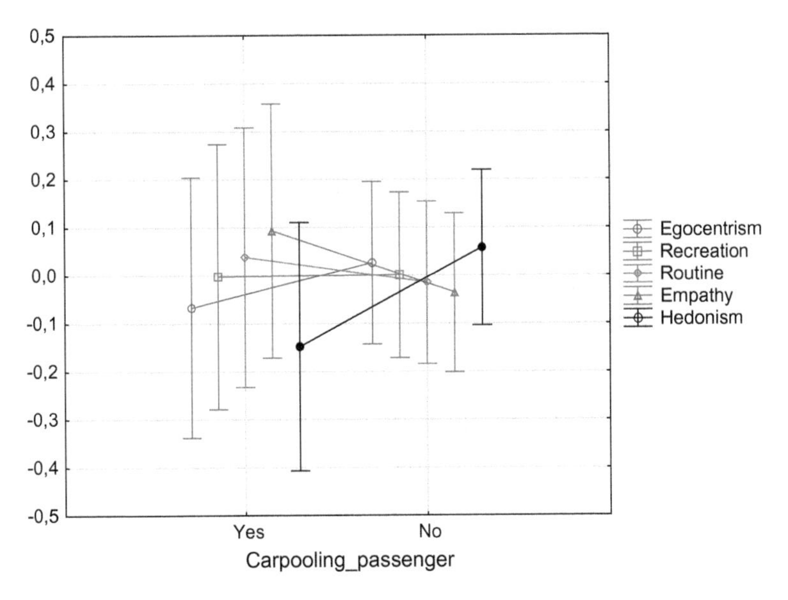

Fig. 4 Differences in attitudes of people who use carpooling as a passenger. *Source* Own estimation

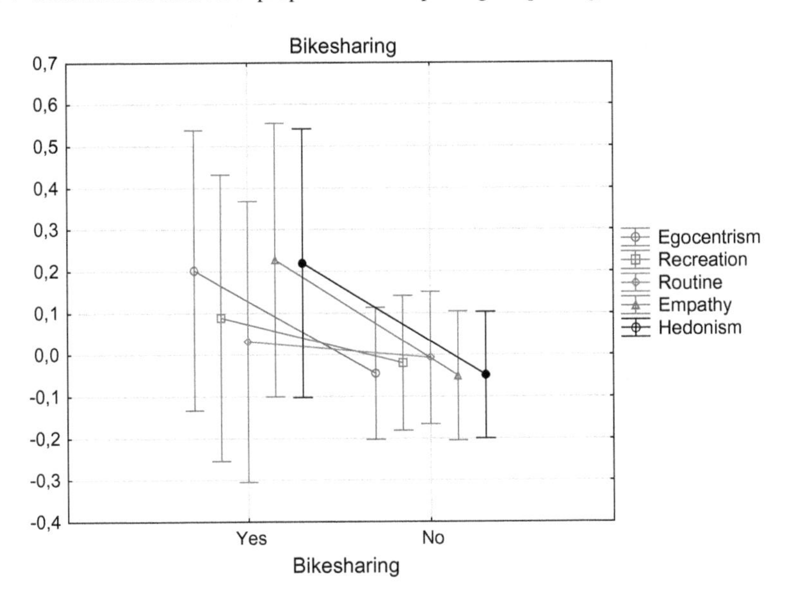

Fig. 5 Differences in attitudes of people who use carpooling as a driver. *Source* Own estimation

References

1. Hebel, K.: Transport behaviours as a type of consumer behaviour. Res. J. Univ. Gdańsk Transp. Econ. Logist. **41**, 167 (2011)
2. Hebel, K.: Zachowania transportowe mieszkańców w kształtowaniu transportu miejskiego. Fundacja Rozwoju Uniwersytetu Gdańskiego, pp. 32 (2013)
3. Curtis, C., Perkins, T.: Travel behaviour: a review of recent literature, impacts of transit led development in a new rail corridor working paper No. 3, pp. 7–17 (2006)
4. Goodwin, P.: Policy incentives to change behaviour in passenger transport. In: OECD/International Transport Forum on Transport and Energy: The Challenge of Climate Change. Leipzig, Germany, pp. 2 (2008)
5. Ferguson, E.: The rise and fall of the American carpool: 1970–1990. Transportation **24**, 349–376 (1997)
6. Watts, R.: Increasing carpooling in vermont: opportunities and obstacles. A report by the University of Vermont Transportation Research Center, pp. 4 (2010)
7. Benkler, Y.: Sharing nicely. on shareable goods and the emergence of sharing as a modality of economic production. Yale L. Rev. **114**, 281–285 (2004)
8. Rose, G.: Providing premium carpool parking using a low-tech ITS initiative. Inst. Transp. Eng. J. **71**(7), 32 (2002)
9. Collura, J.: Evaluating ride-sharing programs: Massachusetts' experience. J. Urban Plan. Dev. **120**(1), 32–34 (1994)
10. Li, J., Embry, P., Mattingly, S.P., Sadabadi, K.F., Rasmidatta, I., Burris, M.W.: Who chooses to carpool and why? Examination of Texas carpoolers. Transp. Res. Rec. J. Transp. Res. Board **2021**, 113 (2014)
11. Hebel, K., Wolek, M.: Perception of modes of public transport compared to travel behaviour of urban inhabitants in light of market research. Sci. J. Silesian Univ. Technol. Ser. Transp. **92**, 65–75 (2016)
12. Eurostat. http://ec.europa.eu/eurostat/statisticsexplained/images/c/c6/Table_2_Evolution_COICOP_Total.png
13. European Environment Agency. https://www.eea.europa.eu/data-and-maps/indicators/size-of-the-vehicle-fleet/size-of-the-vehicle-fleet-8#tab-used-in-publications
14. Suchanek, M., Pawłowska, J.: Effects of transport behaviour on public health: a study on the students in the Tricity area. In: Suchanek, M. (ed.) New Research Trends in Transport Sustainability and Innovation. TranSopot 2017. Springer Proceedings in Business and Economics. Springer, Cham (2018)
15. Poraj-Weder, M., Maison, D.: Poszukiwanie uniwersalnego modelu motywacji konsumenckich. Psychologia Społeczna **10**(1) (2015)
16. Voicu, M.C., Babonea, A.: Using the snowball method in marketing research on hidden populations. Challenges Knowl. Soc. **1**, 1341–1351 (2011)

Part II
The Operational and Strategic Challenges of Transport Companies

The Determinants of the Internal Audit in Road Cargo Transport Enterprises

Andrzej Letkiewicz and Ewelina Mandera

Abstract Audit is treated as something obligatory and appropriate for the public finance and large organisations. By definition, the range of audit and the character of auditing activity, it can be pointed out that, in particular concerning the financial, compliance and operational audit, many actions which are preparatory and auditory in character can be and are carried out in road cargo transport enterprises, even if they are not perceived as audit. In the case of financial audit, many entrepreneurs carry out actions specific to audit, by measuring the profitability of sales, assets and equity so as to make better decisions. During the compliance verification, the compliance audit actions are carried out, through the verification of the compliance of activities regarding the bookkeeping and employing, including the verification of the work time of the drivers. Operational audit can be based on many measures describing the effectiveness of the transport performance, which at minimum should include the use of two: the displacement use indicator and the dynamic payload use indicator

Keywords Internal audit · Transport enterprises

1 Introduction

Audit and auditors activity are often associated by the entrepreneurs as a complex combination of activities characteristic and obligatory for the public finance and large entities, in which it is a part of the corporate governance system. It is also associated with high costs connected with the necessity of the use of external entity which performs audits or a necessity of hiring an employee as an internal auditor, which in the case road cargo transport can lead to unprofitability. However, a closer

A. Letkiewicz (✉) · E. Mandera
Chair of Economics and Management of Transportation Companies, Faculty of Economics, University of Gdansk, Sopot, Poland
e-mail: ekoalt@univ.gda.pl

E. Mandera
e-mail: e.mandera@ug.edu.pl

M. Suchanek (ed.), *Challenges of Urban Mobility, Transport Companies and Systems*, Springer Proceedings in Business and Economics,
https://doi.org/10.1007/978-3-030-17743-0_14

look at the goals and determinants of the audit's definition leads to a conclusion that these activities, especially in the internal and operational dimensions of the road transport enterprises, which are mostly small, are constantly carried out as a part of the strive to increase efficiency and effectiveness of economic activity. Thus, they do not require any additional activity but merely the use of proper instruments.

The thesis of the article is: it is relatively easy to build an internal audit system in small- and medium-sized road transport enterprises, based on their potential.

Therefore, the goal of the article is to point out the actions characteristic to the internal audit carried out in road transport enterprises and therefore to indicate the possibility of organising the functionality of the small enterprises based on the audit principles. As a consequence, the article presents indicators which can be used in the enterprises and their levels resulting from research on road transport enterprises carried out in 2017.

2 The Essence and Scope of Audit in Enterprises

The word audit comes from the latin *audire* which stands for interrogating and researching. In the Middle Ages, an auditor was a courthouse member who used to run the investigation. In Poland, since the 1990s, the term audit has started to be associated with the verification of the financial reports [13, p. 15]. There are three main objectives of performing an audit: verifying whether an entity meets the conditions resulting from the legal regulations and the governance norms; verifying whether the expectations and needs of various enterprise' stakeholders are identified and met; assessing the effectiveness of company fulfilling the objectives and allocating the resources [2, p. 100].

This set of goals brings about the conclusion that it is possible to fulfil them continuously or periodically. This leads to a first division of audit into the external and the internal audit. The external audit is carried out by external entities, which are not connected with the audited entity and is therefore carried out when necessary. Such an audit is mostly concentrated on the financial position of the enterprise [13, p. 15]. In Poland, based on the law on statutory auditors, auditor firms and corporate governance, the external financial audit is carried out by statutory auditors. According to the accounting law, some entities are obligated to undergo a financial audit of the financial report. These include, among others, banks, insurance companies, cooperative banks, share issuers and other entities which meet two out of three conditions posted in the law: workforce of over 50 employees, value of assets of over 2.5 mln EUR and sales revenue of over 5 mln EUR [14, art. 64, pkt 1].

Internal audit is performed by a special division in the enterprise or an auditor who is not employed, based on an appropriate contract of auditing services [13, p. 16]. According to the definition of The Institute of Internal Auditors, "internal auditing is an independent, objective assurance and consulting activity designed to add value and improve an organisation's operations. It helps an organisation accomplish its objectives by bringing a systematic, disciplined approach to evaluate and improve the effectiveness of risk management, control, and governance processes" [3, online: 13, March, 2018]. In the case of small and micro-entities, the situation is a bit different, because the internal audit actions, due to a lack of funding, are often inscribed in the functional structures and there is no separate division responsible. They are often treated to be the same as control activities which verify various areas important from the point of view of effectiveness and efficiency.

The comparison between the internal and the external audit shows that the biggest difference is in regard to the objective of the audit. In the case of the external audit, it is to provide reliable information on the asset and financial position to the external entities involved. The internal audit, on the other hand, is supposed to provide an added value for the entity and is often focused on all the areas of the economic activity, not only on the finances. There are also differences in regard to the frequency and the length of the audited period. The external audit is carried out once a year and it focuses on the past events, whereas the internal audit is carried out throughout the whole year and is usually concentrated on current events [13, p. 16].

The most important functions of the internal audit are allocating the resources in the priority areas, researching the priority areas, formulating the conclusions based on reliable and sufficient research proofs, cooperating with the personnel and management in order to maintain an effective internal control [16, p. 45].

The assessment of the entity in which the internal audit is carried out has to be based on the following criteria: legal, economic, purposefulness, reliability, credibility, transparency [13, p. 17]. Despite the fact that the auditor may be employed in the entity—and he usually is—in order for the audit to be purposeful, the auditor has to be provided with a large independence level and information access. If the auditor is not a separate post and some control actions are carried out along the way, this independence is naturally lower, but this is specific to a small enterprise.

Depending on the needs, the audit can be complex or it can focus on a part of the activity. The levels of the complexity can be divided into product/service, process, system [10, pp. 38–40].

Audit of the product/service focuses on whether the product or the service meets the technical requirements and established procedures. The process audit verifies whether the standards of one or many processes are complied with. The audit of the system is concentrated on an enterprise as a whole. Norms, rules and requirements set in the enterprise are verified. In the enterprise system, only proper interactions between these parts allow an efficient activity and a fulfilment of the entity's objectives.

Methodologically speaking, the areas which can be subject to a separate process audit are as follows [9, p. 139]:

- finances—financial audit;
- compliance—compliance audit;
- main activity—operational audit;
- management process—management audit;
- information gathering and transforming—information audit;
- risk—risk audit.

The financial audit focuses on the analysis of the appropriateness of the enterprise financial reports. It focuses among others on the procedures of analysing the accounting proofs, the analysis of the accounting policy and the organisation of accounting system, the analysis of the financial reports and the analysis of the solvency and liquidity of the entity [1, p. 13].

The compliance audit is focused on the verification whether the activity of entity is compliant with the procedures and laws. It mostly focuses on the verification of law compliance of the contracts; compliance with the general rules such as the employment law, the civic code or the company's code; the compliance with the reporting terms [12, p. 53].

The operational audit is focused on the analysis of the savings, efficiency and effectiveness of the systems and organisational units of a given entity. Its goal is to assess the way of planning and controlling the operational activity by the board. It does not only focus on the verification of the formal way of making the decision, but also assess its rationality [5, p. 21].

The management audit is a specific type of audit because it is mostly distinguished by the assessment perspective, not the subject of the assessment itself. It means that the whole activity of the enterprise from the point of view of the manager is analysed [15, p. 22].

The information audit is supposed to assess the effectiveness and the objectivity of the information systems in the enterprise. It should mostly focus on the reliability of the information collected from external sources, the transformation and the flow of information between the departments of the company.

The risk audit is focused on the risk management in the entity. The risk management process is focused on the risk identification and reacting to it [4, p. 145]. The goal of the risk audit is therefore to verify the effectiveness of the risk management system in the enterprise. However, one has to remember that risk should also be assessed in all other types of audit.

3 Internal Audit in Road Transport Enterprises

Road transport enterprise as an economic entity is meant to efficiently carry out the objective set by the owner, that is to multiply its value, but also to carry out secondary objectives resulting from the economic and operational construction of the entity. As shown in the previous part of the article, the areas which should be specifically audited internally are finances, compliance, main activity, management process, gathering and processing of information and risk [9, p. 139].

Internally auditing every one of these areas has different secondary goals than the main goal focusing on the effectiveness and efficiency of the enterprise goals. Different types of audit use a specific set of indicators of characteristics resulting from the secondary goal of the enterprise and from the effectiveness of the processes which lead to the fulfilment of these goals [9, p. 140]. The goals of the different types of internal audit in a road transport enterprise are presented in Table 1.

Table 1 Areas and goals of the internal audit in a road transport enterprise

Internal audit area	Partial goal
Financial audit	Reliability and clear picture of the enterprise financial position
Compliance audit	Compliance of the actions with the formal and legal requirements
Operational audit	The effectiveness and efficiency of the resources used
Management audit	The assessment of the management control mechanism
Information audit	Effectiveness and objectiveness of the information systems
Risk audit	Assessment of the risk management system procedures

Source Reference [9, p. 140]

Based on the areas of the internal audit presented in Table 1 and due to a large capital disruption of the Polish road transport sector, one has to conclude that in reality, three first audit areas are particularly important: financial, compliance and operational audit. Such a focus of the auditing actions results from the needs of the stakeholders, but also the information needs of the enterprise owner if the management is separated from the ownership.

The financial audit, in the given combination of internal audit stakeholders, is focused on identifying and verifying the degree of effectiveness and efficiency of the main objective of the transport enterprise with the use of profitability indicator, that is the net profit. The clear and reliable presentation of the entity's financial situation, which is the goal of the financial audit, can be achieved through the verification of the procedures used to register the economic events. Due to the bookkeeping character, which is formalised by the accounting law and by tax regulations, the audit verification boils down to the second area of the audit, which is the compliance audit. However, in the effectiveness dimension, it is necessary to use the profitability indicator which brings together the net profit with other economic categories. The profitability indicator, which is the measurement of the effect presents the bottom line of the economic activity in a given period and thus requires a combination of values which allow to make it objective. The spectrum of ratios which can be used during audit is very wide, and it allows to assess various areas of the enterprise. In practice, only a few are used. In a transport enterprise, the ratios are made objective with categories such as sales revenue, assets and owner's equity [8, p. 290]. It should be stated that due to the fact that the accounting methods and regulations regarding the audit are common and well known, they leave no doubt regarding the method. The usability and efficiency of the audit are strategically important to the decisions regarding the economic use of the equity in the activity focused on transport with the use of motor vehicles, in general or within one segment of this market. The results of

Table 2 Basic economic indicators of the road transport enterprises

Specification	Value
Average yearly revenue	2,635,548.00
Average yearly cost	2,532,593.80
Average yearly financial result—profit	86,516.13
Owner's equity	82,806.45
Assets	489,838.71
Return on sale (%)	3.28
Return on assets (%)	17.66
Return on equity (%)	104.47

Source Own estimation based on an empirical analysis of 62 randomly sampled enterprises

the research on the condition of road cargo transport enterprises in Poland in 2017 regarding the average profitability ratios are shown in Table 2.

The second area of the audit is the compliance audit. Within this area, the road cargo transport companies are regulated by many acts of law such as the law on the freedom of economic activity, the trade code, the civic code, transportation law, the labour code, road law. Therefore, they are not necessarily distinct from other enterprises in the transport sector. They are, however, also regulated by other laws dedicated to such enterprises. In particular, one has to mention the road transport law and the driver's labour law. The road transport law sets out the rules of carrying out the national and international transport. It points out that in order to perform haulage, one has to obtain a certain permit or licence, which is regulated by the regulations of the European Parliament. The driver's labour time law regulates the maximal times of driving for the drivers, entrepreneurs and other employees, who carry out the haulage and the obligations of the employer regarding them [11, p. 20].

As mentioned previously, the compliance audit can boil down to the verification of whether certain regulations are met, such as the regulations on the profession accessibility which are defined in the road transport law or the compliance with the accounting law. In general, the compliance of the financial reporting system is analysed by external auditors; however, the internal audit should not be disregarder here, as it is especially important in these enterprises which are not assessed by a statutory auditor.

There is also one area which is left for the transport enterprises to carry out which is verified by the external control. It is connected with the law on hiring employees in general, i.e. the labour code, which covers all the employer--employee relations as well as the driver's time law which makes clear the operational area connected with the haulage.[1] The law on road transport defines the sanctions for breaking the

[1]The road transport law regulates the technical condition of the vehicles. The driver is obliged to monitor the vehicles' technical condition. This leads to a necessity for the internal audit in the transport enterprises due to the possible sanctions in the form of withdrawing the car proof of register and banning the vehicle from traffic.

Table 3 Distance use indicator

Percentage of responses	Distance use indicator (B)
45.2	0.90–1.00
19.4	0.50–0.59
11.3	0.80–0.89
11.3	0.70–0.79
6.5	0.6–0.69
6.5	do 0.49

Source Own estimation based on an empirical analysis of 62 randomly sampled enterprises

regulations, which vary from 50 PLN in the case of small offences to 10,000 PLN (there is a list of possible offences with the corresponding sanctions within the act); however, the maximal penalty imposed on the company during a control is dependent on the number of hired drivers within the 6 months before the controle and ranges from 15.000 PLN for entities with no more than 10 workers to 30.000 PLN for entities hiring more than 250 drivers. This law also makes it possible to ban an enterprise from performing haulage.

The operational audit which is focused on the efficient use of resources is strictly connected with the performance on the contracts generated on the market. Due to the efficiency determinants of the operational activity of transport enterprises, the most important values which are assessed within the internal audit are the mileage and the payload use.

The mileage of the vehicles which are used in transportation allows the assessment of the haulage process and the division into loaded distance and idle distance. The carriers want to maximise the revenue. This means that if they sell to loaded distance, they should aim for a situation in which all of the distance is carried with a load. The indicator which shows the efficient use of stock is the distance use indicator (B = loaded distance/total distance) [7, p. 260]. The performed research has shown that in most of the enterprises, the indicator B has a value between 0.9 and 1.0 (more than 45% of the carriers—Table 3).

Furthermore, the operational audit allows to assess the adaptation of the stock to the carried out transport contracts. The level of this adaptation is verified by the dynamic ratio of the payload use. The dynamic payload use ratio (E) indicates the relation between the performance and the possible performance if the payload was fully used. The unused payload is a lost revenue for the enterprise [6, p. 71]. The research on the road transport enterprises shows that in most of the enterprises (42% of all the answers), the E indicator has a value of between 0.81 and 0.90 (Table 4).

It is worth noting that in many entities (more than 24%), almost the whole payload is used, because the E indicator has a value between 0.91 and 1.0. The first two groups of carriers put together suggest that there is a high efficiency of payload use, because more than 66% of the carriers reach an effectiveness of over 0.8.

Table 4 Average payload use in the vehicles

Percentage of responses	Dynamic payload use ratio (E)
41.94	0.81–0.90
24.19	0.91–1.00
12.90	0.41–0.50
6.45	0.71–0.80
6.45	0.51–0.60
4.84	0.21–0.30
3.23	0.61–0.70

Source Own estimation based on an empirical analysis of 62 randomly sampled enterprises

4 Conclusions

Many of the preparatory and auditory actions can be and in fact are carried out in road cargo transport enterprises, even if they are not perceived that way. In case of the financial audit, many entrepreneurs carry out actions characteristic for the audit by assessing the profitability of sales, assets and equity, when assessing the managerial decisions. The data collected during a research on 62 entities show that the profitability ratios are as follows: the return on sales—3.28%, the return on assets—17.66%, the return on equity—104.47%. During the compliance audit, compliance actions are carried out by the verification of the compliance of the actions in the enterprise with the legal regulations of acts such as the accounting and tax law (within the bookkeeping), labour law or transport law (e.g. by the verification of the driver's labour time). The operational audit can be based on many indicators which present the effectiveness of the enterprise transport contracts, which at minimum boils down the analysis of two indicators, that is the distance use indicator and the dynamic payload use ratio. In the analysed entities, data have been collected, which allowed to calculate these two ratios. The dominant value for the first ratio is between 0.90 and 1.0, and the dominant value for the second ratio is between 0.81 and 0.90. For both indicators, 1.0 is the maximum value.

References

1. Bartoszewicz, A.: Praktyka funkcjonowania audytu wewnętrznego w Polsce, CeDeWu, Warszawa (2011). ISBN 978-83-7556-402-0
2. Ciechan-Kujawa, M.: Wielowymiarowy audyt biznesowy—wartość dodana dla organizacji i interesariuszy. Wydawnictwo Naukowe Uniwersytetu Mikołaja Kopernika, Toruń (2014). ISBN 978-83-231-3171-7
3. https://na.theiia.org/translations/PublicDocuments/IPPF-Standards-2017-Polish.pdf
4. Janasz, K.: Proces zarządzania strategicznego ryzykiem w przedsiębiorstwie. w: Studia i Prace WNEiZ nr 12/2009 pod red. Władysława Janasza Wydawnictwo Naukowe Uniwersytetu Szczecińskiego Szczecin (2009). ISSN 1640-6818, ISSN 2080-4881

5. Knedler, K., Stasik, M.: Audyt wewnętrzny w praktyce. Audyt operacyjny i finansowy, Polska Akademia Rachunkowości, Łódź (2007). ISBN 978-83-925032-1-7
6. Letkiewicz, A.: Benchmarki operacyjne, zatrudnienia i ekonomiczno-finansowe w przedsiębiorstwach transportu drogowego. w: Letkiewicza, A. (Pod redakcją) Przegląd Naukowy Nr 8. Wyższa Szkoła Społeczno-Ekonomiczna w Gdańsku, Gdańsk (2009). ISSN 1730-7074
7. Letkiewicz, A.: Benchmarki w przedsiębiorstwach transportu drogowego. w: Brdulak, H., Duliniec, E., Gołębiowskiego, T. (Praca zbiorowa pod red.) Partnerstwo przedsiębiorstw jako czynnik ograniczania ryzyka działalności gospodarczej. Szkoła Główna Handlowa w Warszawie, Warszawa (2009). ISBN 978-83-7378-484-0
8. Letkiewicz, A.: Finansowe czynniki konkurencyjności polskich przedsiębiorstw transportu drogowego na rynku europejskim. w: Rucińskiej, D., Adamowicz, E. (pod redakcją) Transport a Unia Europejska. Polski transport w europejskiej perspektywie. Zeszyty Naukowe Uniwersytetu Gdańskiego, Ekonomika Transportu Lądowego, Zeszyt nr 33. Fundacja Rozwoju Uniwersytetu Gdańskiego, Gdańsk (2006). ISSN 0208-4821
9. Liszek, J., Letkiewicz, A.: Audyt wewnętrzny—nowa formuła funkcjonowania polskiego transportu w Unii Europejskiej. w: Rucińskiej, D., Adamowicz, E. (pod redakcją) Nowa jakość polskiego transportu i logistyki po akcesji do Unii Europejskiej. Zeszyty Naukowe Uniwersytetu Gdańskiego, Ekonomika Transportu Lądowego, Zeszyt nr 34. Fundacja Rozwoju Uniwersytetu Gdańskiego, Gdańsk (2006). ISSN 0208-4821
10. Łuczak, B., Kuklińska, D.: Audi/yty i audi/ytowanie, Wydawnictwo Wyższej Szkoły Bankowej, Poznań (2007). ISBN 978-83-7205-249-0
11. Majecka, B., Mandera, E.: Regulatory conditions of functional efficiency and stability of road transport enterprises. In: Letkiewicz, A., Suchanek, M. (eds.) Research Journal of the University of Gdańsk Transport Economics and Logistics Determinants of Functional Effectiveness and Durability of Road Transport Enterprises, vol. 73. Gdańsk University Press, Gdańsk (2017). ISSN 2544-3224 e-ISSN 2544-3232
12. Skoczylas-Tworek, A.: Audyt we współczesnej gospodarce rynkowej. Wydawnictwo Uniwersytetu Łódzkiego, Łódź (2014). ISBN 978-83-7969-395-5 [9E]
13. Szymańska, H.: Ogólne zagadnienia audytu wewnętrznego. w: Kiziukiewicz, T. (Praca zbiorowa pod red.) Audyt wewnętrzny w strukturze kontroli zarządczej. Difin, Warszawa (2013). ISBN 978-83-7641-868-1
14. Ustawa z dnia 29.09.1994 r. o rachunkowości, Dz. U. z 2018 r. poz. 398 [4E]
15. Winiarska, K.: Audyt wewnętrzny 2008. Difin, Warszawa (2008). ISBN 978-83-7251-873-6
16. Winiarska, K.: Teoretyczne i praktyczne aspekty audytu wewnętrznego. Difin, Warszawa (2005). ISBN 83-7251-570-0

Non-aeronautical Services Offer as an Airport Revenue Management Tool

Dariusz Tłoczyński

Abstract Non-aeronautical revenues are one of the elements of generating inflows from airport operations. Therefore, revenue management plays an important role. The use of this instrument allows air operators to conduct a flexible policy of shaping prices, including prices for renting commercial space. Therefore, it can be stated that the selection of diversified non-aeronautical services provided at the terminal influences the pricing policy of airports. There are effect traffic in market. Airports wanting to be competitive, diversify the offer of non-aeronautical services, using various models of cooperation. The availability of non-aeronautical services in all ports is similar, which is why non-aeronautical services provided at the Airport in Gdańsk have been presented hereby. However, in order for supply to be able to comprehensively meet the needs of consumers, it is advisable to systematically conduct surveys assessing the quality of services provided by infrastructure operators. This is also the purpose of this work. It was carried out based on primary research carried out at all airports.

Keywords Revenue management · Airport · Duty free · Non-aviation charges

1 Introduction

Airports as economic entities operate on two levels. In addition to the function of generating revenues from aviation activities, infrastructure operators strive to obtain additional funds for the management of terminal and landside. Revenue Management policy implemented by ports is not a relatively new method. However, the literature on this subject lacks a broad analysis of the effectiveness of the models used. It should be noted that in the study of airport economics, the problem of charges is widely analyzed.

D. Tłoczyński (✉)
Department of Transportation Market, Faculty of Economics, University of Gdańsk, Gdańsk, Poland
e-mail: dariusz.tloczynski@ug.edu.pl

M. Suchanek (ed.), *Challenges of Urban Mobility, Transport Companies and Systems*, Springer Proceedings in Business and Economics,
https://doi.org/10.1007/978-3-030-17743-0_15

The main prerequisite of undertaking research on revenue management in relation to the commercial area of airports, taking into account passenger satisfaction, is the lack of research in this area. The main objective of the article is to assess non-aeronautical services by passengers using the airport. For this purpose, the concept of revenue management was analyzed and three groups of entities operating in the commercial zone of Duty Free, Retail, F&B were selected to present selected passenger opinions in the final part. The main method used during the research was the comparative and expert method as well as the selection of particular secondary data. In addition, the study presents the author's results of marketing research in relation to this area, carried out at Polish airports.

2 Revenue Management in Airports

In the early 1970s, the revenue management method was developed. The implementation of this model was associated with the phenomenon of civil aviation deregulation and the beginning of transport activity by low-cost carriers. The rapid development of the transport services market and the limited number of seats on board contributed to the interest in the issue of the maximum use of its transport potential.

Revenue Management is defined as a process of skillful, careful and considerate management, control and management of production capacities and sources of revenues within the existing demand and supply [1]. Ivanov et al. points to the technique of maximizing revenues by adjusting the available production capacity of enterprises to defined market segments, at an optimum price [2–4].

The common feature of using RM is the market orientation of the company and the possibility of precise segmentation of its market due to the price sensitivity and customer purchase preferences [5]. The allocation of limited enterprise resources is made through the management of extremely unstable prices addressed to individual recipient markets, which depend on very variable demand over time [6]. Maximizing income from resources, with a given price structure, is the essence of revenue management.

Revenue Management is a process of extensive analysis and determination of future consumer behavior at the level of the micromarket, as well as the optimization of availability and price of the product, to maximize revenue growth. The primary goal of Revenue Management is to sell the right product to the right customer and at the right time at the right price and with the right packaging [7].

Initially, the revenue management method was implemented by air carriers. In the initial period, air operators applied segmentation of the passenger market, taking into account their sensitivity to price and shopping preferences. Price differentiation started depending on the standard of the offered service (business class or economic class) and on the time of advance purchase compared to the service delivery deadline. Currently, the RM method is also used by other entities operating in the air transport sector [8–10].

Two groups of revenues (air and non-aeronautical) are distinguished in the shaping of the airport's financial policy. The literature describes the sources of revenues of the entities managing the air transport infrastructure in a very detailed manner. This issue is analyzed by Tłoczyński, Graham, Doganis [11–14].

Optimization of airport revenue sources often results from functions performed by entities managing air infrastructure. The literature in the field of transport economics distinguishes the following functions: transport: in a branch system, inter-industry; economic: related to the promotion and development of the region, income generation; social: generating jobs, training staff, ensuring safety with the help of sanitary, fire and military aviation.

The implementation of these functions results directly from the model and strategy of operation and development. It is determined by the aspirations of entities that own assets. The assumed goals are to contribute not only to the growth of air traffic and the importance of the port in the transport system, but also to achieve the financial dimension. Tłoczyński mentions various methods of airport management, whose overriding goal is to generate profit for the company [15].

When analyzing the operation of airports, it is necessary to indicate two profiles of the conducted activity, one directly related to air transport, the other indirect. At the airport terminal, passengers use the infrastructure related to check-in, security control, gate, or airport bridge, but also from a variety of shops, services and dedicated points.

There are several models of cooperation between ports and concessionaires (Table 1).

If however, commercial exploitation of the airport is decided upon, a number of operational policy decisions must be made. First, a decision must be made on the model of operation. Five different modes are common; these are operation by [16]:

- a department of the airport authority directly,
- a specially formed fully owned commercial subsidiary of the airport authority,
- a commercial subsidiary formed by the airport authority and the airlines,
- a commercial subsidiary formed by the airport authority and a specialist commercial company,

Table 1 Models of commercial area cooperation with the air transport infrastructure manager

Operator/consisterian		Air infrastructure manager
Independent entities	←	Airport
Airport	←	Airport
Company isolated from the airport	←	Airport
Independent entities	←	Consisterian
Airport	←	Consisterian
Company isolated from the airport	←	Consisterian

Source Own research

- an independent commercial enterprise.

Regardless of the cooperation model, a number of different forms of settlements must be indicated:

- commission on turnover, but no less than a fixed amount (when it is easy to find a relation between income and passenger traffic),
- fee per 1 m^2 (when it is difficult to find a relation between income and passenger traffic),
- rental thresholds depending on the volume of passenger traffic.

Diversified forms of cooperation and settlements, and, above all, meeting passenger expectations are the basis for cooperation between infrastructure managers and concessionaires.

3 Commercial Facilities in Polish Airports

Bearing in mind the desire to increase revenues from non-aeronautical activities, airports effectively use their infrastructure. Over the last dozen or so years, significant changes have occurred in the Polish market related to the dynamics of air traffic. Airports as a result of investment activities increased the space needed to service passengers, expanding terminals, however, keeping in mind the desire to provide high-standard services. The large supply of space in the terminal for business operations makes the ports strive to offer their clients a whole package of additional services.

In turn, landlords wanting to satisfy the expectations of passengers are considering decisions regarding what to sell, when to sell, who to sell and for how much. For this reason, they want to know the aspirations and economic opportunities of travellers, adapting their offer to them in order to achieve the assumed level of revenues.

The selection of the contractor is based on the determination of the demand for the assortment and through the execution of tender procedures.

The main prerequisites for the selection of contractors are:

- passengers' needs;
- employees' needs;
- financial advantages;
- improving the quality of passenger service.

It should be noted that the relation between air traffic and the number of service points located in the terminal, the number of contractors. The growth of air traffic and the diversity of business models of air carriers at airports determines the number and variety of the range at airports (Fig. 1).

In small airports with very small traffic (Radom, Olsztyn) the offer of services is very poor. A completely different model exists in Modlin, where operates only one air carrier—Ryanair. This is an airport dedicated to handling low-cost traffic. In large

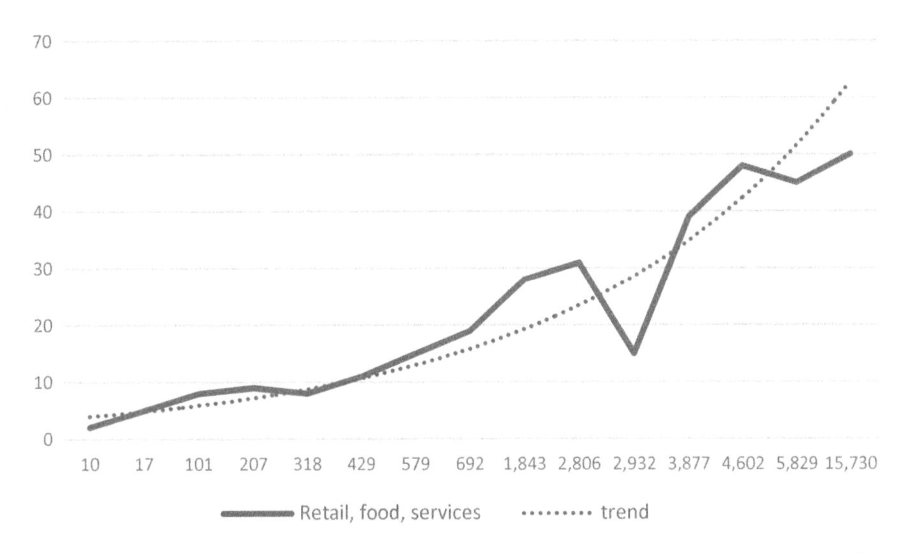

Fig. 1 Relation between air traffic and the number of points located at airport terminals at Polish airports. *Source* Internal materials of Polish airports and ULC

airports such as Gdańsk, Kraków, Warsaw, air infrastructure as a result of air traffic, causes an increase of supply of services provided at terminals, creating a diversity of stores and catering points.

Although large airports have a large number of counterparties competing with each other, the main segments of services are offered at all airports. Based on the airport in Gdańsk, the diversified offer of non-aeronautical services provided in the immediate vicinity of the terminal was presented (Table 2).

The largest retail area of over 5000 m^2 has Aelia Duty Free (one of the world's largest duty free sellers). The decor of the shops refers to the architecture of the city. The Airport of Gdansk has concluded a contract with Lagardere Travel Retail for a period of 10 years, until 2018 with regard to investments carried out by the Group of approximately PLN 30 million. Passengers using the store have the opportunity to buy in the special Sense of Place zone, local products of the highest quality (including cold meats, bread, fish and cheese, as well as alcohol) [17].

The most characteristic commercial element in the airport of Gdansk is the creation of a free-of-charge Aelia Duty Free shop, a "walk-through" type, as well as the construction of a Premium zone for particularly demanding passengers. The axis of this zone is the shops: Le Connaisseur (a premium place with an offer of alcoholic beverages and the possibility of tasting them) and Premium Food Gate designed as a culinary "postcard of the city of Gdańsk". In addition, passengers flying from Gdansk can do the shopping at the Victoria's Secret store. The Victoria's Secret store impresses with its unique design and high-quality service. The offer is addressed not only to the segment of passengers who appreciate the high quality of cosmetics [18].

Table 2 Tenants of retail space at the Gdańsk Airport [17]

Tenant	Industry	Location
Aelia Duty Free	Retail	Airside
Chef's	F&B	Airside
Coffee corner	F&B	Airside
Emanuel Berg	retail	Airside
Executive Lounge	F&B	Airside
Flying Bistro	F&B	Airside
Food gate	F&B	Airside
Furore	F&B	Airside
I love Poland	Souvenirs	Airside
Imperial Club	F&B	Airside
ING	Insurance	Airside
McDonald's	F&B	Airside
Paul	F&B	Airside
Relay	Press	Airside
The flame	F&B	Airside
Toys Star	Retail	Airside
Victoria's Secret	Retail	Airside
Virgin	Press	Airside
1 min	Retail	Landside
Business Shark Pub & Restaurant	F&B	Landside
Discover Poland	Retail	Landside
Tourist information point	Services	Landside
Exchange	Services	Landside
Chapel	Services	Landside
Post office	Services	Landside
Press4you	Press	Landside
Passenger service points of selected carriers and handling agents	Services	Landside
Car rentals: Avis, Budget, Enterprise Car Rental, Europcar, Express Rent a Car, Hertz, Panek, Sixt, Yourrent	Services	Landside
S&A Jewellery Design	Jewellery	Landside
So Coffee	F&B	Landside

For a selected segment of travellers, people with Priority Pass, Dragon Pass, HON Circle cards on flights of LOT and Lufthansa carriers, Golden Star Alliance cards travelling on flights of LOT and Lufthansa carriers as well as passengers:

- business travellers travelling with LOT, KLM and Lufthansa,
- having LOT invitations to the Executive Lounge for LOT flights,
- who have concluded appropriate agreements with the salon manager—Gdańsk Airport,
- holding Benefits and Privileges Cards issued by Bank Handlowy w Warszawie SA, Diners Club cards, Lounge Key cards and Oneworld Emerald and Oneworld Sapphire cards for Finnair and Flying Blue flights (gold and platinium) for KLM—Air France flights.

They can take advantage of the Executive Lounge in Gdańsk [17].

In 2013, a hotel located in the immediate vicinity of the terminal was commissioned for potential passengers. The Hampton by Hilton Gdańsk Airport hotel is located on a plot belonging to the Gdańsk Airport. It offers guests 116 rooms decorated in a style characteristic of the Hampton by Hilton brand. The hotel rooms include conference rooms, business centers and parking for 60–80 cars. The hotel was designed in a style adapted to the needs of people travelling for business and leisure purposes and offers best-in-class service and convenience, such as a 24-hour bar [19].

The management of the network of car parks located in the vicinity of the passenger terminal is handled by Interparking Polska. Passengers have the option of parking the car in several zones adapted to the expectations of travellers. For those leaving and waiting there is a Kiss and Fly zone, for passengers carrying out business flights, single- or multi-day there is a short-term zone and several long-term parking lots.

The main task of entities operating in the terminal and landside area is to provide additional, diversified services, which from the point of view of the airport constitute a group of non-aeronautical revenues [17].

4 Prerequisites for the Selection and Assessment of Non-aeronautical Services by Passengers at Polish Airports

For some passengers shopping in the commercial area of the airport is an integral part of air travel. At selected airports, passengers can order products at a specified time before departure. Passengers travelling with Thomas Cook can order all products using the Airshoppen service. This service creates the opportunity for customers of carriers to plan their purchases well in advance, avoiding hurry at the airport and ensuring the availability of products. Heathrow Airport offers passengers its own "boutique" service, which allows each passenger to book products online, then confirm three days before the trip, pick up and pay for the purchased goods at the airport on the day of departure [20, 21].

At some airports, passengers can book goods online and then collect and pay for them in the appropriate flight storage from one month to 24 h before departure.

At the Gdansk airport, the largest duty-free operator—Aelia Duty Free offers the option of placing an online order and a 10% discount at the same time.

Based on marketing research at airports, passengers use duty free shops due to low prices (39%) (Fig. 2).

The price level of products sold in duty-free stores in relation to prices in the city center remains a debatable issue. Certainly in duty-free shops products at prices exempt from customs duties are sold (e.g. perfumes and accessories for skin care, wine, champagnes, sweets, articles—fashion and electronics).

Most passengers travelling on international routes used the Dufry operator [255.6 million pax and Heinemann (179 million pax)]. It is estimated that one passenger spends an average of about 37 min shopping in duty-free shops, 55% arrive earlier to the airport to buy anything in the commercial zone. The most common (72%) passengers buy gifts for a partner. The main purchases in the commercial zone include: perfumes, alcohol, groceries, newspapers/magazines/books, drinks, wine, beer, cafe, local food [22, 23].

Based on DKMA research, satisfied passengers spend 10% more time in the commercial zone, unlike passengers who do not like to use retail, duty-free and

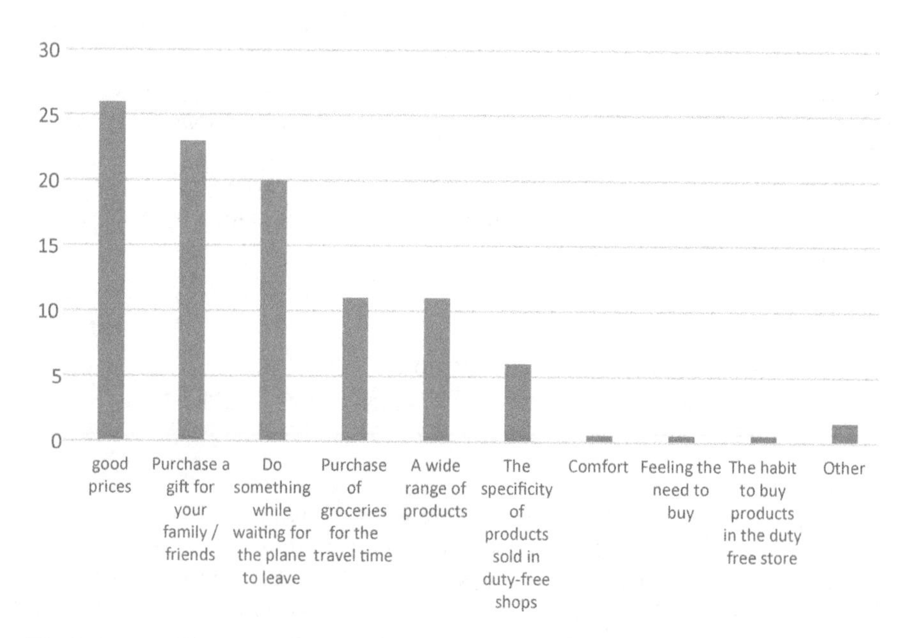

Fig. 2 Reasons for shopping in duty free stores. *Source* Internal materials of one of the European airports

Table 3 Factors affecting the level of satisfaction from the commercial zone at airports [24]

	F&B		Duty-free	
Rank	Most important to satisfaction	Least important to satisfaction	Most important to satisfaction	Least important to satisfaction
1	Menu selection	Friendliness of staff	Selection of products	Speed of service
2	Selection of restaurants	Cleanliness of facilities	Atmosphere of shops	Friendliness of Staff
3	Quality of food/drinks	Availability of seating	Selections of shops	Ease of finding shops

F&B. The most important factors affecting the level of satisfaction are presented in Table 3 [24].

There is a wide variation in the use of the commercial zone at airports, taking into account:

- frequency,
- nature of travel,
- economy,
- time to boarding.

52% of business passengers spend less than 75 min at airports, and 67% of leisure passengers spend more than 75 min at airports. This means that leisure passengers spend more time in commercial zones and are more likely to go shopping. In 52% they use F&B services, 31% of duty-free, and 16% purchase in stores. However, business passengers spend more on F&B and retail, and leisure on duty-free. Passenger purchases satisfaction surveys in commercial zones determine the average level of expenses made by passengers in two travel segments (Table 4) [24].

Based on marketing research carried out at Polish airports, 48% people positively assessed the quality of services provided at the terminal and the level of catering services (Fig. 3).

In many foreign airports, non-aeronautical revenues account for up to 60% of all revenues, in Polish airports this share is much lower, on average 40%. The exception is the Gdańsk Lech Walesa Airport, where in 2015 non-aeronautical revenues accounted for 45%. A large share of revenues from renting and leasing (18% of all revenues) and commissions (23%) in the Gdańsk Airport [25] enforces the management of the implementation of the quality policy of passengers served also in the

Table 4 Passenger expenses in the commercial zone of airports in USD [24]

Business	Profile	Leisure
16.5	F&B	15.5
43.2	Retail	41.5
96.8	Duty-Free	100.6

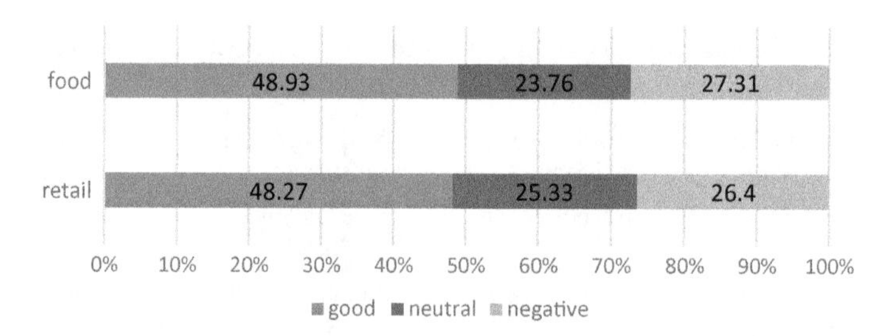

Fig. 3 Evaluation of commercial and catering services by passengers using the services of Polish airports. *Source* Own research

commercial zone, including by concluding a long-term contract with Aelia Duty Free for building and operating a walk-through store, concluding long-term contracts with McDonald's and other operators. Activities involving the implementation of procedures related to passenger service, taking into account the network of connections from the airport contribute to the increase of non-aeronautical revenues of the airport. In 2009, revenues from rental space per 1 PAX amounted to PLN 3.89, while in 2015—PLN 6.34.

5 Summary

The revenue management field combines data examination with research strategy, understanding customer behavior and cooperation with sales departments. Therefore, revenue management must be analytical and detail oriented, and at the same time capable of strategic thinking and managing relationships in the sales process.

The variety of services provided at airports is part of the aviation revenue stream management. But the diversity of commercial points at airport terminals depends primarily on air traffic. In addition to the trading points, the quality is also influenced by catering points, taxi, point distribution, service at check-in points and the way security controls are performed, as well as the instrumentation of providing information about air services. The opinion of Polish passengers in the light of the conducted research in Polish airports is highly satisfactory.

Unfortunately, many airports do not conduct any research on the relationship between the offer of services provided in airports and the satisfaction of passengers. The report prepared by DKMA formulates conclusions that [24]:

- 64% of airports have no research or let concessionaires manage retail,
- 26% are strongly involved in retail development,
- 6% do some research,
- 5% stopped research.

Unfortunately, these conclusions apply to all Polish airports. Therefore, the operators should consider the issue of passenger satisfaction surveys using the commercial zone of airports. For this purpose, it is necessary to formulate the concept and main objectives of the research. Certainly, the implementation of such research will contribute to the increase in non-aeronautical revenues.

References

1. Talluri, K.T., van Ryzin, G.J.: The Theory and Practice of Revenue Management, p. 1. Springer (2005)
2. Ivanov, S.: Hotel Revenue Management: From Theory to Practice, pp. 8–9. Varna (2014)
3. Adams, D., Burgess, C., Kelly, J., et al.: Revenue Management, p. 5. Wentworth Jones Limited, Bournemouth (2013)
4. Talluri, K.T., van Ryzin, G.J., An Introduction to Revenue Management, pp. 144–147. Tutorials in Operations Research, INFORMS (2005)
5. Rücker, M.: Revenue Management Integration. The Financial Performance Contribution of an Integrated Revenue Management Process for Hotels, p. 13. Diplomica Verlag GmbH, Hamburg (2012)
6. Daudel, S., Vialle.: Yield Management: Applications to Air Transport and Other Service Industries. Institute of Air Transport (1994)
7. Huefner, R.J.: Revenue Management: A Path to Increased Profits, pp. 79–81. Business Expert Press, New York (2011)
8. Shaw, S.: Airline Marketing and Management, pp. 207–209. Ashgate, Oxfordshire (2011)
9. Holloway, S.: Straight and Level: Practical Airline Economics, pp. 115–116. Ashgate, Burlington (2003)
10. Vasigh, B., Fleming, K., Tacker, T.: Introduction to Air Transport Economics. From Theory to Applications, pp. 280–281. Ashgate, Hampshire (2008)
11. Tłoczyński, D.: Non-aviation factors shaping competitive position of airports—Gdańsk Airport case study. In: Transbalitica 2013, pp. 235–240. Vilnius Gediminas Technical University, Vilnius (2013)
12. Tłoczyński D.: Charges for access in aviation infrastructure. Case Study: Poland. In: Suchanek, M. (ed.) New Research Trends in Transport Sustainability and Innovation, pp. 173–174. Springer, Cham (2018)
13. Graham, A.: Managing Airports. An International Perspective, pp. 71–76. Elsevier, Burlington (2008)
14. Doganis, R.: The Airport Business, pp. 53–58. Routledge, New York (2002)
15. Miecznikowski, S., Tłoczyński, D.: Privatization of entities in the air transport sector—airports. In: Macioszek, E., Sierpiński, G. (eds.) Recent Advances in Traffic Engineering for Transport Networks and Systems, pp. 170–180. Springer, Cham (2017)
16. Ashford, N., Stanton, H.P.M., Moore, C.A.: Airport Operations, p. 223. McGraw-Hill, New York (2012)
17. Gdańsk Airport. www.airport.gdansk.pl
18. Aelia Duty Free. www.aeliadytyfree.pl
19. Hampton Hotel. www.hiltonhotels.com
20. How to by duty-free goods before reaching airport. The Telegraph. www.telegraph.co.uk
21. Airshoppen. www.airshoppen.thomascookairlines.com
22. Heinemann Duty Free. www.heinemann-dutyfree.com
23. Dufry Duty Free. www.dufry.com
24. Passenger satisfaction. The key to growing non-aeronautical revenue. In: Trends From the 2013 Airport Retail and F7B Survey. DKMA (2004)
25. Financial Report. Gdańsk Airport (2009, 2015)

Rationalization of Raw Materials Supply in a Manufacturing Company—Case Study

Konrad Niziołek and Katarzyna Boczkowska

Abstract The primary objective of the article is rationalization of raw material supply to the steel warehouse of a manufacturing company. The study was conducted at a company manufacturing stainless steel components. The seat of the company is the Łódzki region. The total headcount is in excess of 50. The study analyzed the movement of the raw material with the highest turnover and largest dimensions. That material was steel. The goal of the study was to collect information required for analysis and evaluation of the existing inventory procurement and warehousing processes for all types of steel used in production. The study covered the period of five months, from August to December, when the turnover of the studied raw material is the greatest. Seventeen types of steel were included in the analysis. In this article, based on the analysis of the selected raw material, the existing manner of raw material purchase order placement and methods of the establishment of new rules and standards for the same are discussed. Simulations of inventory control based on the proposed modifications were also performed. The results of the research show that organizational activities such as review of reordering points of each type of stock, review of deliveries with regard to time and quantity, examination of production orders for each type of stock, and monitoring compliance with the established rules and standards by the workers will result in optimal utilization of the company's warehouse space.

Keywords Inventory control · Rationalization of storage locations · Supply management

K. Niziołek (✉) · K. Boczkowska
Department of Production Management and Logistics, Faculty of Management and Production Engineering, Lodz University of Technology, Lodz, Poland
e-mail: konrad.niziolek@p.lodz.pl

K. Boczkowska
e-mail: katarzyna.boczkowska@p.lodz.pl

M. Suchanek (ed.), *Challenges of Urban Mobility, Transport Companies and Systems*, Springer Proceedings in Business and Economics,
https://doi.org/10.1007/978-3-030-17743-0_16

1 Introduction

Warehouse management is defined as planned operations encompassing a repertory of measures, organizational and technical activities, and economic tasks related to the storage of goods (inventory) [3]. Warehouse management exists at small, medium-sized as well as large enterprises. It is an inextricable part of any type of business activity, although it is doubtless of utmost significance for manufacturing companies. Warehouse management is a process inseparably connected with organizational management and with the way all logistic processes that may possibly be involved in it proceed because the warehouse is a focal point in the logistic network of connections and relationships where goods are temporarily stored or directed to a different route that intersects the network of relationships. Further, it could be described as a link connecting procurement with production and sale, and procurement with the market [1]. In other words, the warehouse is the nerve center in the logistic supply chain, connected with transport, manufacturing, and distribution of goods [3]. According to Niemczyk, warehousing is a group of processes related to receiving, stacking, storing, picking, moving, maintaining, documenting, controlling, and releasing stock (goods and materials with a precisely specified location, defined in units of quantity and/or units of account) [5], which involves the entire warehouse infrastructure. The task of warehouse management in logistic supply chain is to receive raw materials and materials from suppliers, store them, and release them for production in an organized manner [4].

Inventory, held in the supply chains because of uncertain supply and demand for certain goods, is one of the fundamental elements of logistic processes. Inventory is maintained because of seasonal variation in the supply of raw materials, length of production cycles, delivery delays, level of efficiency and precision of production planning, its control and organization, current trends in fashion, etc. [6]. Although there are evident benefits of holding inventory as well as economic gains that result from it such as a reduced number of shipments and the possibility of saving on purchasing larger quantities of materials [2], there are also costs related to inventory procurement and maintenance. Furthermore, organizations should be mindful of the fact that holding an inventory of raw materials means tying up money which can only be recouped, hopefully with a profit, once the raw materials have been processed into finished goods, distributed, and sold. Therefore, effective inventory control is critically important. The primary methods of stock control, otherwise known as models of replenishment or purchasing policy, are the reorder point method and the reorder cycle method. For the former, two decision rules for inventory control need to be specified [7]:

- a reorder point which determines when it is necessary to prepare an order and place it with the supplier,
- a quantity of the order.

Going back to the cost of maintaining inventory, it mainly involves the cost of the warehouse space for the storage of goods, especially raw materials for manufacturing companies because the larger the space, the bigger the cost. The cost is connected with the investment made in the construction or renting of the warehouse space and, subsequently, fitting it with warehouse infrastructure, the cost of heating, lighting, maintenance, technical and organizational processes, etc. Many organizations, in particular those that have been in the market for longer periods, are faced with the problem of a shortage of storage space in the warehouse. Frequently, it is the task of warehouse managers as well as of warehouse workers to stack raw materials on each available surface. It is not unusual for stock units to be stacked on top of other stock units, in traffic routes, passageways, and other undesignated spaces, which causes problems primarily with locating a specific item of stock, hinders movement of people and equipment in the warehouse, and is a hazard to the health or even life of the workers. Another solution that management may decide to use is to expand the warehouse space by redevelopment: conversion of a low-storage warehouse to a high-storage warehouse, extension to the existing warehouse or construction of a new one, each of which entails high investment costs and requires time for preparation and completion.

A different approach to optimization of the warehouse space involves organizational activities such as analysis of stock reorder points, analysis of raw material deliver with regard to time and quantity, keeping track of production orders for particular stock, and monitoring worker compliance with the rules and standards in the organization. Conclusions from the analysis and possible implementation of new rules usually lead to optimized raw material stock control, which, in turn, brings about optimization of the warehouse space.

Therefore, the following hypothesis can be put forward: Review and changes to inventory control will result in optimal warehouse space utilization by the enterprise and benefits in regard to internal transport.

2 Analysis of Inventory Control in the Studied Enterprise

2.1 Research Object, Subject and Methods

The study was conducted in a manufacturing company specialized in manufacturing stainless steel components. The seat of the company, which has been in the market since 2008, is in the Łódzki region. The headcount is in excess of 50. The object of the study was the movement of the raw material with the highest turnover rate and the largest dimensions in the company warehouse. The goal of the study was to collect information required for the analysis and evaluation of the processes of movement and storage of all types of steel used for production at the company. The required information was collected during unstructured interviews with workers in the logistics and production planning departments. Data concerning stock turnover rates and reorder points for each type of sheet metal were obtained from the company's business software applications. A five-month period from August to December of 2014 was considered in the analysis. Observation, inventory, and accounting records and reports were additional sources of information. As a result of the study, information related to warehouse turnover specifying quantities of inputs and outputs in kilograms was acquired.

The study was comprised of several stages. The first stage involved analysis of the warehouse and the processes connected with the storage and movement of raw materials, especially steel. The next stage involved classification of stock by turnover rates in the analyzed period. The next step was to collect data on raw materials' turnover, which allowed the authors to compile a summary of movement in the warehouse (inbound/outbound), information indicating current and required storage locations, and critical areas related to the moment of placing an order for raw materials, their receipt at the warehouse, and release to production for each kind of stock.

2.2 Analysis and Evaluation of Raw Materials Storage Locations

The warehouse area is 385 m^2, whereas its height is 7 m. According to classification, the warehouse has a U-shaped warehouse flow layout. Products, materials, and finished goods are typically stored in rows of freestanding racks. Block stacking type of storage is used for some products and for raw materials of larger size if no storage locations are available on the racks.

The warehouse has zones and areas allotted for the storage of different types of materials such as steel, mineral wool, cardboard boxes, connectors, clips and fasteners, intermediate products, gaskets, and final goods.

The raw material (steel in sheet form) and final goods storage areas take up the largest space in the analyzed warehouse. The raw materials and final goods are also the stock with the highest turnover rate and largest dimensions of all stock warehoused by the company. Sheets of steel are moved and stored on pallets.

Steel on pallets should be stored in the warehouse on racks fitted with properly adapted and spaced load beams. Use of racking systems facilitates easy pickup and placement of pallets in a specific location and reduces the time needed to locate and retrieve a required type of steel. Steel sheets should be stored in one of the eight racks. Based on the analysis, there are 36 locations for the storage of steel because only one pallet with steel can be placed on a single beam. Each upright frame has the load capacity of 7920 kg, whereas the load capacity of each load beam is 1500 kg. The width of the column is 2.7 m. The minimum distance between racks where steel is stored is 3.5 m. The maximum allowable weight of steel on a single pallet is 1200 kg. An empty pallet frees space up for other deliveries and should be removed from the warehouse once all the sheets have been picked. The dimensions of each pallet are proportional to the dimensions of the sheet of steel which, in turn, depend on the type of steel. Descriptions of steel including its type and quantity are given on paper slips stapled to the pallet. Seventeen types of steel are stored at the examined company's warehouse where the dimensions of most of them are 2500 mm × 1250 mm, while sheet thickness ranges from 0.5 to 4 mm.

At the studied organization, storage of raw materials, mainly steel, has been challenging. The reasons for that are limited warehouse space, large dimensions of the material, and its high but variable turnover. Therefore, all kinds of losses have been incurred by the company.

2.3 Analysis and Evaluation of the Warehouse Turnover

The warehouse turnover in the studied organization involves the following stages:

- ordering raw materials from suppliers,
- receiving raw materials,
- putting raw materials away in specific locations in the warehouse,
- releasing raw materials to production,
- receiving final products in the warehouse.

Steel is ordered by a logistics–production planning specialist. The decision to order particular steel stock is triggered when its inventory level drops to the reorder point defined in the company's software system. However, the final decision to order a raw material is taken by the worker intuitively based on the predicted increase in customer orders (due to an early onset of the peak season) and monthly material consumption to date. Materials are usually ordered seven days prior to the planned delivery and immediately after an order is placed by a customer in order to replenish stocks whose levels have dropped to the reorder point. The reorder points for each type of steel have been defined as the level of the average monthly usage of the steel.

Sheets of steel, before they are stored at the warehouse, are accurately weighed, registered in the company's system, and then moved to an available storage location. However, it is not infrequent that the location is in a totally different zone than the zone designated for the type of material. The reason is a lack of free space in the designated zones.

In the remaining part of the foregoing article, typical warehouse turnover of the steel with the highest turnover will be presented. The steel is marked with the symbol KRV-05. From August to December 2014, a total of 29,496.53 kg of KRV-05 were used. Monthly picks for production were as follows:

- August—3926.35 kg
- September—5647.75 kg
- October—7137.61 kg
- November—7869.20 kg
- December—4915.62 kg

The inventory level which triggers reordering the material (determined by the enterprise according to the principle discussed above) is 3000 kg. However, as the analyzed data showed, on August 1, 2014, there were 3599.62 kg of the steel in the warehouse, whereas on December 31, 2014, there were merely 1789.09 kg of the material, which is a clear deviation from the rule established by the company. In the examined period, there were eleven deliveries of the sheet metal (based on the warehouse receiving reports).

- Delivery of 2458 kg in August. The stock level of the material prior to the delivery was 3599.62 kg and 5032.22 kg after the delivery.
- Another 2454 kg of the steel was also delivered in August. The stock level on the day of placing the order was 3714.62 kg, whereas after the order was received, the stock level rose to 5407.08 kg.
- In September, 2456 kg was delivered. On the day of the order placement, there was 4131.21 kg of the material in the warehouse, whereas after the delivery, the inventory level of the stock reached 5193.44 kg.
- Another 1516 kg was delivered in September. On the day of the order, the inventory level was 2934.71 kg; after the receipt, the inventory level reached 3388.29 kg.
- In October, 1228 kg was delivered. There were 3388.29 kg of the sheet metal prior to the receipt and 3953.75 kg of the raw material after the receipt.
- Another 2584 kg was also delivered in October. On the day of ordering, the inventory level was 3401.94 kg and rose to 4670.98 kg once the material had been received.
- The third delivery in October increased the item's stock level to 7047.04 from 4004.54 prior to the delivery. A total of 3876 kg of the material was received.
- A delivery of 2486 kg of the steel also took place in October. On the day of the order, there were 5610.86 kg of the sheet metal in the warehouse and 6674.85 kg after the receipt.
- As a result of a November delivery, 2520 kg of the material was received. The inventory level on the day of the order was 3866.63 kg, which means that it reached 5821.78 kg after the delivery.
- Another 3852 kg was also received in November. On the day of the order, the stock level was 3664.24 kg, and after the delivery, it increased to 6485.26 kg.
- A total of 2256 kg of the sheet metal was delivered in the last delivery included in the study. It took place in December. The inventory level on the day of the order was 3729.80 and 4659.84 kg once the delivery was received.

Due to the fact that picks for production were not taken into account in the above analysis, the most active turnover period of the stock item will be presented (Fig. 1) with orders and receipts taken into consideration along with releases to production.

In the chart (Fig. 1), the time of the order from the production planning specialist is marked in orange, whereas the time of the delivery from the supplier is marked in red. The chart shows that varying quantities of the steel were delivered. The orders and deliveries appear to have been rather random (an order would be placed with the supplier one day after the delivery of the previous order or, in some cases, even on the day of the delivery of the preceding order; such occurrence took place, e.g., on September 25). The chart also reveals that the worker who placed the orders did not comply with the reordering point defined for this stock as 3000 kg. Further, it follows from the analysis of the collected data that the decisions to order were not

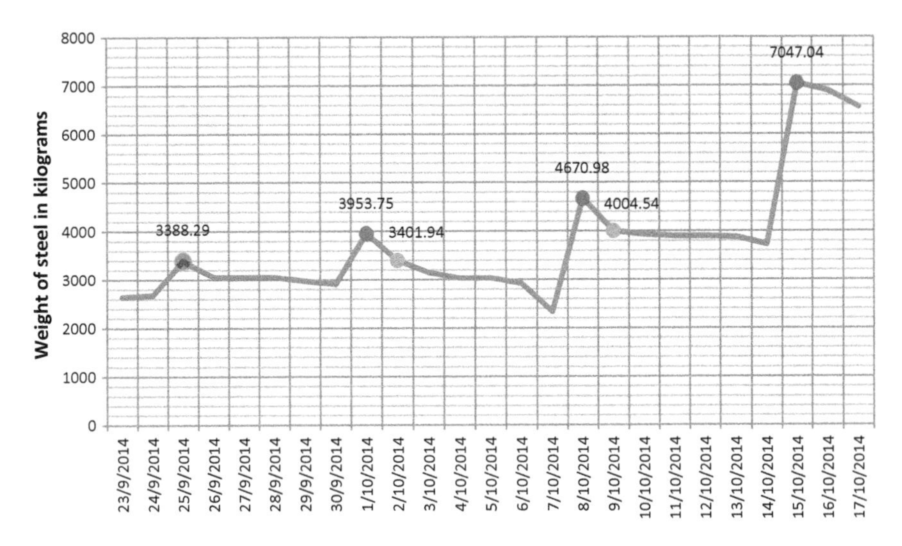

Fig. 1 KRV-05 steel turnover in the analyzed period. *Source* Own analysis

well though through because, for example, an order for a single delivery of 1516 kg of the steel was placed and, at another time, for 1228 kg on separate pallets where 2000 or even 2400 kg of the sheet metal in a single delivery could have been ordered. These decisions resulted in a situation where the same quantity of the stock was stored but required more space in the warehouse. In effect, in the analyzed period, the number of locations taken by the steel ranged from four to the maximum of eight.

3 Simulation of Optimized Warehouse Turnover

To optimize warehouse space utilization with regard to the analyzed raw material, the first step was to determine the inventory level that would trigger the decision to replenish the material. To do that, KRV-05 picks for production in the analyzed period needed to be examined (Table 1).

The weight of the raw material picked for production per week was the basis for determining the reordering point for the stock. The weekly time interval was chosen because it was the longest purchase order lead time for this stock. Table 1 shows significant variation in the consumption of the material. Out of 25 analyzed time intervals, in 9 of them, the weight of the material picked for production ranged from 0 to 1000 kg, whereas in 8 of the intervals, the weight of the picked material did not exceed 1500 kg; in another 4, the weight of the picked material was below 2000 kg, and in 3, it was below 2500 kg. The largest quantity of the steel was released to production in the week from November 23 to 29 and amounted to 2543.83 kg (highlighted bold values in Table 1). Although this quantity was picked for production

Table 1 Weight of KRV-05 steel picked for production per week

Time intervals		Weight of the steel picked (kg)	Time intervals		Weight of the steel picked (kg)
1	08-01/08-07	846.28	14	10-18/10-24	1142.85
2	08-07/08-13	1430.68	15	10-24/10-30	1634.20
3	08-13/08-19	165.77	16	10-30/11-05	2133.78
4	08-19/08-25	1109.24	17	11-05/11-11	656.00
5	08-25/08-31	374.38	18	11-11/11-17	2230.74
6	08-31/09-06	1398.73	19	11-17/11-23	559.03
7	09-06/09-12	1606.63	20	11-23/11-29	**2543.83**
8	09-12/09-18	1006.99	21	11-29/12-05	824.54
9	09-18/09-24	349.79	22	12-05/12-11	1696.85
10	09-24/09-30	1285.61	23	12-11/12-17	2010.36
11	09-30/10-06	1227.18	24	12-17/12-23	383.87
12	10-06/10-12	1597.76	25	12-23/12-29	0.00
13	10-12/10-18	1281.44			

Source Own analysis

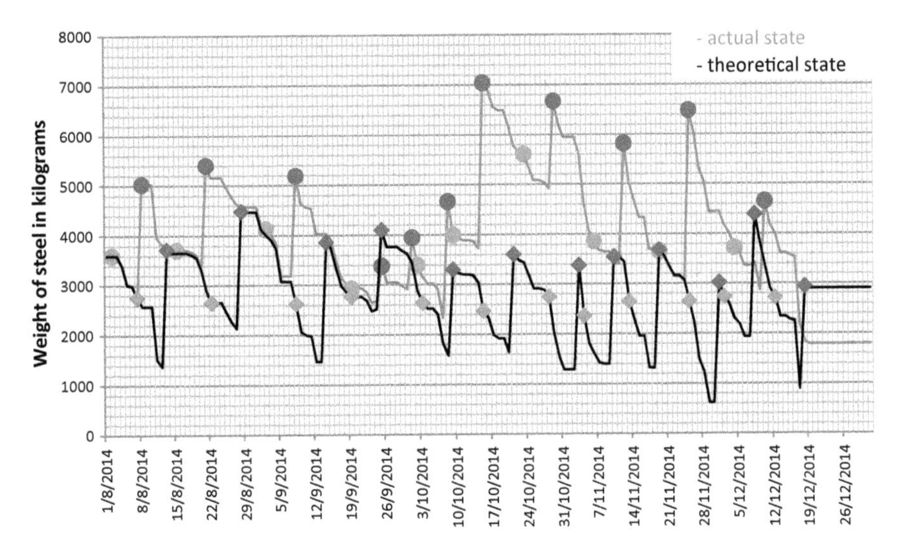

Fig. 2 Actual and simulated turnover of KRV-05 steel in the analyzed period. *Source* Own analysis

only at one time during the analyzed period, the reorder point was hypothetically determined at the level of 2500 kg. This quantity was lowered by 500 kg from the reorder point set by the enterprise. Furthermore, it was established that any single order would be a multiple of 1200 as 1200 kg is the weight of the maximum allowable load on a pallet and on a pallet rack beam. Once the inventory control procedures for the analyzed stock were defined, a simulation of the warehouse turnover was performed. A chart illustrating the simulated KRV-05 steel turnover against the actual turnover is presented in Fig. 2.

As in Fig. 1, order placements are marked in orange and deliveries in red. In the simulated turnover (marked in black in Fig. 2), there are 12 deliveries, whereas, as previously discussed, in the analyzed period, there were actually 11 deliveries of the steel, which means that there was one order and one delivery more. Even at first glance, it appears obvious that the simulated turnover line is more streamlined—the differences between the minimum and maximum inventory levels are smaller compared to the line illustrating the actual behavior of the raw material in the analyzed period. In the simulated turnover of the KRV-05 steel, ca. 600 kg was the lowest value to which the inventory level dropped, whereas 4400 kg was the highest value the inventory level reached. For the actual turnover, the maximum inventory level value was 7000 kg (Fig. 1, on October 15, 2014). It follows that the reorder point set at 2500 kg could be lowered even further by ca. 500 kg to the level of 2000 kg. However, a simulation would need to be performed for these assumptions, preferably, for a longer period of time.

The need to achieve more storage space for the selected raw material was the primary reason for the analysis and simulation of the stock turn. To this end, an analysis of the number of locations where KRV-05 was actually stored in the examined

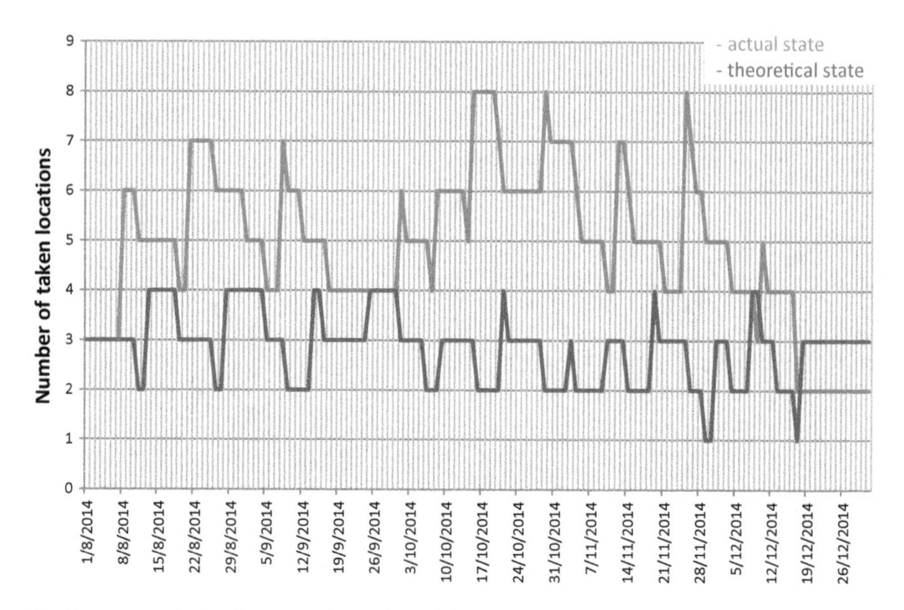

Fig. 3 Actual and simulated number of KRV-05 steel storage locations in the analyzed period. *Source* Own analysis

period and a simulation of locations where KRV-05 would be stored with the implementation of the theoretical procedures were performed. The results are presented in Fig. 3.

As the simulated KRV-05 turnover shows, 1 to the maximum of 4 locations would be required (Fig. 3). For the analyzed raw material, the difference is striking because it actually required up to 8 storage locations. A more balanced demand for storage locations is a consequence of a more rational simulated warehouse turnover. In the simulation, only 2–3 storage locations were usually required, which did not occur in the case of the actual turnover. Figure 3 illustrates that there were times when within a single week the number of the required locations dropped from 7 to 4 or spiked from 4 to 7. It also happened (between November 28, 2014, and December 12, 2014) that the number of location taken by KRV-05 dropped from 8 to 3 within a fortnight.

4 Analysis of External and Internal Transport

After the warehouse turnover rationalization, carried out in the previous part of the paper, an analysis of external and internal transport for the number of transport operations was carried out.

4.1 External Transport

As proven before (Fig. 2) in the analyzed period (from the beginning of September until the end of December), the number of KRV-05 steel deliveries increased from 11 (for the previous variant) to 12 (simulation variant). The number of external deliveries for all the 17 types of sheet metal in the warehouse is presented in Fig. 4.

While analyzing the results presented in Fig. 4, it should be noticed that providing adequate raw material supply, which also results in efficient utilization of warehouse spaces allocated to steel, would require 65 deliveries of the raw material to the enterprise, a growth of 8.3%. It should be also noticed that in case of 9 types of sheet steel, the number of deliveries would not change, and in 4 cases (in the simulated version), it would even decrease. We can observe an increase in the number of deliveries regarding 4 types of steel, the most significant being KRT-04 steel.

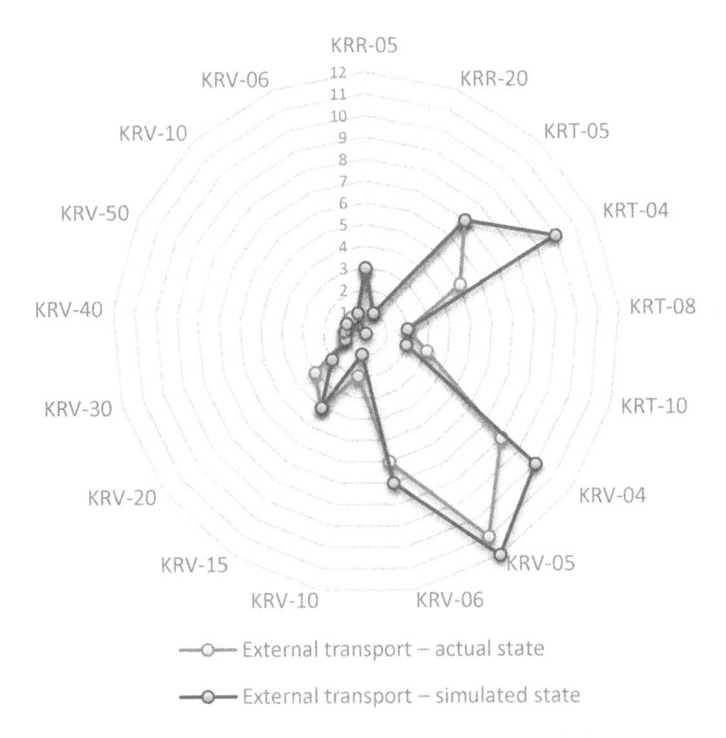

Fig. 4 External transport for 17 types of sheet metal. *Source* Own analysis

4.2 Internal Transport

Analysis of internal transport was divided into two sections: transport from the raw material delivery spot to the storage place (delivery rack) and transport from the storage spot to the production hall (rack production). In the examined company, because of the mass being transported, dimensions, production demand, storage method, and safety during steel delivery, a means of transport can only move one pallet at a time.

In the in-depth analysis of the internal transport of KRV-05 steel, it should be noticed (Fig. 3) that there were 27 transits from the steel delivery spot to the warehouse rack and 28 transits from the raw material storage spot to the production hall. After a simulation of the deliveries, which optimized the utilization of the steel storage space, an improvement of the internal transport was observed—from 27 to 20 transits for delivery rack and from 28 to 20 for rack production.

A similar analysis was carried out for all of the 17 types of sheet steel for internal transport, in both the version actually realized in the examined company and the simulated version. The results are presented in Fig. 5.

As it can be seen (Fig. 5) in respect to all 17 types of steel, the proposed raw material delivery organization solution also positively affects the number of transits. The simulation proved that in case of 9 types of steel a reduction in the number of transits would be observed; in 5 cases, it would not change; and only in 3, the number of transits would increase by 1. Globally, a reduction in the delivery-rack transit of 18.2% was achieved with a 17.2% decrease for the rack-production transit.

Decrease in the number of transits is clearly connected with the decrease in transport costs, especially by decreasing the: transport time, mileage, wear and tear of the means of transport, the engagement of the employees and in transit work, number of loading and unloading operations. The risk of unexpected emergency situations (e.g., accidents at work and damage to property) also decreases.

5 Conclusions

In the study, 17 types of sheet metal were analyzed. However, the foregoing paper discussed the results of changes proposed for the type of steel marked with the symbol KRV-05 as its turnover was the highest. Based on the analysis of this raw material, it can be concluded that a minor change in inventory control, implementation of simple rules, and adherence to them may bring about major advantages. Excessively high reordering point values and a concurrent lack of strict compliance with the established rules and procedures for stock procurement leads to a situation where there is a shortage of storage space in the warehouse. The simulations performed for all 17 types of sheet metal showed that in the periods of the highest demand for storage space, the number of locations required would drop to ca. 30 in contrast to 48 locations in which the material was actually stored in the analyzed period.

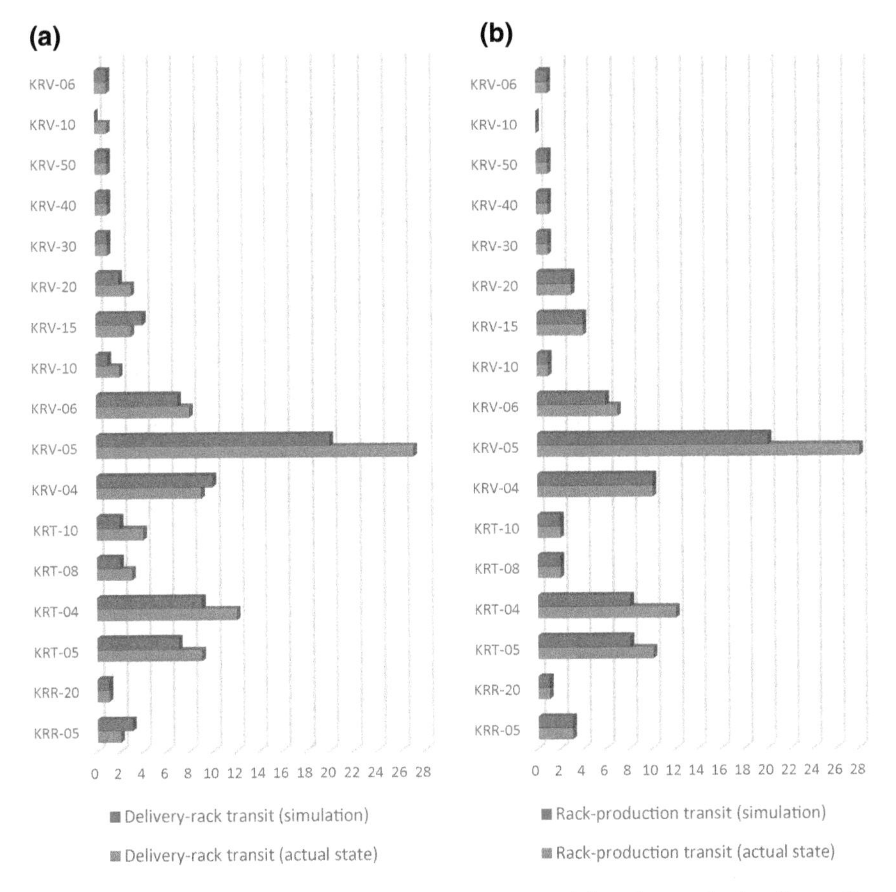

Fig. 5 Internal transport for 17 types of steel, **a** delivery to rack, **b** rack to production. *Source* Own analysis

Therefore, the results of the research and simulations confirm the initial hypothesis that review and changes to inventory control will bring about optimal warehouse space utilization by the enterprise and benefits in regard to internal transport with relatively small increase in the costs of external transport.

References

1. Bendkowski, J., Radziejowska, G.: Logistyka zaopatrzenia w przedsiębiorstwie. Wydawnictwo Politechniki Śląskiej, Gliwice (2005)
2. Coyle, J.J., Bardi, E.J., Langley, C.J.: Zarządzanie logistyczne. PWE, Warszawa (2010)
3. Dudziński, Z.: Poradnik organizatora gospodarki magazynowej w przedsiębiorstwie. Polskie Wydawnictwo Ekonomiczne, Warszawa (2012)

4. Dudziński, Z.: Vademecum organizacji gospodarki magazynowej. Ośrodek Doradztwa i Doskonalenia Kadr, Gdańsk (2008)
5. Niemczyk, A.: Zapasy i magazynowanie. Tom II Magazynowanie. Instytut Logistyki i Magazynowania, Poznań (2008)
6. Pisz, I., Sęk, T., Zielecki, W.: Logistyka w przedsiębiorstwie. Polskie Wydawnictwo Ekonomiczne, Warszawa (2013)
7. Skowronek, C., Sarjusz-Wolski, Z.: Logistyka w przedsiębiorstwie. Polskie Wydawnictwo Ekonomiczne, Warszawa (2012)

Study of the Relationship Between the Potential of Road Transport Enterprises and Socio-economic Development of Poviats in Pomorskie voivodeship

Rafał Szyc, Agnieszka Pobłocka and Krystyna Wojewódzka-Król

Abstract Road transport has been dominating the transport sector for many years and has a significant impact on socio-economic development in Poland and the EU (e.g. in Poland in 2017, 303,000 employees were employed in road transport, which was 41% of employees in all modes of transport). Road transport is also of great importance for the development of local government units, which also results from easy access to the profession of a road transport operator. The aim of the empirical research was:

- to estimate the potential of enterprises related to commercial road transport of goods in poviats of Pomorskie voivodeship (for this purpose an original indicator describing this variable was developed),
- to research on the impact of socio-economic development of the region and the potential of the analysed enterprises.

This paper uses methods of statistical description. In effect of the conducted research, it was shown that the development of road transport enterprises is more intense in economically weaker regions.

Keywords Road transport enterprises · Economic development · Regional development

1 Introduction

Road transport plays an important role in the development of modern economy and its market mechanisms. In the transport system, we can single out subsystems which correspond to each of the transport branches, i.e. road, rail, air and inland waterway

R. Szyc (✉) · K. Wojewódzka-Król
Faculty of Economics, University of Gdańsk, Gdańsk, Poland
e-mail: ravszyc@gmail.com

A. Pobłocka
Faculty of Management, University of Gdańsk, Gdańsk, Poland

M. Suchanek (ed.), *Challenges of Urban Mobility, Transport Companies and Systems*, Springer Proceedings in Business and Economics,
https://doi.org/10.1007/978-3-030-17743-0_17

[1] transport. Road transport in this structure plays a dominant role both domestically and throughout the European Union. In 2014, 49% of loads were transported by road, and if you take the land transport itself, this value goes up to 72%. This branch is also characterised by a great ability to improve its instruments of action [2].

Among all modes of transport, road transport generates the largest external costs; however, it also dominates in terms of generated external benefits, especially those related to employment. In the EU, in 2014, 47% of registered enterprises in the transport sector were road transport companies (563,598 enterprises). They employed almost 28% of employees (2.9 million employees) working in the transport sector and this was the highest value. Road transport is even more dominant in Poland. In 2014, from 135 thousand enterprises from the entire transport sector, 77.14 thousand were road transport companies which meant a 57% share in the total number of transport companies. They employed 41% (293.6 thousand employees) working in the transport sector—720.8 thousand employees [3].

Poland's accession to the EU in 2004 and gaining access to the previously protected common market resulted in the dynamic development of road transport companies in Poland. The position of Poland on the market of goods transport services in the EU is special. Polish transport companies made the largest transport work among 28 EU countries. In 2014, this share was at the level of 25%. In the years 2005–2014, Polish road transport companies increased transport performance from 50.9 to 154.3 billion tkm, which is more than a threefold increase compared to the baseline year [4].

Such a strong increase in the position of road transport in the national economy also has a positive impact on regional development, because it means:

- reducing the unemployment level,
- increasing budget revenues from local taxes,
- increasing revenues to budgets from participation in CIT and PIT,
- development of cooperating enterprises.

Road transport companies significantly contribute to regional development. However, local government officials frequently lack knowledge about the level of development of this mode of transport in regions such as poviats. This determined the desire to deepen own studies on this subject. It was decided to take a closer look at the road transport of loads at the lower levels of the analysis. The poviat level was selected for the study, as it was noted that there are very few statistic data describing this transport sector in regions smaller than the voivodeships. It was noted that in the poviats of Pomorskie voivodeship there may be significant differences in the potential development of this branch of transport.

The main hypothesis has been put forward that says that in poviats of the Pomorskie voivodeship there are significant differences in the number of trucks used. In addition, a detailed hypothesis has been put forward in which it is believed that transportation of goods in Pomorskie voivodeship has most developed in those poviats that are the least economically developed.

Data from the Transport Departments of Poviat Starosties, Chief Inspector of Road Transport in Warsaw and Central Statistical Office [GUS] databases were analysed.

Due to the lack of reliable data on the potential of transport companies, expressed in the number of trucks used the potential of enterprises was determined by developing a proprietary indicator—the number of total transcripts of domestic and EU licences per 1000 inhabitants of the poviat. In addition, an attempt was made to explain the relationship of the indicator developed with other socio-economic indicators.

2 Justification for Test Method Selection

At the beginning of the 1970s, the Statistical Office of the European Communities (Eurostat) specified the "Nomenclature of Statistical Territorial Units", later called Nomenclature of Territorial Units for Statistics to establish a uniform, coherent system for the distribution of the territories of EEC Member States to produce regional statistics for the Community. Their goal was to provide a comprehensive overview of the changes taking place at the regional level across Europe. At the beginning, 6 NUTS levels were held [5]. NUTS 0—covered the whole country, NUTS 1, 2 and 3—the regional level, and NUTS 4 and 5—the local level [6]. Unfortunately, there are very few data on transport from which data could be derived at the poviat level (NUTS 4).

The default database of registered vehicles in Poland is the Central Register of Vehicles and Drivers (CEPIK), the records of which include detailed data and information on road motor vehicles and their owners. The Central Statistical Office (GUS) also uses the data from CEPIK database to compile statistical yearbooks and other reports. Generating data directly from the CEPIK database, or indirectly from the CSO data, with the assumed research objective, would not present a true picture of the road transport industry in the studied regions.

One of the ways to internalise external costs in road transport is a tax on transport means imposed by municipalities on each HGV, tractor, trailer and semi-trailer [7]. The legislator provides for a significant range of amounts, within which municipalities may move, setting the amount of the said tax annually [8]. Almost twofold differences in the amount of tax on transport means induce enterprises to register branches of their enterprises (this applies in particular to commercial law companies) in towns where, by the decision of the Commune (*Polish: Gmina*) Council, this tax was set at a low or a minimum level. This means that a vehicle registered in one of the communes with a low tax on means of transport may be used by an enterprise registered in a completely different commune, poviat or even a voivodeship.

A similar problem occurs in the case of leased vehicles (this does not apply to vehicles bought on credit). Throughout the operating lease period, the lessee (the vehicle user) is not the owner; it is the leasing company (the lessor). In order to gain a competitive advantage, leasing companies have the possibility to open branches in communes where the tax on transport means is at a low or minimal level provided for by the law. In the case of leasing companies, the factor interfering with the research could be significant as these enterprises may register a significant number of vehicles in a small commune.

The lack of precise division into vehicle groups in statistical sources is also a significant factor. In the CSO and CEPIK publications, besides commercially used HGVs, also non-commercially used HGVs are presented, and these are not taken into account in the study.

In order to estimate the number of vehicles and enterprises in the road transport of goods, it was decided to use the legal requirement to license companies and vehicles used for commercial transport of goods [9].

Pursuant to the provisions of Regulation (EC) No. 1071/2009 of 21 October 2009, every carrier wishing to carry out transport activity in the country with heavy goods vehicles with a maximum total vehicle weight or a unit exceeding 3.5 tons must obtain a permit to practise as a road carrier. In addition, for every used motor vehicle, regardless of the form of its ownership, the entrepreneur must obtain a transcript of the permit. The authority issuing the permit and transcripts of the permit is the starosta of the poviat, competent for the place where the enterprise is registered. For an entrepreneur wishing to conduct a transport activity in the EU, a Community licence is issued after obtaining a permit to perform the profession of a carrier. Also in this case, the entrepreneur is obliged to apply for a transcript of the Community licence for each motor vehicle used in transport, or a unit with a total vehicle weight exceeding 3.5 tons. The competent authority to grant the Community licence is the Chief Inspector of Road Transport in Warsaw. The regulation also regulates the requirement to have only one transcript for each vehicle. This condition is beneficial for the needs of research, because it allows to state that one transcript corresponds to one vehicle. In addition, the regulation provides that unused transcripts are to be returned, in the event of sale or scrapping of the vehicle.

The types of issued related documents are presented in Table 1.

The number of vehicles used in commercial road transport of goods in Poland and the EU was estimated on the basis of information on the number of permits (up to 2013 licences) and transcripts permits for such activity.

3 Results of Empirical Research

In order to estimate the potential of enterprises related to domestic and international commercial road transport of goods in poviats of the Pomeranian voivodeship in Poland, a proprietary indicator was developed—the number of total transcripts of domestic and EU licences per 1000 inhabitants of the poviat. The number of combined extracts from national and EU licences describes the actual number of heavy goods vehicles used to transport goods in the country and in the EU.

The statistical data used in the study are secondary data. Statistical data on domestic transport were obtained personally (or by correspondence) from the transport departments of poviat administrative offices of the surveyed units. Statistical data on foreign transport were generated on the basis of postal codes from the IT system of

Table 1 Documents required for carrying out road transport activities

	Domestic transport	International transport
Name of the document issued	Permit to practise as a road transport operator (until 2013, the licence for domestic road transport of goods and transport of people)	Community licence, issued after obtaining a permit to practise as a road transport operator first (until 2013, international transport licence)
Issuing authority	The transport department of a poviat administrative office, competent for the place of registration of business and International Transport Office, at the Chief Road Transport Inspector Office	International Transport Service Office, at the Chief Road Transport Inspector Office
Name of the document issued for each vehicle	Transcript of the permit to practise the occupation of a road transport operator (until 2013, transcript of the licence for domestic road transport of goods)	Transcript from a Community licence (until 2013, transcript of international transport licence)

Source Own elaboration

the International Transport Service Office operating at the Chief Transport Inspector Office in Warsaw in Poland. All data were aggregated at the end of the third quarter of 2017.

Statistical data for Pomorskie voivodeship include all 16 poviats and 4 cities with poviat rights (total survey).

Four indicators available in the CSO [GUS] studies were selected to present the level of social and economic development in the poviats studied. These are:

- the percentage of inhabitants of rural areas,
- the average gross salary in PLN in 2016,
- labour productivity in industry in Pomorskie voivodeship in PLN per employee in 2012.

Due to the lack of consent for publication, the collected data are presented in Table 2.

In Pomorskie voivodeship in 2017, the average number of total transcripts of licences (domestic and EU) per 1000 inhabitants in poviats amounted to 12.49 with a standard deviation of 5.41. The spread of the total number of national and EU transcripts in the poviats of Pomorskie voivodeship is measured (spread $R = 20.92$ of the total number of domestic and foreign transcripts per 1000 inhabitants, coefficient of variance $v = 43.3\%$)—Fig. 1.

- the smallest number of total domestic and foreign transcripts per 1000 inhabitants in poviats was in Sopot (5.58), the largest in the poviat of Gdańsk (26.5),

Table 2 Summary of the data obtained

Item	Poviat	Population in 1000 inhabitants	Potential of road transport of goods—-transcript of EU/domestic licences per 1000 inhabitants	Residents of country areas in %	Average gross salary in PLN in 2016	Labour productivity in industry in Pomorskie voivodeship in PLN per employee in 2012
1	The city of Gdańsk	462.996	7.96	0	5118.59	1450.5
2	The city of Gdynia	247.329	11.82	0	4798.54	307.9
3	The city of Słupsk	92.17	7.26	0	5104.38	237.5
4	The city of Sopot	37.089	5.58	0	3620.36	304.2
5	Bytów poviat	78.805	20.77	64.67	3602.02	221.8
6	Chojnice poviat	96.697	14.74	42.5	3275.78	233.2
7	Człuchów poviat	56.807	8.80	55.02	3480.17	264.5
8	Gdańsk poviat	110.666	26.51	72.41	3950.76	323.3
9	Kartuzy poviat	129.996	17.92	82.74	3477.69	272.4
10	Kościerzyna poviat	71.738	20.11	66.65	3285.04	187.7
11	Kwidzyn poviat	83.457	14.38	42.98	3890.63	578.7
12	Lębork poviat	66.163	8.19	40.43	3705.38	283.4
13	Malbork poviat	64.189	9.77	32.51	3496.46	646.6
14	Nowy Dwór Gdański poviat	36.107	8.28	68.51	3563.66	251.7
15	Puck poviat	83.433	12.75	57.58	3779.75	373.7
16	Słupsk poviat	98.249	11.31	79.34	3642.86	246.4
17	Starogard Gdański poviat	127.339	15.18	50.68	3643.85	383.7

(continued)

Table 2 (continued)

Item	Poviat	Population in 1000 inhabitants	Potential of road transport of goods—transcript of EU/domestic licences per 1000 inhabitants	Residents of country areas in %	Average gross salary in PLN in 2016	Labour productivity in industry in Pomorskie voivodeship in PLN per employee in 2012
18	Sztum poviat	42.284	7.97	62.46	3512	231.1
19	Tczew poviat	115.581	11.13	34.62	3960.08	620.8
20	Wejherowo poviat	210.374	9.27	41.17	3554.31	267.8

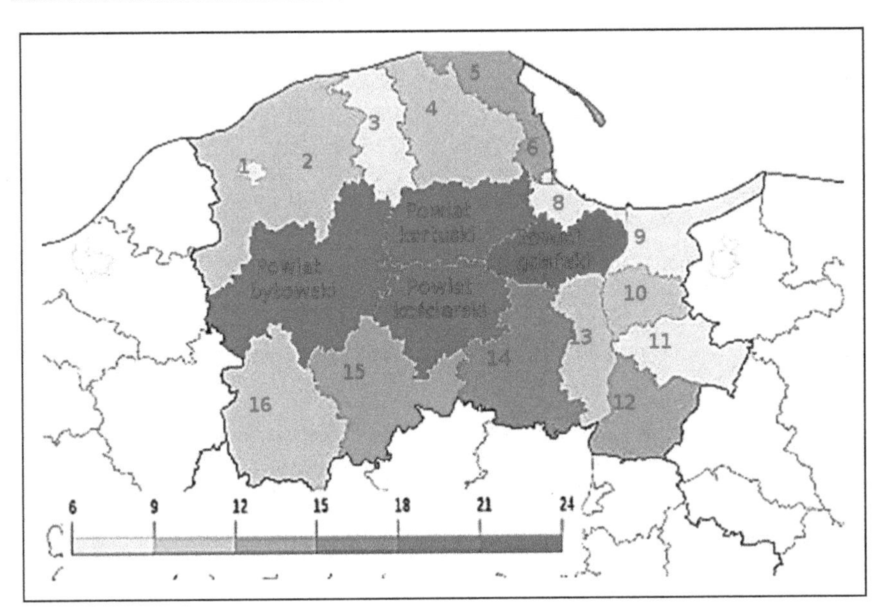

Poviats: 1 - Słupsk City, 2 - Słupsk, 3 - Lębork, 4 - Wejherowo, 5 - Puck, 6 - Gdynia City, 7 - Sopot City, 8 - Gdańsk City, 9 - Nowy Dwór Gdański, 10 - Malbork, 11 - Sztum, 12 - Kwidzyn, 13 - Tczew, 14 - Starogard Gdański, 15 - Chojnice, 16 - Człuchów.

Fig. 1 Number of HGVs used in domestic and the EU transport per 1000 inhabitants in poviats of Pomorskie voivodeship. *Source* Own elaboration, based on research

- in Pomorskie voivodeship in 50% of the poviats (Sztum, Lębork, Nowy Dwór Gdański, Człuchów, Wejherowo, Malbork, Tczew and the cities of Gdańsk, Słupsk, Sopot): the total number of domestic and EU transcripts per 1000 inhabitants in poviats was not more than 11.22 (Q2 = 11.22); in other poviats, it was not less than 11.22 (Słupsk, Puck, Kwidzyn, Chojnice, Starogard Gdański, Kartuzy, Kościerzyna, Bytów, Gdańsk and the city of Gdynia).

As shown in Fig. 1, the number of trucks used in domestic and foreign transport was the highest in the poviats: Kościerzyna, Kartuzy, Bytów and Gdańsk.

In Pomorskie voivodeship in 2017, the average share of inhabitants of rural areas in poviats was 44.71% with a standard deviation of 26.92%. The spread of the share of inhabitants of rural areas in poviats of Pomorskie voivodeship is big ($R = 82.74\%$ points, coefficient of variance $v = 60.21\%$):

- the smallest number of inhabitants was in cities with poviat rights (0%), the largest in Kartuzy poviat (82.7%),
- in 50% of poviats (the city of Gdańsk, the city of Gdynia, the city of Słupsk, the city of Sopot, Malbork, Tczew, Lębork, Wejherowo, Chojnice and Kwidzyn), the total number of inhabitants of rural areas was 46.83% (Median, Q2 = 46.83%); in other poviats, it was no less than 46.83% (Starogard Gdański, Człuchów, Puck, Sztum, Bytów, Kościerzyna, Nowy Dwór Gdański, Gdańsk, Słupsk and Kartuzy)—Fig. 2.

As shown in Fig. 2, the number of people living in rural areas was the highest in poviats: Słupsk, Kartuzy, Nowy Dwór Gdański, Bytów and Kościerzyna.

In Pomorskie voivodeship in 2016, the average level of gross remuneration in poviats was recorded at PLN 3823.12 with a standard deviation of PLN 545.34. In 2016, the average gross remuneration in the poviats of Pomorskie voivodeship was characterised by low spread (spread R = PLN 1842.81, coefficient of variance $v = 14.26\%$):

- the lowest level of gross remuneration was recorded in Chojnice poviat; it amounted to PLN 3275.78; the highest was in Gdańsk—PLN 5118.59,
- in 25% of poviats (in 5 poviats: Chojnice, Kościerzyna, Kartuzy, Człuchów, Malbork), the level of gross remuneration was not higher than PLN 3504.23 (Q1 = PLN 3504.23),
- in 50% of poviats—in 10 poviats (Chojnice, Kościerzyna, Kartuzy, Człuchów, Malbork, Sztum, Wejherowo, Nowy Dwór Gdański, Bytów, Sopot), remuneration not higher than 3631.61 PLN was recorded (median, Q2 = 3631.61 PLN); in other poviats, it was not less than PLN 3631.61 (Słupsk, Starogard Gdański, Lębork, Puck, Kwidzyn, Gdańsk, Tczew, Gdynia, Słupsk, Gdańsk)—Fig. 3.

As shown in Fig. 3, the lowest level of average gross remuneration was recorded in poviats: Kartuzy, Sztum, Chojnice, Kościerzyna, Człuchów.

In Pomorskie voivodeship in 2012, the average labour productivity in industry per capita in poviats was PLN 384.35 thousand with a standard deviation of PLN 284.28 thousand.

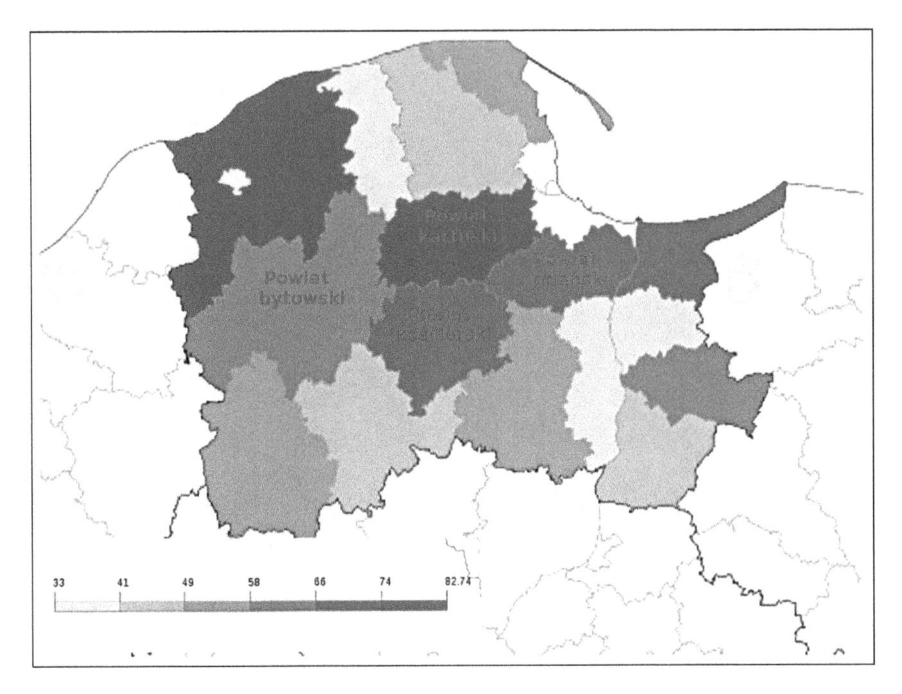

Fig. 2 Percentage of population living in rural areas in Pomorskie voivodeship in 2016. *Source* Own elaboration based on data from the Central Statistical Office, Branch in Gdańsk, Sustainable Development Indicators

In 2012, the labour productivity in industry per one employed in the poviats of Pomorskie voivodeship was characterised by a big spread (R = PLN 1263.5 thousand, coefficient of variance $v = 73.97\%$):

- the lowest labour productivity in industry per capita was recorded in Kościerzyna poviat—188 thousand PLN, and the largest labour productivity in industry per capita was recorded in the city of Gdańsk—PLN 1450.0 thousand,
- in 25% of poviats (Kościerzyna, Bytów, Sztum, Chojnice, Słupsk), the reported productivity in industry per capita was not higher than 242 thousand PLN (Q1 = 242 thousand PLN),
- in 50% of poviats (Kościerzyna, Bytów, Sztum, Chojnice, Słupsk, Nowy Dwór Gdański, Człuchów, Wejherowo, Kartuzy and Słupsk), the average productivity in industry per capita was 277.9 thousand PLN (Median, Q2 = 277.9 thousand PLN); in other poviats, it was not less than 277.9 thousand PLN (poviats: Lębork, Sopot, Gdynia, Gdańsk, Puck, Starogard, Kwidzyn, Tczew, Malbork, Gdańsk)—Fig. 4.

As shown in Fig. 4, poviats: Kościerzyna and Bytów are distinguished by the lowest labour productivity poviats in Pomorskie voivodeship.

The next analysed variable was the level of GDP per capita of a poviat in Pomorskie voivodeship. In Poland, the Central Statistical Office (*Polish*: GUS) publishes the GDP for the whole country and broken down into macro-regions of the country.

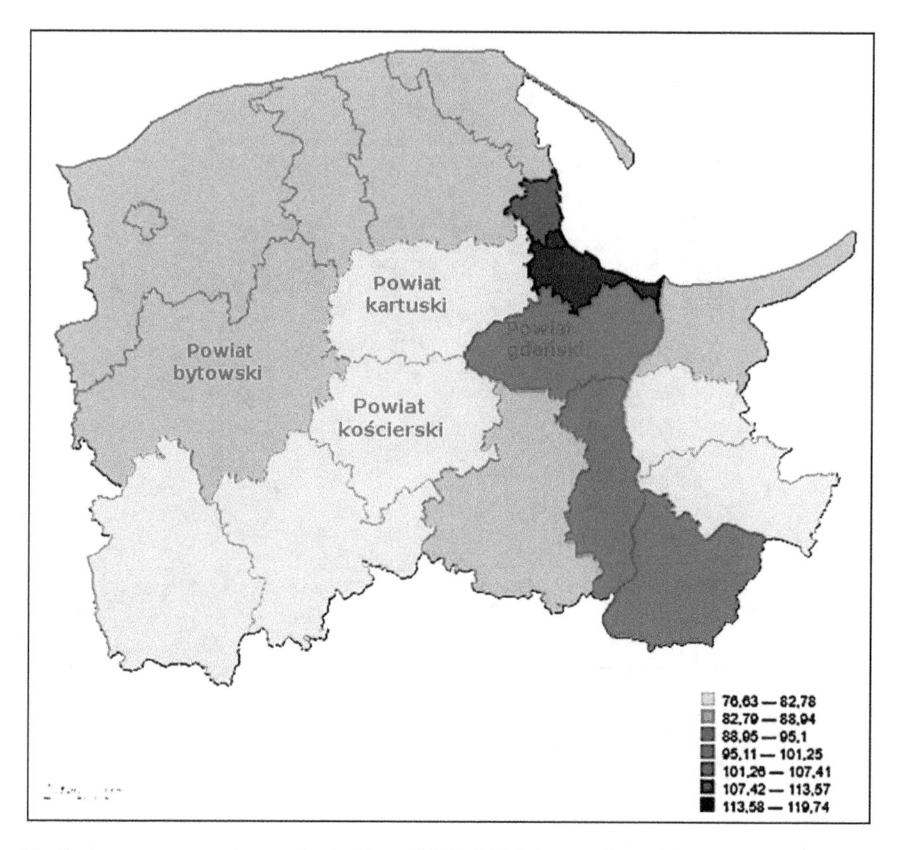

Fig. 3 Average gross salary, voivodeship = 100%, 2016. *Source* Own elaboration, based on data from the Central Statistical Office

In the study, the 2015 level of GDP at the poviat level estimated by D. Ciołek, J. Zaucha, 2015 was used. In Pomorskie voivodeship in 2015, the average level of GDP per capita in poviats was recorded at 39.8 thousand PLN with a standard deviation PLN 16.11 thousand. During this period, the level of GDP per capita in the poviats of Pomorskie voivodeship was characterised by a measure spread (R = PLN 63.36 thousand; coefficient of variance $v = 40.49\%$):

- the lowest level of GDP per capita was recorded in Nowy Dwór Gdański poviat and amounted to 27 thousand PLN,
- the highest GDP per capita was in Sopot, amounting to 91 thousand PLN,
- in 25% of poviats—in 5 poviats (poviats: Nowy Dwór Gdański, Bytów, Kościerzyna, Kartuzy and Człuchów) GDP per capita was not more than 30.26 thousand PLN (Q1 = 30.26 thousand PLN); in the other poviats, GDP per capita was not less than 30.26 thousand PLN,
- in 50% of the poviats—in 10 poviats (poviats: Nowy Dwór Gdański, Bytów, Kościerzyna, Człuchów, Kartuzy, Chojnice, Lębork, Puck, Wejherowo) GDP per

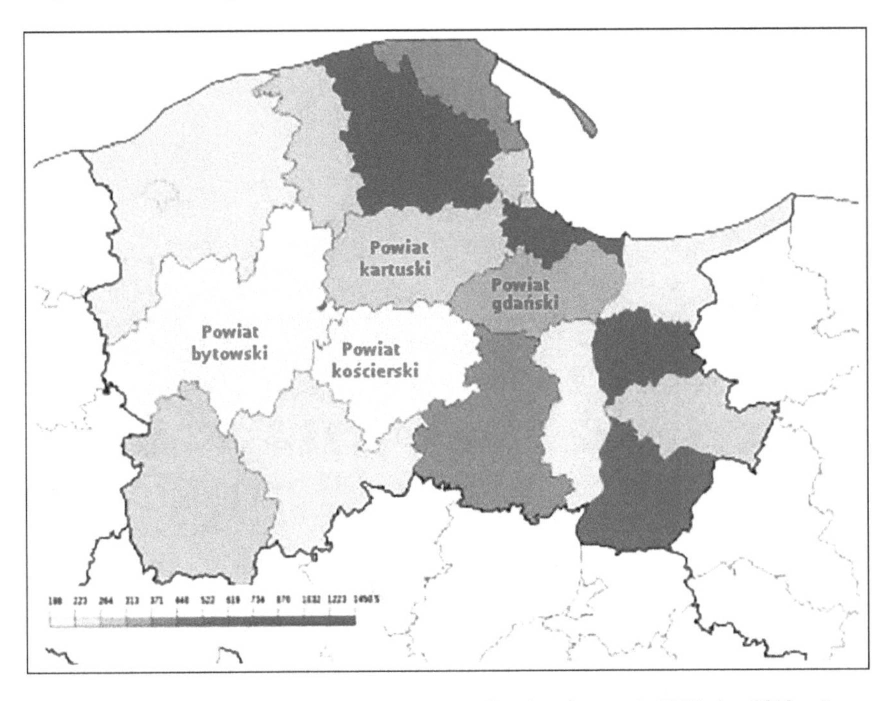

Fig. 4 Labour productivity in industry per capita in thousand PLN in 2012. *Source* Industry in Pomorskie voivodeship in 2009–2012, Central Statistical Office, Branch in Gdańsk, 2014, http://gdansk.stat.gov.pl/publikacje-i-foldery/przemysl-budownictwo/przemysl-w-wojewodztwie-pomorskim-w-latach-2009-2012,1,2.html [March 10, 2018]

capita was not more than 33.88 thousand PLN (Median, Q2 = PLN 33.88 thousand); in the other poviats, GDP per capita was not less than 33.88 thousand PLN (poviats: Słupsk, Starogard Gdański, Kwidzyn, Tczew, Gdańsk, Malbork, Słupsk City, Gdynia City, Gdańsk City and Sopot City).

As shown in Fig. 5, the lowest ratio of GDP per capita in Pomorskie Voivodeship was recorded in poviats: Kościerzyna, Bytów Kartuzy, Chojnice, Lębork and Człuchów.

4 Relationship Between the Potential of Road Transport Enterprises and the Level of Socio-economic Development

The diversification of the potential of road transport companies in the poviats of Pomorskie Voivodeship is related to socio-economic development. The greatest focus is on poviats: Kościerzyna, Kartuzy, Bytów and Gdańsk (Fig. 1). Road transport of goods, expressed in the number of HGVs used, had the highest value. Therefore, the

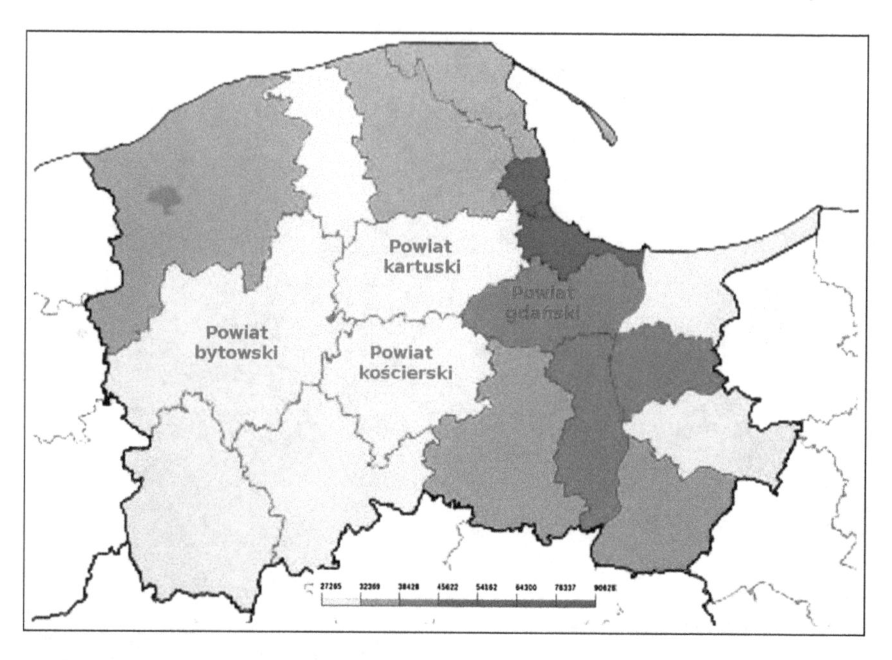

Fig. 5 Ratio of GDP per capita in PLN in Pomorskie Voivodeship in 2015. *Source* D. Ciołek, J. Zaucha. Estimation of GDP per capita at the poviat level. Faculty of Management, University of Gdańsk, 2015

largest number of road transport companies of goods can be noticed in the poviats in which the percentage of inhabitants of rural areas was the highest. It can be assumed that in the 1990s, during the dynamic changes in the structure of agriculture in Poland, it was a complementary activity for farm owners or the direction of retraining. This was partly supported by similar characteristics of the activity, consisting of basic preparation in the field of machine mechanics and the possibility of adjusting the buildings and land to repair and park the vehicles.

A favourable factor for the development of road transport companies in poviats, Bytów, Kościerzyna, Kartuzy, Gdańsk of Pomorskie Voivodeship, was clearly lower labour costs (Gdańsk poviat was an exception—Fig. 3). This factor gained additional significance after opening the market of transport of goods by road transport in the EU in 2004. Among others, low labour costs allowed to develop a competitive advantage on the market of road transport in the EU and the development of export services related to this industry.

In the poviats of Pomorskie Voivodeship, in which the road transport of goods is developed the most, there is also low labour productivity in industry (Fig. 4). In these poviats, there is the lowest number of industrial enterprises in the whole voivodeship, and the few that are do not have technologies that enable them to achieve medium or higher productivity in the voivodeship.

A similar relationship is visible in relation of GDP per capita, which is an indicator of the level of economic development (Fig. 5). It reached the lowest value in those poviats in which the road transport of goods was developed the most. (The exception was Gdańsk poviat.)

5 Summary

Road transport, unlike other branches of transport, is characterised by a relatively low barrier of entry into the profession of a carrier. This makes companies registered in virtually every commune in the country. The potential of car transport enterprises is unevenly distributed in various poviats in Poland. In the case of Pomorskie Voivodeship, the largest development of road transport enterprises can be observed in the least developed poviats in relation to the average in the voivodeship. The low level of economic development of poviats is reflected in such indicators as: labour productivity in industry, GDP per capita and average gross remuneration.

The reasons for the high development of road transport enterprises in Kościerzyna poviat should also be seen in the increased demand for transport services at the back of two seaports—Gdańsk and Gdynia—and the numerous open-cast aggregate mines in Kościerzyna, Kartuzy and Bytów poviats.

The activity of road transport companies is significant, sometimes crucial for the development of the lowest level territorial units—poviats and communes. According to observations and interviews, a significant part of poviat and commune [gmina] managements in Pomorskie voivodeship does not have sufficient knowledge regarding the economic potential of road transport companies operating in their area and their importance in the economic life of the region. Therefore, there is a need for further research regarding the impact of road transport companies on the development of these economically weakest territorial units.

It can be assumed that development trends in road transport, concerning the implementation of intelligent transport system (ITS) and the introduction of autonomous vehicles, will have a major impact on the labour market in Kościerzyna, Kartuzy, Bytów and Gdańsk poviats, i.e. those poviats of Pomorskie voivodeship, in which the potential of road transport companies has developed the most.

References

1. Grzelakowski, A.S., Matczak, M., Przybyłowski, A.: Transport Policy of the European Union and its Implications for Transport Systems of Member Countries, pp. 9–10. WAM, Gdynia (2008)
2. Wojewódzka-Król, K., Załoga, E.: Transport. New Challenges, p. 96. PWN, Warsaw (2016)
3. EU Transport in Figures, Statistical Pocket Book, pp. 24–25. Luxembourg (2016)
4. EU transport, op. … p. 40

5. Central Statistical Office: http://stat.gov.pl/statystyka-regionalna/jednostki-terytorialne/klasyfikacja-nuts/historia-klasyfikacji-nuts/, 21 Feb 2018
6. Regulation (EC) No 1888/2005 of the European Parliament and of the Council of 26 October 2005 amending Regulation (EC) No 1059/2003 as regards the establishment of a common classification of Territorial Units for Statistics (NUTS) due to the accession of the Czech Republic, Estonia, Cyprus, Latvia, Lithuania, Hungary, Malta, Poland, Slovenia and Slovakia to the European Union, EU Official Journal L 309 (2005)
7. Act of 12 January 1991 on Taxes and Local Fees, Dz. U. item 1785 (2017)
8. In Monitor Polski [The Polish Monitor] of 9 August 2017, item 800, issued on the basis of Art. 20. 2 of the Act on Taxes and Local Fees, the maximum and minimum tax rates for 2018 are specified. For a tractor or a weighted tractor adapted for use together with a semi-trailer or trailer with a maximum total weight of a combination of vehicles equal to or higher than 12 tonnes and over 36 tonnes—from PLN 1,684.91 up to PLN 3,130.90. For trailers or semi-trailers, which together with a motor vehicle have a permissible total weight equal to or higher than 12 tons, except for those above 36 tons related solely to agricultural activity carried out by the tax payer of agricultural tax—from PLN 921.50 up to PLN 24,199.98
9. These rules are governed by Regulation (EC) No. 1071/2009 of the European Parliament and of the Council of 21 October 2009 laying down common rules concerning the conditions to be complied with to practise as a road transport operator and repealing Council Directive 96/26/EC; and Regulation (EC) No. 1072/2009 of the European Parliament and of the Council of 21 October 2009 concerning common rules for access to the international road haulage market

Transport of Works of Art—Challenges and Opportunities—Case Study

Katarzyna Caban-Piaskowska

Abstract The aim of the article is to show what opportunities and challenges stand before specialised transport, i.e. the transport of works of art. The study was divided into theoretical and practical parts. The theoretical part describes the reason for dealing with the issues and the essence of transporting works of art and the reasons for further development on the market of transporting works of art. The practical part of the article aims at presenting the results of research in the field of transport of works of art. The case study formulated on this basis presents the characteristics of the transportation of works of art, including challenges and opportunities for this type of activity.

Keywords Work of art · Transporting works of art · Specialised transport

1 Introduction

The art market in Poland and the world is developing very dynamically. It is a global market, and as a result, specialised transport services covering the entire globe are increasingly needed for the transport of works of art. On the one hand, operating in the field of transporting art is very lucrative, but on the other, it is connected with high risk of possible destruction or theft of goods carried, because they tend to be delicate, have high value and are often oversized. The ability of the company to ensure the security of transported goods (works of art) determines the achievement of a competitive advantage and its further success on the market.

K. Caban-Piaskowska (✉)
Faculty of Industrial and Interior Design, Strzemiński Academy of Art
in Łódź, Łódź, Poland
e-mail: k.caban-piaskowska@2.pl

M. Suchanek (ed.), *Challenges of Urban Mobility, Transport Companies and Systems*, Springer Proceedings in Business and Economics,
https://doi.org/10.1007/978-3-030-17743-0_18

2 Transport on the Market of Works of Art

2.1 Reasons for Developing the Market of Works of Art

The value of a work of art seems to be arbitrary as a rule as it is founded in subjective opinion [1, 2]. The basic reason for coming into contact with a work of art (which effects in the desire to own it) is its subjective reception. However, if we look at how a work of art obtains its value, we will see a number of operations performed by different participants of this process. And, respectively, the creator defines what is the work of art, the owner of a gallery exhibits the work of art, critics and curators interpret a particular work of art in comparison with other works of art, formulating diagnoses, and as a result, the audience discusses not only the work of art itself, but also the meaning that was created in the above-mentioned process. Contemporary multitude of alternative tools for constructing the process (e.g. author's Internet gallery) does not change the essence of the process itself. The work of art—as indicated by Boltanski—is characterised first of all by its uniqueness, singularity; even if it is multiplied and copied, the issue of originality has fundamental meaning for its value [1].

Works of art can be seen as emotional assets, which are more and more often perceived as investments [see also 3, 4]. Moreover, history and contemporary practice prove that works of art are infallibly luxurious goods [5].

After the Second World War, there was an increase in turnover and development of the market of works of art on an unprecedented scale. The interest in buying works of art not only for collecting but as a form of investment rose, as the rate of return from the investment from a work of art can be compared to the one from shares and sometimes it is even higher [6]. The situation is similar in Poland. Together with the increase in the level of wealthiness, Polish people choose various ways to invest. They do not only buy shares of companies operating on the stock market, but also more and more often invest in art (paintings, sculpture or artistic craftsmanship). Research conducted by KPMG indicates that among people who earn 7.1–10 thousand PLN gross, only 5% invest in works of art currently. However, among Poles who have higher income (over 20 thousand PLN gross), the percentage increases up to 18%, while 36% plan such investment in the nearest future [5].

Sociologists who analyse the market of art [1, 7, 8], notice that more and more often the market of art, through the mechanisms which appear, seems to say a lot about the nature of contemporary capitalism, and certainly the market has become the area of intensified activity of investors. And we are not only talking about collectors of works of art, owners of fortunes or art dealers, but about institutional investors, who more and more often treat investing in works of art as one of the so-called alternative investments [3, 4].

Big interest in works of art caused a large number of art auctions. Many auction houses have numerous branches, especially the biggest ones: Sotheby's, Christie's, Philips—they have from a dozen or so to several dozen branches and representative offices all over the world. The first one has 20 branches abroad on four continents and as many as 75 representative offices around the globe [6].

The situation is similar in case of trade fair events (such as Tefaf in Maastricht, Art Cologne, Fiac in Paris and first of all Art Basel in Basel and Miami), which are developing in an unheard-of-before way nowadays, the proof of which is the growing number of exhibitors and presented pieces [6].

2.2 *The Reason for Focusing on the Issue of Transporting Works of Art*

There are a number of reasons why transporting works of art is an important and interesting area for research. First of all, currently we can observe increase in the turnover on the market of art on an unprecedented scale and development of the market connected with it. This has led to significant growth of prices for works of art and in the same time forced the creation of a system for specialised transport for them [see 6].

Secondly, on the market of works of art next to the past buyers from the USA, Western Europe and Middle East, there appeared the so-called new players, i.e. vendors from China, India and Russian oligarchs. For this reason, art travels from Europe to the USA, to Asia and back, which requires transport between continents on an uncommon scale [6].

Thirdly, the number of auction companies together with their branches has risen significantly all over the world. The situation forces the owners of auction houses and organisers of such auctions to master the difficult skill of transporting works of art to perfection and constantly improve the solutions for logistic problems that arise as they organise bidding and sell art almost all over the globe [3, 9].

Fourthly, the increase in the number of trade fair events in the world makes the issue of transporting works of art on large distance not to be a marginal concept any more [6].

Fifthly, the development of Internet art trading (online and live auctions), where pieces from all corners of the world are sold, poses a challenge for logistic companies [3, 10].

Sixthly, there is a huge interest in organising and looking at exhibitions of art as well as contests in Poland and in the whole world [1, 11].

2.3 *The Essence of Art Transportation*

The main customer for companies which specialise in transporting works of art is museums; nevertheless, the meaning of private collectors has grown—starting with

2006, when the average level of prices for works of art on auctions and in dealer trade increased, as much as 30% of offers were addressed to them [6, 12].

There are around 19,500 companies which can potentially offer services of transporting art [13]. In practice, there are three important players in Poland.

Two companies render services first of all within the scope of public tenders [14]. According to the data from this portal, analysed on the basis of Public Order Newsletter/Biuletyn Zamówień Publicznych (BZP), Renesans Trans from Warsaw has won 21 tenders with 12 commissioning parties since 2013. The estimated value of the orders is 2,354,974 PLN. Art Logistic from Łódź, which has belonged to the German concern Hasenkamp Group since 1996, won 19 tenders organised by 13 entities. The estimated value of those contracts is 4,072,047 PLN [12]. The third main player on the market of transporting art—Pol-Artu addresses its offer directly to artists and private collectors. It is a company which focuses mainly on packing and collecting [15].

Transporting works of art is characterised by the following needs [6, 10–12, 16]:

- proper security of the transported things,
- proper packaging,
- climate conditions (humidity and temperature),
- strictly defined time of realisation,
- insurance,
- compliance with legal regulations.

Margery Gordon, the author of columns dedicated to the market of works of art and issues connected with their transport emphasises that works of art are the most vulnerable to damage during transportation [6]. As a result, the companies which transport works of art adapt their vehicles so that they could deliver the load in the safest possible way. The cars have hard container-like space with isotherm, soft suspension, padded load space and a special system of handles and belts for securing the load against shifting [12].

Robert Noortman, who deals in art in Maastricht, admits that the most valuable pieces are kept in two types of chests. The outer one is equipped with a sensor which signals any damage or shifting of a given shipping. If he finds proof that the shipping was not treated appropriately, he opens it assisted by a heritage conservationist and a representative of the shipping company, in order to avoid making further mistakes when handling the piece and possible misunderstandings [16].

Works of art should be provided with optimum atmospheric conditions (humidity and temperature) during transport. As a result, the vehicles have air-conditioned load space with the possibility to maintain permanent temperature of 20 °C and permanent humidity of 45–55% [17].

An incredibly important issue when transporting works of art is the fixed period of time allocated for the delivery. It is connected with the fact that pieces dedicated for a traditional auction, i.e. a hammer auction, must be exhibited on a given day at a fixed time or—in case of trade fair events—must be presented at a trade fair on a given day. Especially because pieces on auctions and trade fairs come from various, often remote places on Earth [6].

It is a standard procedure when transporting art, to include insurance to the value of 100 thousand PLN, but it is possible to organise transports with a higher declared value of the shipped pieces [10, 12].

The most important Polish legal acts connected with transporting art are the following: Regulation of the Minister of Culture and National Heritage of the 15 May 2008 on the conditions, methods and ways of moving museum exhibits [18] and of 2 September 2014 on securing museum collections against fire, theft and other dangers which might cause damage or destruction [1, 19], Act of 23 July 2003 on protecting and taking care of relics [20] and the Act of 22 June 2017 on amending the act on protecting and taking care of relics [21] and some other acts. In the second of the above-mentioned, the crucial thing is §22. According to the paragraph, protecting collections during transport means adjusting the type of vehicle, the kind of convoy and the number of guards to the type of collection and its value. The destination and the length of the journey should also be taken into account. We should also provide the attendance of a museum courier while transporting collections and meeting the requirements in the area of organising protection for exhibits in transportation defined by appendix 3 to the Regulation. Transporting museum collections, where appropriate temperature, humidity or the slightest vibrations can influence their condition, is done with the use of means of transport which guarantee such environment. The museum courier can refuse to transport art if the vehicle does not fulfil the requirements [11].

3 Research Results

3.1 Research Methodology

In order to verify the hypothesis that transporting works of art has characteristic features and its specific character depends on the receiver for whom the service of transport is rendered, the author decided to gather data by conducting deepened interviews and observation analysis of documents and reports as well as interviews with the former Director of the National Museum in Gdańsk, with heritage conservationist in Gdańsk, with an ex-employee of Dessa and with a graphic artist who used the service of art transportation between the years of 2017 and 2018 in Poland.

The aim of the analysis was to answer the question: what are the characteristic features of transporting works of art from the point of view of various groups of customers.

In order to obtain empirical conclusions from the data, a thorough and multifaceted analysis was used, which gave the chance to obtain a possibly accurate and multifaceted picture of cases of a given phenomenon. The choice of method resulted from the initial stage of the research, the unique phenomenon, as well as the will to make the analysis wide enough and the issue to be understood better.

3.2 The Criteria for Selecting the Sample

Choosing four people who represent different groups of customers of companies which transport works of art was deliberate and was supposed to present the same phenomenon from various points of view.

3.3 Results

The research conducted indicates that the analysed companies and creators characterise with the following features:

According to the research conducted, the heritage conservationist and the museum representative chose specialised companies for transporting works of art, and there are two of them in Poland.

However, according to the representative of an art gallery and a young artist for the above-mentioned purpose, he always seeks a company which is cheap and provides secure transport or he handles the pieces personally. Very often, when sending small works of art, e.g. graphic designs, as many people as possible are gathered to send their pieces in one package in order to share the cost of transport; thus, the parcel contains from 5 to even 30 works. The transport is then equally shared among the participants who send their pieces. There are exhibitions which offer guaranteed financing of transport—but they are an exception, and then, a specialised company is chosen.

If an artist or a gallery uses the services of a courier, according to the representative of a gallery, it is important to add parcel tracking on your order on the courier's website. It is significant, as the customers tend to be impatient and like to ask questions about where the parcel is, especially the Americans.

If the package is not over 2 kg and the sum of all the sides does not exceed 90 cm, it can be sent by a registered letter. The cost is much lower. Nevertheless, the tracking of letters has been disabled in the USA recently. That is why sending small packages to the USA is better with the option of Global Express (Poczta Polska), even though it is more expensive. Tracking your parcel is essential. Without online tracking, we can expect problems with delivery or the package can go missing.

In case of young artists, packages are not insured due to financial reasons, and parcels, especially those sent abroad, are marked as "not valuable". It is connected with the duty which must be paid for the declared high value of a parcel. It is assumed that the duty should be paid for by the receiver; however, organisers of contests condition in the rules that the artist is responsible for delivering the works, financially as well. Problems appear when a customs office of a given country decides to check what is in the parcel.

Very often, organisers of contests determine in the rules of the contest that the works are returned at the cost of the artist as well. Hence, young creators decide to leave the works with the organiser. This happens when the possible price for which the pieces could be sold later is lower than the cost of transport; in such situation, getting the works back does not pay off. This is often the case with graphic designs.

Transporting art by the artist him/herself often stems from two things: it does not pay off to transport the works; this concerns for example sculpture, where due to its weight (and sometimes the weight of the box made of OSB or wood too) or atypical size, the price is very high or the courier company/post office refuses to handle the parcel. Thanks to transporting art on your own; according to our respondents, they can be sure that no one will be throwing the package.

As reported by our interviewees, a company which transports works of art characterises with properly adapted vehicles. Trucks can be big or small, depending on your needs. They are equipped with special belts for strapping chests, often with an AC unit, which provides appropriate conditions inside during long stops, ramps for unloading, etc. another example of a special feature of a company which carries art is the packaging. The chests are padded with suitable materials, which make it possible to safely (i.e. with secured surfaces and edges) transport various objects in vertical position. Stability of the objects as well as minimising the influence of vibrations on the way is crucial. Another example is the specially trained employees. Very often, there is security when transporting art, especially on international distances. The conditions of transport are defined by the standards of galleries and museums, as well as the insurance. Courier companies know them and are able to fulfil them; they have certificates to prove it. The criterion for choosing a given company in a tender for transporting works of art of high value from museums and other public institutions is having certificates for such type of transport. Another criterion might be the experience—i.e. what institutions selected the company and how many times it happened.

A characteristic feature of specialising in transporting art is the significance of the carried loads, which goes beyond their financial value.

According to the respondents, the differences between a company which transports works of art and a company which carries other things are: the people, the equipment, the certificates, experience and insurance of the art up to the value of 100 thousand PLN, which is included in the price of service. It is possible to order transport with a higher declared value of the objects.

Companies which specialise in transporting works of art have the competitive advantage over other similar companies, thanks to the quality and quantity of equipment and the skillfulness of the qualified workers, who pack the works and take this duty off of the gallery or the museum; there is always conservationist supervision during packing, and each object is accompanied with the so-called opinion, i.e. a condition report, with photographs, where in compliance with a special scheme, the state of objects before they are put inside the chests is marked in the form of pictures and descriptions. When the chests are open, the reports are checked by: a courier conservationist from the institution from which the pieces were sent and the conservationist from the receiving institution. The price is also a competitive factor: what kind of insurance is offered, how many cars they have and how big they are, how many drivers—i.e. how far they can go (changing the driver every 8 h—tachometer). However, such transport is only affordable for museums and rich institutions.

Companies which handle art transport are out of reach for young artists due to high prices.

As our respondents say, the direct competition of companies which carry art in Poland is not huge—there are only two such companies. In European companies, they met those two and two more companies. The services are specialised and expensive; they generate a lot of costs; thus, there are not many companies in the sector.

According to the respondents, transporting works of art is done with the use of specialised courier companies due to the regulations and acts on protection of relics and museology. There is also the question of insurance. These are often objects of very high (not only) material value, and they must not be damaged. Various ways of securing the pieces are supposed to protect them from getting broken while being transported and in case of an accident; nevertheless, it is also important to maintain the same temperature and humidity during the journey for example in case of paintings. Changes of these parameters during the day are very harmful for most of museum objects, which are historical and valuable.

As the interviewees report, specialised transport companies can carry any type of art which an institution agrees to borrow to another institution for an exhibition—those that can be bought as well. Private customers rarely use such services because of the price; however, auction houses also respect those rules.

A specialised company which ships works of art, as the respondents claim, is responsible, if it decides to do it, for proper packaging of flat objects, paintings, ceramics, stone, glass, etc.—different for each type of things, for appropriate securing of the objects on the car, for AC and respecting other special recommendations included in a contract, etc.; for safety during transportation, including guards, as well as for watching the load during stops if there are no guards. Overnight stops are scheduled in places designated by the person who plans the route.

4 Conclusions

The direct cause for getting interested in the issue of transport on the market of art for an economist or a market analyst is the increase in its global significance. The road from the market of art to the capital market has shortened greatly; the distance between the market of art, not very fluxional and considered arbitrary, and a capital market has shortened in terms of financial and capital connections, but also, due to institutional conditions, in Poland both markets are still distant from each other [4].

For transporting works of art, those valuable ones and belonging to national heritage specialised transport companies should be chosen. However, only public institutions, museums, big galleries and auction houses can afford them. Other institutions and less-known artists (or those who are at the beginning of their career) decide to use courier companies or the post office, because of lower costs. Unless, the sent object is extremely valuable or the cost of transport is covered by the receiver of the package.

Proper level of transport is offered only by specialised companies. In Poland, there are two companies which fulfil public orders and the third one which realises private orders only, but it is a very expensive service.

For smaller players on the market of art, sending pieces seems to be quite problematic not only because of financial reasons. It often happens that the corners of the works are folded or some other damage is done to them. Few people decide to send a sculpture as it is easy to damage it, especially if it is made of fragile material such as plaster.

Insuring works of art up to 100 thousand PLN is included in the price of the courier service; nevertheless, it is possible to organise transport with a higher declared value of the transported objects. Such insurance is only accessible for big and medium players; smaller players declare that the shipped pieces have no value or decide to deliver the pieces on their own.

References

1. Boltanski, L., Thevenot, L.: On Justification: Economies of Worth. Princeton University Press, Princeton (2006)
2. Benjamin, W.: Twórca jako wytwórca. Wydawnictwo Poznańskie, Poznań (1975)
3. Cichorska, J.I.: Rozwój rynku sztuki jako przykład inwestycji alternatywnych. Ann. Univ. Mariae Curie-Skłodowska Sect. H Oecon. **49**(4), 69–83 (2015)
4. Lewicki, M.: Wycenić ideę. Z rynku sztuki na giełdę iz powrotem. Analiza procesów wartościowania i wyceny na przykładzie rynku sztuki. Psychol. Ekon. **22**(7), 5–24 (2015)
5. Raport KPMG Rynek dóbr luksusowych w Polsce Edycja 2016
6. Korzeniowska-Marciniak, M.: Pakowanie, transport i magazynowanie dzieł sztuki (cz. 1). Logistyka 3/2007, 29–32 (2007)
7. Hutter, M., Throsby, D.: Beyond Price: Value in Culture, Economics, and the Arts. Cambridge University Press, Cambridge (2008)
8. Karpik, L.: Valuing the Unique: The Economics of Singularities. Princeton University Press, Princeton (2010)
9. Korzeniowska-Marciniak, M.: Międzynarodowy rynek sztuki. Univertas, Kraków (2001)
10. Niewiarowski, W.: Artysta u progu kariery zawodowej. Galaktyka, Łódź (2017)
11. Matassa, F.: Organizacja wystaw. Podręcznik dla muzeów, bibliotek i archiwów. Univertas, Kraków (2015)
12. Szczodra, A.: Sztuka w transporcie. Truck Bus. **55**, 29–31 (2017)
13. clicktrans.pl/transport/dziel-sztuki
14. przetargi.egospodarka.pl
15. http://eurologistics.pl/wydania_pdf/TBP%202_2017_NET.pdf
16. Korzeniowska-Marciniak, M.: Pakowanie, transport i magazynowanie dzieł sztuki (cz. 2). Logistyka 4/2007, 42–44 (2007)
17. http://dynatrans.pl/pl/transport-dzie%C5%82-sztuki.html
18. Dz. U. 2008 nr 91 poz. 569 Rozporządzenie Ministra Kultury i Dziedzictwa Narodowego: z dnia 15. maja 2008 roku w sprawie warunków, sposobu i trybu przenoszenia muzealiów

19. Dz. U. 2014 poz. 1240 Rozporządzenie Ministra Kultury i Dziedzictwa Narodowego z 2 września 2014 roku w sprawie zabezpieczania zbiorów muzeum przed pożarem, kradzieżą i innym niebezpieczeństwem grożącym ich zniszczeniem lub utratą
20. Dz. U. 2003 nr 162 poz. 1568 Ustawa z dnia 23. lipca 2003 roku o ochronie zabytków i opiece nad zabytkami
21. Dz. U. 2017 poz. 1595 Ustawa z dnia 22 czerwca 2017 r. o zmianie ustawy o ochronie zabytków i opiece nad zabytkami

Knowledge as a Factor of Process Maturity Growth in Polish Logistic Organizations

Aleksander Dreczka and Piotr Sliż

Abstract The purpose of this article is to assess the process maturity of logistics organizations in Poland. The article describes the management of the knowledge resource, defined as a key factor in the implementation of the process approach in the organization management. Two research questions (RQ) have been formulated based on the problem discussed. RQ1: What is the level of process maturity in the surveyed group of logistics organizations in Poland? RQ2: What determinants of knowledge diffusion in the studied units affect their level of process maturity? As a result of the literature review, the desirable characteristics of the knowledge employee in the process-oriented organization are presented. Subsequently, the results of the empirical study were presented, using a survey opinion poll carried out on a non-random attempt to organize logistics in Poland. The empirical approach was carried out using a multidimensional model of process maturity evaluation (MMPM). The study units were evaluated in two dimensions: short term and long term. In the vast majority, the examined objects were classified at the first level of process maturity. This means that the dominant management formula in the units under examination is a functional approach with single symptoms, demonstrating the implementation of process solutions.

Keywords Logistic organizations · Knowledge transfer · Process maturity · MMPM model · Process management

1 Introduction

The management staff of modern organizations operating in a turbulent market environment faces the challenge of designing highly flexible structures that enable

A. Dreczka (✉) · P. Sliż
University of Gdańsk, Sopot, Poland
e-mail: aleksander.dreczka@op.pl

P. Sliż
e-mail: piotr.sliz@ug.edu.pl

M. Suchanek (ed.), *Challenges of Urban Mobility, Transport Companies and Systems*, Springer Proceedings in Business and Economics,
https://doi.org/10.1007/978-3-030-17743-0_19

dynamic response to the variability of the environment as well as factors generated within the organization. Process approach is one of the responses to these challenges proposed in the discipline of management science. This means that the ability to dynamically respond to changes requires the design of process-based activities. This in turn implies the composition of the organization in the market environment in the space of processes. Jokiel [1] indicates the contribution of process management to many management concepts, such as quality management, logistics, marketing, production management, controlling and knowledge management. At this point, it is necessary to emphasize the importance of knowledge in modern organization, referred to as its most important resource. This, in turn, forces the managerial staff to have new competences in identifying intellectual potential and transferring knowledge between all employees in the organization. A departure from the desired role of the employee in restorative proficiency in favour of the presence in two roles at the same time is also significant: the task implementer and stimulator of improvements in the organization [2]. This is confirmed by Grajewski, according to whom organizations operating in a turbulent environment must generate information and knowledge, not only efficiently processing it [3].

2 The Importance of Knowledge in the Course of Transformation Towards a Process Organization

During the first decade of the twenty-first century, business process management has become a mature part of the global achievements in the discipline of management science. Methods and tools combining knowledge in the field of information technology, management theory and industrial engineering were developed to improve business processes. One of the aspects of the world scientific achievements is to determine the success factors of business process management. In one of their papers, Karagiannis and Woitsch [4] distinguished six basic success factors of business process management, based on the literature research, analysis of three models of process management maturity, international research with the Delphi method and a case study, including [4]:

- strategic connections,
- methods,
- IT technologies,
- personnel, including skills and expert knowledge as well as knowledge on process management,
- organizational culture.

This study focuses on the role of knowledge as a component of the personal factor of the success of business process management. The mentioned authors define the importance of knowledge in process management through [4]:

- skills and expert knowledge as focused on versatility and detail in the light of the requirements of given processes,
- knowledge about process management as combining explicit and implicit knowledge about BPM practices and assumptions. It contains a level of understanding of BPM, including knowledge of methods, information technology and their impact on the results of processes.

Process knowledge can also be captured in two dimensions—as an identification of the direction in which the organization and the regenerative capabilities of this knowledge are headed. This approach is presented by Yaxiong et al. [5] specifying one dimension of knowledge as an indication of the direction of the process and the other as the ability to create new knowledge in the processes of the organization. These authors distinguish two types of knowledge about processes in the organization. The first type is the basic process knowledge which focuses on activities in the process and their relations. It is a combination of knowledge related to individual activities organized according to principles. It covers four aspects: resources, rules, staff and purpose [5]. Process knowledge regarding resources contains information about entities that perform activities and objects involved in these activities. The rules describe the time, conditions and limitations of the process activities, for example, the earliest/latest start times, the earliest/last time the activity ended. Knowledge about people involved in the process is an indispensable component of effective management. The objectives and system of process evaluation must be included in process knowledge as a basis for management actions. The second type is creative process knowledge based on the assumption that regeneration is one of the two basic attributes of knowledge, that is, the expansion of knowledge available through deduction and search for cause-and-effect relationships. Knowledge supports the current process progress and contributes to the creation of new knowledge about processes. There are three fundamental approaches to creating new knowledge from the perspective of business process management [5]:

- change of the process assumptions;
- adding knowledge on the handling of unforeseen exceptions when they arise;
- analysis of the process history recorded in the process statistics.

Understanding the importance of knowledge in contemporary process-oriented organizations, the issue of its propagation in the organization's space becomes an important issue. The very definition of propagating knowledge in an organization in the literature is recognized in a different way. Apart from the concept of "spreading knowledge", "transfer", "exchange", "diffusion" or "sharing" knowledge is also encountered. Although the difference between them may seem out of focus, one needs to indicate their semantic differences. Transfer is a purposeful and one-directional action consisting in communicating knowledge, that is, to find a specific application. According to the definition of Argote and Ingram, transfer of knowledge is a process that subjects the individual to the experience of another [6], while sharing knowledge is a multidirectional action without a specific purpose and takes place between equivalent recipients/senders [7]. In turn, dissemination of knowledge applies only to the planned activities of organizations that are to guarantee the access and use of knowledge. This term refers to facilitating access to knowledge, in particular to coded knowledge [8]. In contrast to the sharing of knowledge that is directed to specific recipients, which is important in the context of protection against unauthorized access, the spreading of knowledge creates a publicly accessible resource [9].

Considering the usefulness of the concepts of "distributing", "transfer", "sharing" and "disseminating" knowledge in process organizations, it should be recognized that disseminating will be the most appropriate form of knowledge transfer. This results from the purposefulness of this task, the universality of knowledge made available in the organization (which exhausts the assumption of the complexity of the process organization) and the default encoding of knowledge that can be included in maps and process descriptions.

3 Knowledge Employees in the Process Organization

The transformations that took place in the twentieth century in the economic environment, the progressive automation and computerization of work resulted in the creation of a new concept of an employee—a knowledge employee. Table 1 presents the selected characteristics of the knowledge worker.

When considering the characteristics of a knowledge worker in the context of process management, attention should be paid to its participation in knowledge processes,[1] its independence and responsibility.[2] As a result of decreasing dispersion in the education of employees, traits such as formal education or professional specialization do not reflect the employee participation in process management precisely

Table 1 Knowledge employee distinguishing features

Author	Knowledge employee distinguishing features
P. F. Drucker	Formal education
N. Beck	Specific professions
Central Plan Bureau	Formal education
T. Davenport	Specialist knowledge, participation in knowledge processes
M. Morawski	Specialist knowledge, independence
E. Skrzypek	Participation in knowledge processes
K. Perechuda	Independence, network interactions
D. Morello, F. Caldwell	Independence, responsibility

Source Reference [10]

[1]The concept of "knowledge processes" includes two issues related to knowledge—its acquisition and codification.

[2]The importance of independence and responsibility of employees in the context of process management is emphasized by [1].

any longer. This means that the dynamic environment implies the need to give broad prerogatives that enable employees to solve problems during the implementation of processes. This, in turn, depreciates the usefulness of the restoration skills of the staff. Independence, responsibility and individuality are other terms for the validation of process contractors and the network interactions a cooperation with the other participants of the process regardless of their functional location. According to Brajer-Marczak, improvement of processes is related to the development of individual knowledge of employees, the use of experience to solve problems and teamwork skills [11], thus presenting the attributes of the knowledge employee.

4 Results of the Study of the Process Maturity Assessment of Logistic Organizations in Poland

4.1 Assessment of Process Maturity Using a Multidimensional Model of Process Maturity Assessment

Process maturity of the organization is defined as "the state of the system, in which the benefits derived from the level of advancement of the applied process solutions are discounted in a conscious way" [12]. It has a gradual character, and specific patterns are used to evaluate it, which are referred to in the literature as models for the assessment of process maturity. In turn, the level of process organization maturity identifies the degree, to which processes are formally defined, managed, flexible, measured and effective [13].

The study characterized in this article was carried out using a multidimensional model of process maturity assessment of the MMPM organization [2, 14]. At this point, it should be emphasized that the level of process maturity in the surveyed organizations has been identified in two dimensions: short and long terms. In this article, the levels of maturity in the short-term assessment have been marked as follows: L1—management organization functionally showing weak symptoms of implementation of elements of the process approach, L2—identified and formalized processes, L3—measurement processes, L4—managed processes and L5, the highest level—improved processes. In addition, the evaluation criteria for individual levels have been extended by three long-term dimensions, in order to analyse the implementation of elements of the process approach in the long run. The following were qualified in them: development in the implementation of process solutions, stagnation, defined as a state in which the organization remains at the current level of maturity and atrophy, understood as the cessation of implementation of process solutions and oriented towards a functional approach to management [2, 14].

4.2 The Structure of Empirical Proceedings and the Structure of the Studied Population

The empirical proceeding was carried out using the opinion poll method with a non-probabilistic sampling technique [15]. At this point, it should be emphasized that the selection of the sample was deliberate. The choice of the research method and sampling technique proved appealing to general opinion polling centres in Great Britain [16]. At this point, it should be emphasized that the sample selection used disallows the analysis of statistical results. The selection of units was the size of the organization and the location. On this basis, the organization group was selected. The direction of further research and described issues is to confirm the conclusions formulated on the basis of this quantitative research, based on the empirical facts examined. The empirical proceedings were carried out in the third and fourth quarters of 2017 using the CAWI technique[3] on a sample of 44 organizations operating in the transport, storage and courier sectors [17]. All organizations included in the study were located in Poland. The structure of the surveyed population was determined on the basis of the employment in the surveyed organizations [18]. They were classified according to the criterion of the number of employees employed. On this basis, the following were distinguished: micro-organizations—employing up to 9 employees (43.18%), small organizations employing from 10 to 49 employees (34.09%), average organizations from 50 to 249 employees (11.36%) and large organizations employing more than 250 employees (11.36%). In addition, the results obtained in the empirical proceeding made it possible to differentiate the respondents according to the criterion of the employment position. The largest percentage share in the surveyed organizations was the group of CEOs and owners (50.00%), then the directors and executives (20.45%), specialists (15.91%) and managers and coordinators (13.64%). More precisely, the research questionnaire was addressed only to the middle and senior management.

4.3 The Results of the Assessment of Process Maturity of the Examined Objects

Summing up the results of the empirical study presented in Table 2, it should ne stated that the majority of the examined objects in Poland were qualified to the first level of maturity. In addition, it has to be emphasized that in the long-term dimension, 34.09% of the facilities were placed at the L1 E+ level. According to the assumptions of the MMPM model used, the identified level in which, despite the dominant formula of the functional approach in management, single symptoms can be observed that indicate the initial stage of implementation of process solutions [2, 14].

[3]CAWI (ang. computer-assisted website interview)—a method based on the implementation of a quantitative survey via the Internet. The study used a tool available at www.webankieta.pl.

Table 2 The results of the process maturity assessment of logistic organizations in Poland

Level	Designation	Number of organizations	Percentage (%)	Accumulated percentage (%)
Level 1	L1 E−	0	0.00	0.00
	L1 E	14	31.82	31.82
	L1 E+	15	34.09	65.91
Level 2	L2 D−	0	0.00	65.91
	L2 D	1	2.27	68.18
	L2 D+	3	6.82	75.00
Level 3	L3 C−	0	0.00	75.00
	L3 C	3	6.82	81.82
	L3 C+	2	4.55	86.36
Level 4	L4 B−	0	0.00	86.36
	L4 B	0	0.00	86.36
	L4 B+	3	6.82	93.18
Level 5	L5 A−	3	6.82	100.00
	L5 A	0	0.00	100.00
	L5 A+	0	0.00	100.00
Total		44	100.00	100.00

Source Own study

Of the research objects qualified for the second level of maturity (9.09%), most of them were qualified for the dimension—development (D+). This means that organizations showed symptoms indicating a possible transition to the third level in the long-term dimension. Then, among the five examined units qualified for the third level of maturity, none indicated any symptoms proving the atrophy of the implementation of the elements of the process approach in favour of functional solutions. In addition, only three organizations were qualified to the fourth level (L4 B+). Three units meeting the criteria of the fifth level of process maturity according to the MMPM model were identified in the sample of the organizations examined (Table 2).

4.4 The Analysis of the Selected Criteria for Process Maturity Assessment of the Multidimensional Organization Process Maturity Assessment

The selected process maturity assessment criteria presented in this section refer to the epistemological part of this article. This means that in addition to presenting the summary results of the process maturity of the group of organizations under investigation, the results of partial questions regarding the subject of knowledge management are described.

Among the respondents' answers regarding the desired role of the employee in the organization, the most frequent response was the role of an employee being an independent member of the interdisciplinary team executing tasks and stimulating improvement throughout the organization, desirable from the perspective of the process organization determinants, indicated by 34.09% of units. The share of the remaining answers was as follows: the role of multitasking implementer in the area of the selected department of the organization (22.73%), the formula of the contractor-assigned tasks and the initiator of improvements at the position held (22.73%), the role of the expert performed of other tasks (6.82%). At this point, it should be emphasized that 6.82% marked the answer "none of the above".

The next question asked the respondents to indicate the type of training implemented in the examined group of objects. The distribution of responses is as follows: the organization carries out internal training (50.00%), they are carried out by specialist external organizations (45.45%), training is carried out according to the planned training cycle (13.63%), additional training resulting from the current needs, %) and supplementary training, which do not require participant attendance, are implemented on the principle of Internet training (e-learning) (13.63%) [2, 14].

Then, the respondents were asked about the desirable nature of the training being carried out. In the vast majority, in the analysed organization, the trainings: create the direction of learning new ways of acting (50.00%), are one of the elements of the incentive system (50.00%) and create the development of competence (47.72%). The minority share of responses: inform about the benefits and threats resulting from the planned changes (25.00%) and are the element of organizational strategy (18.18%). At this point, it should be emphasized that the following answers were not marked in any of the analysed organizations: trainings allow employees to exchange views [2, 14].

Another element related to the area of diffusion of knowledge in the organization was the assessment of the implementation of internal training in the studied group of units. In 43.18% of such facilities, this type of training results from the needs of implementing new employees, in 38.64% it concerns the implementation of current changes in the organization, in 22.72% of the facilities, internal trainings are implemented in a planned and cyclical manner, whereas in 15.91% they are the result of employees' inventiveness in transferring knowledge obtained during external training. At this point, it should be emphasized that in 29.54% of the surveyed organizations, internal training is not implemented [2, 14].

In the further part of the study, an attempt was made to assess the desired and expected status of a leader in the surveyed units. In 38.64% of the surveyed group of organizations, a response was indicated pointing to the role of the coordinator of the tasks of the subordinate subdivision, division of department, solving problems during the process realization. From the perspective of the process organization, the desired response was to indicate the leader responsible for the transfer of knowledge between employees, intervening when the implemented personnel activities depart from the established assumptions. This position was indicated by 18.18% of the respondents.

The remaining answers were as follows: coordinating the tasks of the subordinate subdivision, division or department (6.82%), specialist in the implementation of tasks in the selected department (20.45%) and none of the answers proposed in the questionnaire (15.91%) [2, 14].

In one of the questions included in the research tool, the respondents were asked about the nature of the improvements made in the whole organization. In the largest number of units, they are planned based on the customer requirements (38.36%). It was further confirmed that improvements are generated by all employees (15.91%), are carried out during the implementation of the process (36.36%), are conducted on the basis of the cost analysis of actions in individual processes (34.09%), are planned on the basis of the identified external and internal threats (e.g. crisis) (25.00%), start with the planning the course and the date of implementing the improvement (22.72%) and are designed by the process planning centre (6.82%) [2, 14].

5 Summary

The results of the process maturity assessment of organizations operating in the economy sectors: transport, warehouse management and courier services in Poland, presented in the article, show that in the majority of surveyed organizations, despite high awareness of process identification, they are not formalized, metered and managed. Moreover, only 6.82% of the organizations were classified as the fifth, the highest level, while 6.82% at the fourth level, 11.37% at the third level, 9.09% at the second level and 65.91% at the first level.

Three general conclusions were formulated on the basis of the analysis of the results obtained.

First, defining the term "knowledge" as "information about reality, stored in the memory of the subject of the action" [19], the thesis of Krzakiewicz should be quoted, according to which "all economic and managerial relations are based on knowledge, the preferences of the participants of these relations are revealed and the exchange process becomes understandable" [20]. It should be understood that, in the assessment of process maturity, it is necessary, after symptoms, to perform the analysis of symptoms concerning the attempts to acquire knowledge in the surveyed organizations and its transfer between employees.

Secondly, the relationship between the size of the examined units and the level of process maturity was confirmed. This means that in the surveyed group of organizations employing up to 9 employees, in 84.21% the units were classified to the first level of maturity (L1). The low degree of scale of activity, which does not require documentation of identified processes, may be the reason for this situation, in which processes are not formalized, metered and managed in the micro-organizations [21].

Thirdly, it should be noted that in 63.63% of the organizations surveyed, the role of the employee was determined in accordance with the determinants of the process organization. Moreover, only in 21.42% of organizations in this group it pointed to the desirable role of the leader responsible for the transfer of knowledge between employees, intervening when the implemented personnel actions deviate from the established assumptions. Such a result may indicate that despite high awareness of the employee's role in process-oriented organizations, the role of the leader remains in the functional formula of the department's work coordinator, solving problems during the process [21, 2, 14].

Balancing the above, it should be noted that due to the non-probabilistic sample selection technique, the conclusions formulated in this summary cannot apply to the entire population of logistic organizations, but only to the group of objects under study.

References

1. Jokiel, G.: Podejście procesowe w zarządzaniu—geneza i kierunki rozwoju koncepcji. EU research works in Wroclaw, no. 52 (2009)
2. Sliż, P.: Concept of the organization process maturity assessment. J Econ & Manage, Katowice (2018)
3. Grajewski, P.: Organizacja procesowa. PWE, Warszawa (2007)
4. Karagiannis, D., Woitsch, R.: Knowledge engineering in business process management. In: vom Brocke, J., Rosemann, M. (eds.) Handbook on Business Process Management, vol. 2. Springer, Heidelberg (2010)
5. Yaxiong, T., et al.: A research on BPM system based on process knowledge. In: Proceedings of IEEE Conference (2008)
6. Argote, L., Ingram, P.: Knowledge transfer: a basis for competitive advantage in firms. In: Organizational Behavior and Human Decision Processes, vol. 82 (2000). In Rudawska, A.: Dzielenie się wiedzą w organizacjach—istota, bariery i efekty. Organizacja i kierowanie **4**(157) (2013)
7. King, W.R.: Knowledge sharing. In: Schwartz, D.G. (ed.) Encyclopedia of Knowledge Management. Herslen, London (2006)
8. Probst, G., Raub, S., Romhardt, K.: Zarządzanie wiedzą w organizacji. Oficyna Ekonomiczna, Kraków (2002)
9. Mikuła, B.: Zadania organizacji w zakresie zarządzania wiedzą. E-mentor **5**(17) (2006)
10. Kowalski, T.: Pojęcie i cechy pracownika wiedzy, vol. VII. Lubusz Studies, State Higher Vocational School in Sulechów, Sulechów (2011)
11. Brajer-Marczak, R.: Doskonalenie procesów w organizacji z perspektywy zasobów ludzkich. ZN WSB University in Wroclaw, no. 8 (46) (2014)
12. Grajewski, P.: Procesowe zarządzanie organizacją. PWE, Warszawa (2012)
13. Grajewski, P.: Organizacja procesowa. PWE, Warszawa (2012)
14. Sliż, P.: Dojrzałość procesowa współczesnych organizacji w Polsce, Wydawnictwo Uniwersytetu Gdańskiego, Sopot (2018)
15. Sturgis, P., Nick, B., Mario, C., Stephen, F., Jane, G., Jennings, W., ... Patten, S.: Report of the Inquiry Into the 2015 British General Election Opinion Polls (2016)
16. Etikan, I., Musa, S.A., Alkassim, R.S.: Comparison of convenience sampling and purposive sampling. Am. J. Theoret. Appl. Stat. **5**(1), 1–4 (2016)
17. Serwis internetowy dla przedsiębiorcy. https://www.biznes.gov.pl/tabela-pkd. Accessed 05.03.2018
18. The act on freedom of economic activity. J. Laws, no. 173, item 1807 (2004)
19. Pszczołowski, T.: Mała encyklopedia prakseologii i teorii organizacji. Ossolineum, Wroclaw (1978)

20. Krzakiewicz, K.: Zarządzanie wiedzą—czy nowe wyzwanie dla nauk o zarządzaniu? w: Błaszczyk, W., Kaczmarka, B. (eds.) Przeszłość i przyszłość nauk o zarządzaniu. University of Lodz, Łódź (2001)
21. Sliż, P.: Dojrzałość procesowa organizacji—wyniki badań empirycznych. Prace Naukowe Akademii Ekonomicznej, Wroclaw (2016)

Analysis of Attitude Differences of Professional Drivers in Light of Occupational Change Intention

Joanna Fryca-Knop, Beata Majecka, Michał Suchanek and Dagmara Wach

Abstract The chapter presents the results of the research on professional drivers in the context of their professional behaviour. Based on a development of the Ajzen's Theory of Planned Behaviour, a questionnaire is constructed, in which four groups of variables which shape the drivers' professional behaviour are analysed. The research sample consists of 120 professional drivers. Collected data undergo the validity and reliability analysis in order to identify the latent attitudes. The results indicate that there are a number of latent attitudes in every group of variables. A proper questionnaire can then be used in further research on the relations between the attitudes and the professional behaviour.

Keywords Professional behaviour · Occupational change attitudes · Professional driver · Road transport

1 Introduction

For a few years, the Polish labour market is characterised by a qualitative and quantitative imbalance which applies to the professional drivers in both the cargo and the passenger transport. It is hard to assess the precise scale of the imbalance due to a lack of data which would allow to identify professionally active drivers. Based on the information on the number of attendees at regular professional training courses (which every professional driver has to do every five years) and preliminary pro-

J. Fryca-Knop (✉) · B. Majecka · M. Suchanek · D. Wach
Faculty of Economics, University of Gdansk, Sopot, Poland
e-mail: j.fryca@ug.edu.pl

B. Majecka
e-mail: ekobma@ug.edu.pl

M. Suchanek
e-mail: m.suchanek@ug.edu.pl

D. Wach
e-mail: dwach@ug.edu.pl

M. Suchanek (ed.), *Challenges of Urban Mobility, Transport Companies and Systems*, Springer Proceedings in Business and Economics,
https://doi.org/10.1007/978-3-030-17743-0_20

fessional qualifications, PwC estimates that there were around 600–650 thousand professional drivers in Poland in 2016. Based on the difference between the growth of transport performance and the growth of employment in the years 2004–2014, one can assess that there is a shortage of 20% [1], which has increased since 2014 according to the opinions of transport companies representatives. This means that the market still needs 100 thousand more professional drivers [2]. A high rotation of employees (estimated at 10–30% annually) and an insufficient number of employees obtaining the professional qualifications are the additional problems of the road transport labour market [2]. Therefore, it is necessary to identify the reasons for this situation and to propose effective methods of controlling the social behaviour of the professional drivers. They are an important professional group, whose attitudes, intentions to work and the professional behaviour constitute the conditions for an effective functioning of road transport companies.

There is still a shortage of research on the behaviour of professional drivers. Many researchers focus on the drivers' behaviour, but mostly on the people who are not professionals. The research focuses on the traffic safety as a result of drivers' personality [3, 4] or attitudes, especially towards the law [5–8]. The research on professional drivers also mostly focuses on the safety [9]. In this case, mostly the effect of the health of the drivers on their performance is analysed, mostly the safety of the passengers [10–13] or the cargo [14]. The other area that is the reasons behind the occupational change by the professional drivers is much less explored.

In the 80s and 90s of the twentieth century, the high fluctuation of staff in transport companies was of high interest. It was analysed as a result of the level of salary, the inconvenient working conditions, as well as health [15], loyalty [16] or social relations [17]. A number of different strategies which would allow to sustain a stable level of employment were analysed [18]. Later research indicates the importance of job satisfaction as a complex variable which affects the attitude of drivers towards the occupational change or even their resignation [19, 20]. In the USA, a research with the use of structural modelling has been carried out, which proved that the increased fluctuation of the professional lorry drivers is affected by a mix of different factors—professional stressors, responsible for the physical and emotional exhaustion [21].

So far, at least in the identified research directions, there has been no complex analysis of the reasons for the professional drivers behaviour, especially when it comes to occupational change. According to the Ajzen's Theory of Planned Behaviour [22, 23], which is the development of the Fishbein's and Ajzen's Theory of Reasoned Action [24, 25], human behaviour is conditioned by the intentions which represent the desire of a person to engage in a given activity. At their roots, there are all of the motivational components of the human behaviour which constitute the set of attitudes resulting from the personal beliefs. Both theories allow the prediction of diversified social behaviour. Their universality meant that in order to analyse the professional drivers' behaviour the Ajzen and Fishbein models had to be deconstructed. A model of professional behaviour regulation has been proposed based on the systems theory and the regulation theory [26]. It should allow the explanation of decisions which people make in their work, which on the labour market have a form of professional behaviour. It should also allow to identify the rules for the regulation of such behaviour.

An effective technique has to be developed in order to use the model of professional behaviour regulation in the research on occupational change attitudes. The variables which shape the professional behaviour are:

- attitude towards the behaviour,
- subjective norm,
- attitudes towards the profession,
- perceived behavioural control.

If the professional behaviour of the drivers is caused by their intentions, which are in turn caused by their attitudes, then the attitudes themselves should be measured. In this case, the attitudes towards the occupational change are measured, which are internally heterogenous. The goal of the article is to analyse the differences between the attitudes of the professional drivers which are facing the possibility of an occupational change. Bus drivers, which are hired in one of the public transport companies in Gdańsk, Poland, were interviewed. A validity and reliability check was performed on the measurement scales. Afterwards, a factor analysis was performed to identify the underlying latent structure of the attitudes.

2 Materials and Methods

2.1 *The Dataset Characteristics*

Professional drivers[1] who work in the enterprise Gdańskie Autobusy i Tramwaje Sp. z o.o. (Gdańsk Buses and Trams Ltd.—a public transport company in the city of Gdańsk) were subject to the research which goal was to show which of the factors affect the attitudes towards the occupational change. The drivers were interviewed in the April of 2018.

A total of 496 bus drivers are employed in the company, and they are treated as the population for the purpose of this study. All of the drivers who weren't on a sick or holiday leave were asked to fill in the questionnaire. 138 questionnaires were received back, 18 of which were not included in further analysis due to missing or incorrect data. Out of all the respondents, 83 drivers had both the C and D class drivers' permit, while 37 only had the D class permit.

[1]A professional driver is a person who drives a vehicle or is carried in a vehicle so as to be able to start his professional duties of driving if needed [27], who has a C or D driver's licence while also being obliged to fulfil periodic training appropriate for the vehicle type which he drives, every five years from the day he acquired his professional qualifications [28].

2.2 Research Methods

The usefulness of the test in practice was assessed based on the validity and reliability analysis, i.e. the degree in which the results of the test represent the analysed characteristic. The validity and reliability are the formal characteristics of the test measurement. Factor analysis was then used to present the observed relations between the test questions within the attitudes.

The coherence between the positions of the scale was done with the internal coherence analysis method. All of the test positions have more than two possible answers, which is why the Cronbach's alpha was calculated which is believed to be the best estimate of the reliability of a test as well as the indicators of the discriminative power of the position which will be used to eliminate the questions which are the least correlated.

Due to a relatively large number of test positions, factor analysis was used, which based on the principle component method allowed to identify four factors which match the four attitudes in the theoretical model. Afterwards the positions which relate the strongest to the factors were identified.

Validity and reliability analysis was performed in the IBM SPSS 25, while the factor analysis was performed in Dell Statistica 13.1

2.3 Hypotheses

All of the respondents were asked if they plan an occupational change within the next 5 years, based on the assumptions of the model of professional behaviour regulation (question A). It was important that the question was precise, that it focused on one particular behaviour, planned within a given time perspective. Afterwards, in the questionnaire there were multiple questions which goal was to identify the attitudes of professional drivers in the context of the occupational change intention. The questions were divided into four groups which represent the four model types of attitudes: attitude towards the behaviour, subjective norm, attitudes towards the profession, and perceived behavioural control.

The questions were supposed to form a measurement scale. The creators of the original model, Fishbein and Ajzen, say that it is important that the questions allow to assess whether the attitude towards a behaviour is negative or positive (the bipolarity of the assessment). They recommended the use of a 5-point or 7-point Likert scale [29]. In this research, a 7-point Likert scale was used.

In order to identify the variables which affect the professional drivers' attitude towards the occupational change, we've asked them the following question: "Please imagine that you change your occupation within the next 5 years. Would that affect the following aspects of your life in a positive or a negative way? (question B). Six variables representing the following aspects were identified (a bipolar scale—1 signified a negative impact, 7 signified a positive impact):

- (B1) development possibilities,
- (B2) financial situation,
- (B3) the feeling of independence,
- (B4) the feeling of professional safety,
- (B5) life situation (e.g. family situation),
- (B6) well-being.

In order to analyse the internal differentiation of the drivers' attitudes towards the behaviour, a following hypothesis has been assumed:

Hypothesis 1: Latent attitudes can be identified within the attitude towards the behaviour.

A following question has been asked, which allowed to formulate the six initial factors and identify the variables which affect the subjective perception of the norms of the occupational change: To what extent do you agree with the following statements (question C):

- (C1) the expectations for the drivers are too high,
- (C2) I'm not willing to comply with the formal rules of the driver's profession,
- (C3) I'm not willing to comply with the ethical norms of the driver's profession,
- (C4) the pressure from the employers is too high,
- (C5) the investment connected with acquiring the qualifications necessary in the driver's profession doesn't affect the occupational change,
- (C6) my family and friends think I'm wasting my time being a professional driver.

A bipolar scale has been used, in which a negative influence on the attitude has been described as "do not agree" (a value of 1) and the positive influence has been described as "agree" (a value of 7). In order to analyse the internal differentiation of the drivers' subjective perception of the norms, a following hypothesis has been assumed:

Hypothesis 2: Latent attitudes can be identified within the subjective perception of the norms.

A following question has been asked, which allowed to formulate the seven initial factors and identify the variables which affect the subjective perception of the norms of the occupational change: For you, the attractiveness of the driver's profession is a result of (question D):

- (D1) high salary in comparison with other professions,
- (D2) a demand for responsibility (for the cargo or the passengers),
- (D3) a need for extensive knowledge,
- (D4) a requirement of rare skills,
- (D5) a feeling of freedom,
- (D6) the work being interesting,
- (D7) a feeling of prestige.

A bipolar scale has been used, in which a negative influence on the attitude has been described as "true" (a value of 1) and the positive influence has been described as "false" (a value of 7). In order to analyse the internal differentiation of the drivers' attitudes towards the profession, a following hypothesis has been assumed:

Hypothesis 3: Latent attitudes can be identified within the attitude towards the profession.

A following question has been posted to identify the variables which affect the subjective feeling of control of the drivers over their profession. It allowed to specify 6 detailed variables: The possibility of my occupational change within the next 5 years depends on (question E):

- (E1) my desire to change the occupation,
- (E2) the degree of my dissatisfaction with the profession of the driver,
- (E3) the pressure from my family/friends,
- (E4) the ease of adapting to a different profession,
- (E5) the level of salaries in different professions,
- (E6) the environment for occupational change (e.g. the ability to get funding for starting a new company).

A bipolar scale has been used, in which a negative influence on the attitude has been described as "they don't depend on that at all" (a value of 1) and the positive influence has been described as "they definitely depend on that" (a value of 7). In order to analyse the internal differentiation of the drivers' subjective feeling of control over the behaviour, a following hypothesis has been assumed:

Hypothesis 4: Latent attitudes can be identified within the subjective feeling of control.

3 Results

3.1 Validity Analysis

The validity analysis of the scale has been performed individually for the four types of attitudes: attitude towards the behaviour, subjective norm, attitudes towards the profession, and perceived behavioural control.

The Cronbach's alpha can have values between 0 and 1. The higher the value, the more valid the scale. For the B scale, which measured the attitude towards the behaviour, it was equal to 0.91 (Table 1), which proves a high relations of the test positions with the latent construct. High values of the discriminatory power of the positions (0.87, 0.73, 0.72, 0.58, 0.82 and 0.84) prove a high degree of internal consistency of the test

For both the C scale (subjective norm) and the D scale (attitudes towards the profession), the Cronbach's alpha is equal to 0.66. This is a relatively low value, but

Table 1 Cronbach's alphas for the groups of questions

Group of questions	Cronbach's alpha
B	0.909
C	0.662
D	0.664
E	0.748

still acceptable for scientific research [30]. In all of the positions of the C scale, the discriminatory power is higher than 0.2, which proves that the correlations between the positions are strong enough not to discard questions from the scale. In this particular research, it is worth considering discarding the question C5 (discriminatory power 0.24). Such a solution would allow to remove the variability resulting from the specifics of the drivers' employment. In their case, the road to the professional qualifications is not as expensive as it is for other drivers on the labour market, because the employer refunds them to a high extent, provided that the driver agrees to sign a stable employment contract. However, in the case of a comparison research, the use of this variable seems legit.

In the D scale, there is a low level of correlation of the D1 (discriminatory power −0.002) and D2 (discriminatory power 0.15) variables with the other positions on the scale. This is also proven by the reliability analysis of the scale, where the question on the attractiveness of the salary in comparison with other professions (D1) is more correlated with the attitude towards the behaviour than the attitude towards the profession. Discarding this question would increase the reliability of the scale significantly. The same goes for question D2. Discarding both of these questions would allow an increase in the indicator to 0.73.

The Cronbach's alpha for the perceived behavioural control is equal to 0.75. There is a high level of coherence for all the positions on the scale, similarly to the first group. Interestingly enough, discarding any of the questions would result in a decrease in the reliability indicator for this attitude which means that the questions reflect it correctly.

3.2 *Reliability Analysis*

In the analysed model, four attitudes were identified which were the starting point for the analysis. A preliminary factor analysis proved that there are eight correlated groups of questions which explain the 70% of all the variability. For the first three principal components, the eigenvalue based on the Kaiser criterion was higher than 2, which proves that there are at least three position groups strongly correlated with each other.

The alignment of the questions around the four principal components has been checked for the purposes of the analysis. The results are presented in Table 2. The first component is connected with the questions on the attitude towards the behaviour. The second component is concentrated on the questions on the perceived control. Both the first and the second factor groups the questions on the attitude towards the profession which confirms the conclusions from the validity analysis, where these positions led to a decrease in the Cronbach's alpha. In the case of the subjective norms, all the test positions have the highest factor loading within the third component. The fourth component strongly correlates with the positions connected with the attitude towards the profession.

Table 2 The principal component matrix—Varimax rotation

Variable	Factor			
	1	2	3	4
B1	*0.851*	0.189	−0.020	0.073
B6	*0.846*	0.251	0.070	0.126
B2	*0.813*	0.142	0.015	−0.092
B5	*0.812*	0.258	0.122	0.070
B3	*0.785*	0.032	0.113	−0.100
B4	*0.663*	0.005	0.271	0.169
D1	*0.431*	0.118	0.386	−0.374
D7	*0.387*	0.371	0.022	0.336
E2	0.212	*0.704*	−0.074	0.045
E1	0.037	*0.690*	−0.200	00.136
E6	0.054	*0.639*	0.337	0.146
E3	−0.032	*0.626*	0.263	0.118
E5	0.208	*0.581*	0.013	−0.241
E4	0.144	*0.531*	0.088	0.020
D5	0.276	*0.478*	0.181	0.185
C3	−0.128	−0.015	*0.768*	0.152
C2	0.097	−0.006	*0.716*	0.275
C1	0.267	0.077	*0.551*	−0.256
C4	0.319	0.207	*0.516*	−0.293
C6	0.377	0.393	*0.454*	0.084
D3	0.044	0.215	−0.043	*0.711*
D4	0.126	0.116	0.003	*0.639*
D2	−0.144	−0.163	0.068	*0.638*
D6	0.137	0.258	0.436	*0.572*
C5	0.101	0.148	0.205	*0.245*

Values are given in italics for factor loadings higher than 0.5 or lower than −0.5

The validity and reliability analysis proves that there are strong connections between the test positions for the attitude towards the behaviour, subjective norms and the perceived control and a bit weaker connections for the attitude towards the profession. Such results allow to draw true conclusions regarding the attitudes towards the occupational change of the drivers.

3.3 Factor Analysis

Factor analysis was then performed for the four groups of variables (B–D). Firstly, the significance of the eigenvalues was verified by identifying the factors using the principal component method, while the final factor analysis was performed applying the highest reliability method. The varimax rotation was applied to improve the readability of the results.

The attitude towards the behaviour, which is the occupational change, is connected with the perceived results of the effects of a given behaviour. Six variables forming this attitude were analysed.

Only one factor is significant (an eigenvalue higher than 1). This means that all the questions within this group are coinciding and represent only one latent attitude which can be treated as a general development vision in case of the occupational change.

All of the answers are correlated. The total variability is mostly affected by the assessment of the potential: well-being (B6), development possibilities (B1) and life situation (B5) (Table 3).

Another group of factors is connected with the subjective norms, which implicate the professional behaviour of the drivers and determine the difficulty of the job. They are connected with the individual liability to social pressure which enhances or decreases the attractiveness of a given behaviour.

Based on the factor analysis, three factors which represent the difficulty of the job are significant.

The dominant factor (38% of total variability) is created by the questions on the requirements towards the drivers (C1) and the pressure inflicted by the employers

Table 3 Factor loadings for occupational change

Variable	Factor loadings
	Factor 1
B1	−0.868
B2	−0.803
B3	−0.745
B4	−0.599
B5	−0.867
B6	−0.898

(C4). The difficulty of the professions is mostly determined by the determinants which put the drivers under pressure. The second factor (26% of total variability) represents the latent attitude of the willingness to break the rules. The strongest factor loadings were calculated for the willingness to disobey the formal rules (C2) and the ethical rules (C3). There is a high correlation within the factor (the people who are willing to disobey the formal rules are also willing to disobey the ethical rules), but also a high canonical correlation with the first factor, i.e. the stronger the attitude towards the pressure, the stronger the willingness to break the rules. The third factor compiles the other determinants of the drivers' trouble. It is strongly represented by the attitude regarding the inefficiency of the investment in the driver's profession (C5) and the perception of the career by the family and friends (C6). Generally, the third factor represents the latent attitude which presents the way the driver's profession is perceived. What is more, there is a high canonical correlation with the other factors, meaning that the higher the perceived requirements, the higher the willingness to break the rules and the lower the perception of the possibilities for development. The third factor represents the total 17% of the variability of the subjective norms within the driver's profession (Table 4).

The attitudes towards the profession are shaped by the personal beliefs. They are connected with the variables which determine the attractiveness of the profession for the employee. Seven such variables were analysed which were grouped into two significant factors and thus two latent attitudes were identified.

The strongest determinants of the attractiveness of the driver's profession are the perception of the knowledge and skills necessary to perform the job (D3, D4). They are highly correlated and are the basis of the level of professional expertise necessary in the profession. A low value of these determinants would mean that the overall attractiveness is low as well. The second factor is dominated by the perceived prestige connected with the driver's profession (D7). A lower perceived prestige results in a lower attractiveness. A high canonical correlation means that the stronger the perception of the driver's profession requiring a high level of knowledge and skills, the higher the perception of the profession as prestigious. However, these two first factors inly determine the 42% of total variability, which means that there is a high dispersion of the factors. The perceived profitability of the profession is not

Table 4 Factor loadings for subjective norms

Variable	Factor loadings		
	Factor 1	Factor 2	Factor 3
C1	0.811	0.053	0.106
C2	0.048	0.861	0.086
C3	0.018	0.798	0.122
C4	0.903	0.017	0.142
C5	0.073	0.074	0.431
C6	0.316	0.217	0.697

Table 5 Factor loadings for the attitude towards the profession

Variable	Factor loadings	
	Factor 1	Factor 2
D1	−0.163	0.439
D2	0.396	−0.124
D3	0.864	0.124
D4	0.632	0.177
D5	0.253	0.582
D6	0.427	0.446
D7	0.166	0.715

clearly represented in any of the factors, which means that it is correlated with all the other attractiveness determinants (Table 5).

The last group of the variables is connected with the perceived control. The control of the resources and the capacities necessary to act towards the occupational change is described by 6 variables. The factor analysis for these 6 variables proved that there is only one underlying attitude in this case.

All of the loadings within the first factor have the same sign, which means that they all affect the potential occupational change in the same way. The strongest determinant is the suitable environmental conditions (E6), while the weakest is the salary level in the other professions (E6); however, the overall differences are rather small (Table 6).

4 Discussion and Conclusions

The research has proven that the constructed questionnaire is valid and reliable and adequate for the research on the professional drivers' attitudes. The factor analysis has proven that the professional behaviour of the drivers is in fact affected by four groups of variables:

Table 6 Factor loadings for perceived control

Variable	Factor loadings
	Factor 1
E1	−0.607
E2	−0.625
E3	−0.593
E4	−0.529
E5	−0.472
E6	−0.675

- attitude towards the behaviour,
- subjective norm,
- attitudes towards the profession,
- perceived behavioural control.

Therefore, all of the research hypotheses have been proven correct. Additionally, the analysis of the factor loadings allows to state that the attitudes towards the behaviour and the perceived control are coherent and thus represented by one latent attitude. Subjective norms can be decomposed, based on the factor analysis, into three attitudes which represent: the pressure-making determinants, the willingness to break the rules and other difficulties. The attitude towards the profession is represented by two latent attitudes, which are connected with the professional competence level and the prestige of the profession.

The results of the factor analysis allow a true inference in the further research on the professional behaviour of the drivers in the context of possible occupational change. It is particularly advisable to perform a research on the differentiation of the professional behaviour depending on the four identified factors while also including other determinants such as the professional qualifications or the level of professional experience.

References

1. PricewaterhauseCoopers International Limited: Rynek pracy kierowców w Polsce. Raport PwC (2016)
2. Onet Biznes: Raport: w Polsce brakuje kierowców. http://biznes.onet.pl/praca/zawod-kierowca-ciezka-praca-niskie-zarobki-brak-100-tys-osob/qcldlm (2016). Accessed 26 Mar 2018
3. Bucchi, A., Sangiorgi, C., Vignali, V.: Traffic psychology and driver behavior. Proc. Soc. Behav. Sci. (2012). https://doi.org/10.1016/j.sbspro.2012.09.946
4. Useche, S., Serge, A., Alonso, F.: Risky behaviors and stress indicators between novice and experienced drivers. Am. J. Appl. Psychol. (2015). https://doi.org/10.12691/ajap-3-1-3
5. Forward, S.E.: The intention to commit driving violations—a qualitative study. Transp. Res. Part F Traffic Psychol. Behav. (2006). https://doi.org/10.1016/j.trf.2006.02.003
6. Parker, D., Manstead, A.S.R., Stradling, S.G., Reason, J.T., Baxter, J.S.: Intention to commit driving violations: an application of the theory of planned behavior. J. Appl. Psychol. **7**(1), 94–101 (1992)
7. Oviedo-Trespalacios, O., Haque, M.M., King, M., Washington, S.: Influence of road traffic environment and mobile phone distraction on the speed selection behaviour of young drivers. In: Victor, T., Regan, M. (eds.) 4th International Conference on Driver Distraction and Inattention (DDI2015), 9–11 November 2015. ARRB Group Ltd. and Authors, Sydney (2015)
8. Warner, H.W., Åberg, L.: Drivers' decision to speed: a study inspired by the theory of planned behavior. Transp. Res. Part F Traffic Psychol. Behav. (2006). https://doi.org/10.1016/j.trf.2006.03.004
9. Douglas, M.A., Swartz, S.M.: Career stage and truck drivers' regulatory attitudes. Int. J. Logist. Manag. (2016). https://doi.org/10.1108/ijlm-11-2014-0180
10. af Wåhlberg, A.E., Dorn, L.: Absence behavior as traffic crash predictor in bus drivers. J. Saf. Res. (2009). https://doi.org/10.1016/j.jsr.2009.03.003

11. Mallia, L., Lazuras, L., Violani, C., Lucidi, F.: Crash risk and aberrant driving behaviors among bus drivers: the role of personality and attitudes towards traffic safety. Accid. Anal. Prev. (2015). https://doi.org/10.1016/j.aap.2015.03.034
12. Shi, X., Zhang, L.: Effects of altruism and burnout on driving behavior of bus drivers. Accid. Anal. Prev. (2017). https://doi.org/10.1016/j.aap.2017.02.025
13. Vetter, M., Schünemann, A.L., Brieber, D., Debelak, R., Gatscha, M., Grünsteidel, F., Herle, M., Mandler, G., Ortner, T.M.: Cognitive and personality determinants of safe driving performance in professional drivers. Transp. Res. Part F Traffic Psychol. Behav. (2018). https://doi.org/10.1016/j.trf.2017.11.008
14. Zohar, D., Huang, Y.-h., Lee, J., Robertson, M.: A mediation model linking dispatcher leadership and work ownership with safety climate as predictors of truck driver safety performance. Accid. Anal. Prev. (2014). https://doi.org/10.1016/j.aap.2013.09.005
15. Backman, A.-L., Järvinen, E.: Turnover of professional drivers. Scand. J. Work Environ. Health **9**(1), 36–41 (1983)
16. Beilock, R., Capelle, R.B.: Occupational loyalties among truck drivers. Transp. J. **29**(3), 20–28 (1990)
17. Keller, S.B., Ozment, J.: Exploring Dispatcher Characteristics and their effect on driver retention. Transp. J. **39**(1), 20–33 (1999)
18. Stephenson, F.J., Fox, R.J.: Driver retention solutions: strategies for for-hire truckload (TL) employee drivers. Transp. J. **35**(4), 12–25 (1996)
19. Large, R., Breitling, T., Kramer, N.: Driver shortage and fluctuation: occupational and organizational commitment of truck drivers. Supply Chain Forum. Int. J. **15**(3), 66–72 (2014)
20. LeMay, S.A., Johnson, L., Williams, Z., Garver, M.: The causes of truck driver intent-to-quit: a best-fit regression model. Int. J. Commer. Manag. (2013). https://doi.org/10.1108/ijcoma-03-2013-0028
21. Kemp, E., Kopp, S.W., Kemp, E.C.: Take this job and shove it: examining the influence of role stressors and emotional exhaustion on organizational commitment and identification in professional truck drivers. J. Bus. Logist. (2013). https://doi.org/10.1111/jbl.12008
22. Ajzen, I.: From intentions to actions: a theory of planned behavior. In: Kuhl, J., Beckmann, J. (eds.) Action Control. From Cognition to Behavior. Springer Series in Social Psychology, 1st edn. Springer, Berlin (1985)
23. Ajzen, I.: The theory of planned behavior (Teoria planowanego zachowania). Organ. Behav. Hum. Decis. Process. (1991). https://doi.org/10.1016/0749-5978(91)90020-t
24. Fishbein, M., Ajzen, I.: Belief, Attitude, Intention and Behavior. An Introduction to Theory and Research. Addison-Wesley Series in Social Psychology. Addison-Wesley, Reading, MA (1975)
25. Ajzen, I., Fishbein, M.: Understanding Attitudes and Predicting Social Behavior. Prentice-Hall, Englewood Cliffs, NJ (1980)
26. Fryca-Knop, J., Majecka, B.: The research study on principles of drivers' behaviour in the labour market of road transport in Poland. Deconstruction of I. Ajzen and M. Fishbein Model. In: Suchanek, M. (ed.) New Research Trends in Transport Sustainability and Innovation. TranSopot 2017 Conference, vol. 50, pp. 135–148. Springer Proceedings in Business and Economics, Cham (2018)
27. Rozporządzenie (WE) nr 561/2006 Parlamentu Europejskiego i Rady z dnia 15 marca 2006 r. w sprawie harmonizacji niektórych przepisów socjalnych odnoszących się do transportu drogowego oraz zmieniające rozporządzenia Rady (EWG) nr 3821/85 i (WE) 2135/98, jak również uchylające rozporządzenie Rady (EWG) nr 3820/85. In: Dz.U.UE.L.06.102.1 (2006)
28. Ustawa z dnia 6 września 2001 r. o transporcie drogowym. In: Dz.U., art. 39 d, ust. 2 (2001)
29. Fishbein, M., Ajzen, I.: Predicting and Changing Behavior. The reasoned Action Approach. Psychology Press, New York, NY, Hove (2010)
30. Rószkiewicz, M., Perek-Białas, J., Węziak-Białowolska, D., Zięba-Pietrzak, A.: Projektowanie badań społeczno-ekonomicznych. Rekomendacje i praktyka badawcza, pp. 101. Wydawnictwo Naukowe PWN, Warszawa (2013)

Part III
Green and Innovative Transport Systems

Irregularities in Level Crossings and Pedestrian Crossings

Arkadiusz Kampczyk

Abstract Level crossings are single-level crossings; they differ from pedestrian crossings in that the latter are meant only for pedestrian and/or bicycle traffic. Railway line administrators and road administrators are obliged to perform their duly efforts related to ensuring safety on level crossings and pedestrian crossings. This paper focuses on the results of studies that cover irregularities occurring on level crossings and pedestrian crossings. Based on the innovative measuring instrument called magnetic-measuring square (MMS), five steps of the measuring innovation were defined. Also, wireless identification technology (NFC—Near-Field Communication) implemented by the author in the aforementioned type of works was presented. The use of MMS and NFC technologies significantly streamlines the process of maintaining railway infrastructure and carrying out land surveying and diagnostic works. The paper includes proprietary observations and conclusions. The chapter was prepared as part of the AGH statutory research no. 11.11.150.005.

Keywords Level crossing · Level crossing measurement · Pedestrian crossing · Passage · Irregularities in level crossings · Magnetic-Measuring Square, MMS · Near-Field Communication, NFC · Measurement innovations · Railway Surveying Grid, KOG · Railway Special Grid, KOS

1 Introduction

Railway and road crossings are laid out on a single level; they are different from pedestrian crossings as those are intended only for pedestrian, only for bicycle or for pedestrian and bicycle traffic. When a pavement passes through a crossing, then it

A. Kampczyk (✉)
Department of Engineering Surveying and Civil Engineering, Faculty of Mining Surveying and Environmental Engineering, AGH University of Science and Technology, al. A. Mickiewicza 30, 30-059 Krakow, Poland
e-mail: kampczyk@agh.edu.pl; arkadiusz.kampczyk@gmail.com

M. Suchanek (ed.), *Challenges of Urban Mobility, Transport Companies and Systems*, Springer Proceedings in Business and Economics,
https://doi.org/10.1007/978-3-030-17743-0_21

is called a pedestrian crossing. Railway and road crossings and pedestrian crossings should guarantee safe passage through railway lines.

Railway and road crossings should meet plenty of technical requirements, e.g., visibility, lighting, inclination of a road at an approach to the crossing, water drainage, width and length of the crossing, angle values of a railway and road crossing which are described in [1–3].

The performed research was intended to determine any irregularities that occur on the railway and road crossings and pedestrian crossings which were approved during measurements works over crossings geometry. In his research, the author developed some hypotheses related to the occurrence of hazards on railway and road crossings as well as level crossings. Therefore, the following hypotheses were verified:

1. There are some irregularities on the railway and road crossings.
2. Possibility to adopt the NFC wireless identification system in the Railway Surveying Grid and the Railway Special Grid.
3. Possibility to compare the results from railway and road crossings geometry measurements and level crossings obtained in real time with the required parameters.
4. Measurement innovation may include imagining, designing, verifying, analyzing and evaluating.

Papers [1–3] include results from the geometrical condition measurement on the D cat. railway and road crossings, including some author's solutions. However, the performed studies proved that there are also numerous additional irregularities that were presented in this work. The chapter is a part of comprehensive research including measurements of railway and road crossings geometry, employing some author's measuring devices (magnetic-measuring square—MMS—and adapters for installation of surveying prisms on a digital or manual track gauge) and Near-Field Communication (NFC). NFC was utilized in signs of the Railway Surveying Grid (KOG, Kolejowa Osnowa Geodezyjna) and Railway Special Grid (KOS, Kolejowa Osnowa Specjalna). Technical and engineering staff can compare the obtained parameters with required values in real time and communicate them to a database. Based on the innovative measuring device—MMS—five steps of measurement innovation were defined as well. The chapter presents the author's observations and conclusions. The chapter was prepared as part of the AGH statutory research no. 11.11.150.005.

2 Review and Analysis of Selected Literature of Research Subject

Tey, Ferreira and Wallace conclude in [4] that railway crossings pose one of the most complex road safety control systems, especially because of the conflicts that emerge between vehicles–roads and infrastructure–rolling stock.

Salmane, Khoudour and Ruichek argue in [5] that safety on railway and road crossing is a priority issue for the sector of intelligent transport systems. They presented some research results based on a video recording system. This system detects and tracks an object on a crossing via dedicated video sensors.

The author of [1] and [2] concludes that safety on railway and road crossings is a priority. This is a highly dangerous type of level crossings, especially when it comes to D category—meaning the unguarded crossings. At the same time, in [1] the author presents A-F categories of crossings with descriptions and focuses on geometric measurements of visibility conditions of a railway and road crossing. In turn, in [2], the author stresses that safety on road and railway crossings depends on behavior of road users, and correct implementation of tasks by the railway and road infrastructure managers. The tasks under implementation by the railway and road infrastructure managers also cover the scope of surveying works. The measurement innovations in railway infrastructure safety regard not only the railway transport but also railway surveying. The work [3] includes results of research over a D cat. railway and road crossing geometry. It was noted that development of technical and exploitation documents for a railway and road crossing or for a pedestrian crossing—called a railway and road/pedestrian crossing certificate—is obligatory.

Pesonen and Horster state in [6] that the use of mobile devices such as smartphones and tables has been on an increase for some years. An increasing number of mobile devices offer the Near-Field Communication, which is a short-range technology for wireless transmission of data. In their work [6], they referred to the NFC technology in four fields: business models and ecosystems, software and application, safety and hardware, as well as threats and problems. They discussed examples of NFC use in tourist industry. They also concluded that technological advancement and tourism are two inseparably connected fields.

Morak, Hayn, Kastner, Drobics and Schreier in their work [7] presented possibilities provided by combining the NFC technology with mobile phones to obtain some clinical data when it comes to recording and processing of research information in healthcare facilities.

In turn, Feldhofer, Dominikus and Wolkerstorfer refer to Radio-Frequency Identification (RFID) in [8]. They note that wireless communication technologies—RFID in this case—offer some great advantages in terms of capacity, wherever objects must be verified automatically. They stressed the notions related to RFID systems safety and privacy. They presented an authentication protocol that proves the concept of authenticating an RFID tag located on a reader via the Advanced Encryption Standard (AES) as a cryptographic primitive. It was an innovative approach to the implementation of AES which encrypts a 128-bit data block within 1000 clock cycles and consumes less than 9 μA of power in the CMOS 0.35 μm process.

Nasution, Husni and Wuryandari state in [9] that technological development provides access to the digital world without any boarders. This fact was proved by development of technology that allowed to trade and conclude transactions virtually. The process of virtual shopping has Near-Field Communication support. They notice that Google integrated the Android platform with a smartphone in 2011 which was a starting point for introduction of transactions with virtual money. Especially, these e-tickets pose one of the largest services in online trading. They are e-documents, with no paper, commonly employed in case of passenger tickets. This solution proves the use of NFC in daily life, by launching applications for public transport. An app prototype allows to purchase train tickets through smartphones with NFC (with an interface for a passenger and for a conductor).

In [10], Komsta, Brumercikova and Bukova described some possible uses of the NFC technology as a ticketing system in passenger railway transport. They presented some research that evaluated the passengers' interest in this method of purchasing railway tickets in Slovakia. Furthermore, the authors described the planned use of the NFC technology as a ticketing system in passenger railway transport, with consideration of data protection in smartphones, actions in case a smartphone is lost and provision of communication among the system components. They also analyzed the possibility to purchase tickets regardless of a carrier and a mode of transport.

In [11], Weller presents the Swedish Railway Lines SJ that introduced the technology of micro-chip implants scanning that replaces the need to be in hold of real tickets. Micro-chips are implanted under skin of a human hand. He notices that subcutaneous implants have become a common form of "biohacking." People use those implants to lock the doors in their homes or cars.

Coffey claims in [12] that the use of a biometric chip implanted in a hand instead of a paper ticket is an innovative solution. This small chip offers the same technology as the Oyster and contactless bank cards, namely NFC which allows the teams of conductors to scan the passengers' arms. Clients purchase tickets in a regular way, by logging in through a website or a mobile app, and their member ID, which is also a ticket reference code, is linked to the chip. The SJ Railways observe that it takes less time to scan a micro-chip than a travel card; thus, it is possible for the train staff to save some time. However, the main advantage is the fact that this technology places a train in the middle of the digital revolution. The SJ Railways implement certain projects that facilitate digital customer service.

In [13], Kampczyk and Strach presented a concept of RFID implementation in performance of surveying measurements and diagnostic examinations. They referred to two measurement methods adopted in case of track geometry, i.e., indirect and direct.

3 Irregularities on Railway and Road Crossings and on Pedestrian Crossings

Railway lines infrastructure managers and road infrastructure managers are obliged to perform adequate actions that guarantee safety on railway crossings. It was found out on the basis of examinations on condition of the railway and road crossings that they are not marked properly with road signs and railway indicators. The condition of access roads is not fully satisfactory as they limit visibility of the approaching trains because of their poor surface or growing trees and bushes. The railway and road crossings are located on the routes of national, provincial, district, communal, internal, forest, agricultural and other types of roads (including those located at

Fig. 1 "A" cat. road and railway crossing with tracks managed by various entities (own photograph)

suspended railway lines). The applicable category of railway and road crossing is underrepresented in relation to the target category the reason for which is lack of required visibility (geometrical condition of visibility splays). Apart from the crossings recorded in the Railway Crossings Database (records kept by PKP PLK S.A.), there are also some "wild passages." The railway infrastructure manager marks them as well.

The research proved that apart from active railway lines there are also some lines with suspended carriage where crossings are also located, and there may not be any decisions issued to liquidate those lines. There is also a technical inspection performed in spring intended to provide safety on such crossings.

What is more, the railway lines embrace some crossings that apart from the tracks managed by PKP PLK S.A. also include some tracks that belong to other entities (Fig. 1). Such crossings are characterized with a greater number of tracks which in turn is reflected in the crossing length. The length of a railway and road crossing is a road fragment limited with boom barriers on both sides. If there are no boom barriers, it is a road fragment that the extreme points of which lie at a distance of 4 m from each extreme rail. PKP PLK S.A. infrastructure manager undertakes actions intended to liquidate the tracks owned by external entities so the crossing length can be reduced or requires the track owner to repair a part of the crossing that they own or to liquidate the track and lay some road surface.

The most frequent irregularities are:

1. Lack of basic items and adequate data of crossing certificates required by a template included in Attachment 2 to the Regulation [14]. Discrepancies between entries in the certificates and the actual conditions, e.g.,

   - no information about signs informing about the height of an overhead contact line;
   - no entries on location of sings along the whole crossing;
   - incorrect distances from road signs and railway indicators;

- incorrect visibility distances for trains;
- the number and type of signaling units in the field do not correspond to the values entered into the certificate;
- incorrect widths of a roadway;
- traffic measurement information is outdated.

2. Incorrect angle values for the railway and road crossing [1–3].
3. Incorrect statutory visibility distances for trains, in case of 5, 10 and 20 m, respectively.
4. No measurement of current road and railway traffic intensity.
5. Terms entered incorrectly, e.g., "built-up area" instead of a "not a built-up area."
6. No route number and class.
7. No road category and a straight fragment length on each side of the track for both sides of the crossing.
8. Trees and bushes growing at a distance lower than 15 m from the axis of the extreme track nearby the crossing. There are some cases that trees and bushes are located at a distance lower than 15 m from the extreme track axis on the access roads with W-6a "Attention indicator" (indicating that there is a need to emit the Rp1 sound signal "Warning") or W-6b "Attention indicator" (indicating that there is a need to emit the Rp1 sound signal "Warning" numerous times). This distance is incompliant with §1 of the Regulation of the Minister of Infrastructure of 7 August 2008 on requirements regarding distances and conditions permitting localization of trees and bushes, noise protection and ground works performance elements nearby the railway lines, and management and maintenance of snow fences and firebreaks [15]. It should be noted that in such a case, it is necessary to obtain a consent from offices with jurisdiction over the territory for cutting and removing the trees and bushes that grow at a distance lower than 15 m from the extreme track axis nearby the crossing.
9. Lack of calculated traffic exposure factors for A, B, C and D cat. railway and road crossings located on railway lines that are exploited periodically. There are situations where the exposure factors for exploited lines are outdated, and they come from 2009 and previous years.
10. Restraining from increase of the crossings category, adequately to the existing traffic intensity.
11. Lack of periodical technical inspection performed on objects.
12. Incorrect marking of the railway and road crossings, for both rail and road vehicles.
13. Slabs on crossings are distributed unevenly (gaps between particular slabs of several centimeters); they partially miss angle bars on edges and fastenings to the track. There are also some loose bolts [1].
14. Accidents occur most frequently on railway and road crossings – of the lowest categories, i.e., C and D [3].

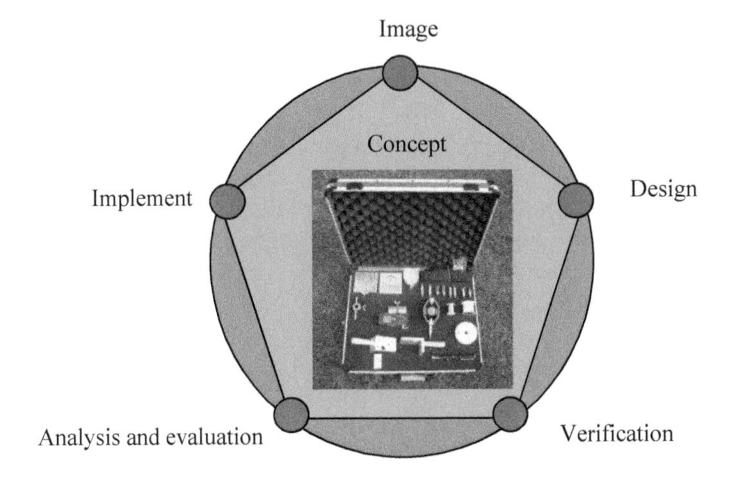

Fig. 2 The measurement innovation in five steps based on an example of MMS (own elaboration)

4 Measurement Innovation in Five Steps

The railway and road geometry measurements were carried out with a measuring device named a magnetic-measuring square (MMS) and adapters for installation of surveying prisms on a digital or manual track gauge [1–3, 16]. At the same time, the term measurement innovation is defined in paper [2]. It was found out that the measurement innovation in the railway infrastructure safety is the manner of both thinking and acting, including selection of measurement techniques and methods, as well as development and utilization of a measuring device that ensures good quality of acquired data. At the same time, it is a "flywheel" in railway surveying, as it means introducing "something" new.

These five essential steps for the measurement innovation were not defined in [2]. The five steps of measurement innovation are (Fig. 2):

1. Image: thinking about new possibilities, talking with other, broadening the existing possibilities and optimization.
2. Design: testing the ideas and concepts, discussing them with clients—practitioners of rail transport surveying and diagnostics, development of prototypes and models.
3. Verification: testing directly in the field—in practice, comparing the obtained results with those obtained from other measuring devices and methods, value of a new product, possibilities to employ it in surveying and diagnostic works.
4. Analysis and evaluation: identification of strengths and weaknesses, benefits, costs, sales market and introduction of constructive changes.
5. Implementation: of a new product—a measuring device, of new knowledge, commercialization of the new product or service.

5 NFC Wireless Identification in Surveying and Diagnostic Works

The NFC wireless identification allows two parallel and smart devices to communicate. This process takes place via a reader reading out the data saved on a tag. The NFC communication uses 13.56 MHz frequency at a distance not exceeding several dozens of centimeters. However, regarding structure of a reader (e.g., a smartphone) and electromagnetic noise, the NFC devices must touch (they need to get in contact), so the communication can be executed. There are three basic operation modes in the case of NFC:

- peer to peer, which allows to communicate minor data such as texts, URL (Uniform Resource Locator) links to websites;
- read and write, which allows to program a tag in such a way that the reader acts according to the included instructions;
- card emulation, a device with an NFC chip employed to carry out a certain action.

The Railway Surveying Grid (KOG) is a systematized collection of surveying points that have a mathematically arranged mutual location and its accuracy. The KOG points refer to all surveying works performed within railway areas [17]. The surveying marks (points) are signs of durable material that have their plane coordinates and altitudes determined within the applicable surveying reference system. The Railway Special Grid (KOS) is a collection of points (that includes sings for track axis configuration, stabilized in trolley poles) with mathematically determined plane and altitude coordinates within the applicable surveying reference system. KOS is included in KOG [17]. The author adopted the NFC PVC NTAG213 tags with 25 mm in diameter as KOS signs, located nearby a railway and road crossing and installed on trolley poles, and NFC PIN TAG (in a form of a pin) in the ground KOG signs.

5.1 *NFC PVC NTAG213 as a KOS Sign*

The adopted NFC PVC NTAG213 tags, as substitutes for the Railway Special Grid signs, are made from PVC and are resistant to water, and their diameter is 25 mm.

The NFC tags are installed on trolley poles as KOS signs nearby a railway and road crossing that underwent the surveying and diagnostic measurements on 01.06.2017. The control measurement was performed in winter conditions on 20.01.2018 (Fig. 3). The tags operate correctly. They allow wireless exchange of information (both read and write). The KOS signs included information about X, Y, H coordinates, offsets in the dy vertical and dx horizontal plane toward the track axis, a location km in the railway line and a pole number (pole location) which is a number of a KOS sign. There were two light-gauge straight lines put on the 25 mm NFC tag that crossed at an angle of 90°.

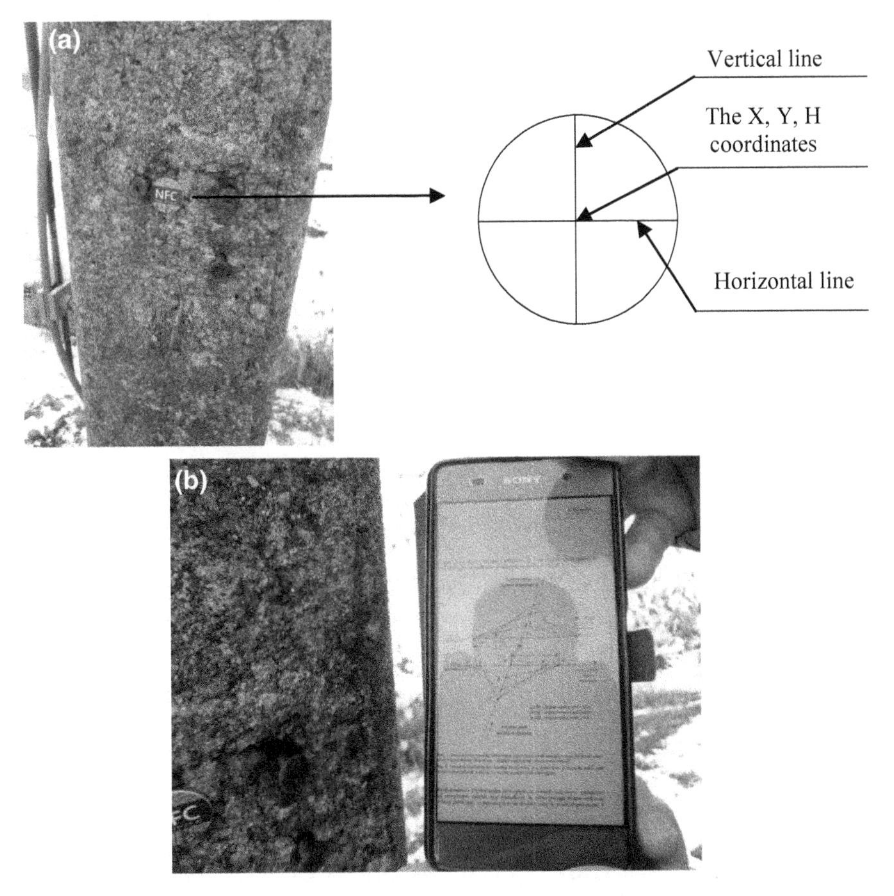

Fig. 3 NFC PVC NTAG213 as a KOS sign (own photograph and elaboration): **a** characteristics of basic data in the NFC tag, **b** readout of the NFC tag via a smartphone—URL link provides measurement data regarding geometry of visibility splays of a railway and road crossing

Horizontal line is a reference point for measurements made in the vertical plane (altitude) toward the upper surface of the running rail. Vertical line is a reference point for measurement made in the line located horizontally to the track axis (i.e., the internal edge of the upper surface of the rail, enlarged by half value of the track gauge). In turn, the point in which the straight lines cross (the middle) determined the *X*, *Y*, *H* coordinates (Fig. 3a). It is enough to bring a smartphone closer to the NFC tag during surveying or diagnostic works, to start communication and receive data. In case of a D cat. railway and road crossing, apart from the data pertaining to the KOS signs, the tag also includes an URL link (password-protected) directing to measurement data on visibility splays geometry, determined on 01.06.2017 (Fig. 3b). Characteristics of NFC PVC NTAG213 are included in Table 1. The NFC tags are compatible with the following standards: ISO 14443A, NDEF, NFC Forum Type 2 Tag.

Table 1 NFC PVC NTAG213 characteristics

Type of characteristics	Description
Tag type: ISO 14443-3A	NXP MIFARE Ultralight (Ultralight C) NTAG213
Available technologies	NfcA, Mifare Ultralight, Ndef
Serial number	04:D5:E8:52:A3:40:80
ATQA	0x0044
SAK Data format Capacity Writeable Possibility to write just in a read-only mode	0x00 NFC Forum Type 2 144 bytes Yes Yes

A NFC tag with the NTAG213 chip is compatible with all common NFC devices equipped with an active NFC module, regardless of the operating system: Android, Windows Phone and Black Berry. The employed tags have 188B of memory (144B at the user's disposal). The NFC tags can be secured against overwriting or removing of data. The tags do not have any internal power source; thus, there is no need to exchange or charge the battery.

5.2 *NFC PIN TAG as a Sign for the Ground KOG*

The employed NFC PIN TAGs have a form of a pin (stud, rivet, nail) put into sings of the ground measuring Railway Surveying Grid (KOG), by fixing them into a hole of a wooden pile. Thanks to the tag's shape, it was possible to place it into a previously performed installation opening for a ground surveying sign (Fig. 4). An important feature of the NFC PIN NAIL TAG is the material it is made of—namely plastic resistant to water and mechanical damage. The employed tag's dimensions are 37 mm in length, 6 mm pin diameter and 8 mm head diameter. They are equipped with 180B of memory, out of which 144B can be used for any data that may be secured from overwriting and removal.

The employed KOG tags do not have any internal power source; thus, there is no need to exchange or charge the battery. They allow wireless information exchange (write and read) at a frequency of 13.56 MHz. A smartphone must be brought closer to the reader to start the communication. Characteristics of NFC PIN NAIL TAG are included in Table 2. The tag adopted as a KOG sign is compatible with the following standards: ISO 14443A, NDEF, NFC Forum Type 2 Tag. The NFC PIN NAIL TAG includes data that are typical for a KOG sign, i.e., X, Y, H coordinates in a surveying reference system, point number and location data resulting from characteristics of the railway infrastructure—based on a railway and road crossing, where the measurement works were performed.

Fig. 4 NFC PIN TAG in KOG applications (own photograph)

Table 2 NFC PIN TAG characteristics

Type of characteristics	Description
Tag type: ISO 14443-3A	NXP MIFARE Ultralight (Ultralight C) NTAG213
Available technologies	NfcA, Mifare Ultralight, Ndef
Serial number	04:C0:C8:4A:FE:4A:80
ATQA	0x0044
SAK Data format Capacity Writeable Possibility to write just in a read-only mode	0x00 NFC Forum Type 2 144 bytes Yes Yes

6 Conclusions

The publication presents irregularities that occur on railway and road crossings and pedestrian crossings, found out during the author's research over measurements of the crossings geometry. Conclusions along with their interpretations were presented on current basis. The key element was the examinations of railway and road crossings and pedestrian crossings, and evidence of irregularities in terms of applicable legal regulations. The performed research proved the presented hypotheses to be correct. The presented results enrich both the theoretical and practical knowledge.

It was found out that the surveying and diagnostic works are significant as regards correctness of railway and road crossings and pedestrian crossings, first of all when it comes to their safety. The author's NFC solutions employed in the Railway Surveying Grid signs significantly facilitate performance of surveying and diagnostic works on the railway infrastructure, the component of which is posed by railway and road crossings. It is recommended to include the KOG signs with NFC into the hand-over protocols for track axis configuration signs and the topographic descriptions of the KOG signs. Therefore, the presented solution can be employed in the civil surveying grid signs. The defined five steps of measurement innovation based on an example of MMS in the railway infrastructure safety determined the development stages of its definition. They are employed in measurement of railway and road crossings geometry; they cover the manner of thinking and acting, including selection of measurement techniques and methods, as well as development and utilization of a measuring device that ensures good quality of acquired data.

References

1. Kampczyk, A.: Pomiar geometrycznych warunków widoczności przejazdu kolejowo-drogowego (Measurement of geometric visibility conditions level railroad crossing). Prz. Komun. **8**, 2–7 (2017)
2. Kampczyk, A.: Measurement innovations in railway infrastructure safety. World Sci. News **89**, 336–347 (2017)
3. Kampczyk, A.: Geometria przejazdu kolejowo-drogowego kategorii D (Geometry of D category road and railway crossing), Czas. Inż. Lądowej, Śr.Arch. (Journal of Civil Engineering, Environment and Architecture. JCEEA) T. XXXIV, 64 (4/II/17), 269–286 (2017). https://doi.org/10.7862/rb.2017.245
4. Tey, L.-S., Ferreira, L., Wallace, A.: Measuring driver responses at railway level crossings. Accid. Anal. Prev. **43** (6), 2134–2141 (2011)
5. Salmane, H., Khoudour, L., Ruichek, Y.: Improving safety of level crossings by detecting hazard situations using video based processing. In: Proceedings of the IEEE International Conference on Intelligent Rail Transportation (ICIRT '13), pp. 179–184. IEEE, Beijing, China (2013)
6. Pesonen, J., Horster, E.: Near field communication technology in tourism. Tour. Manag. Perspect. **4**, 11–18 (2012)
7. Morak, J., Hayn, D., Kastner, P., Drobics, M., Schreier, G.: Near field communication technology as the key for data acquisition in clinical research. In: 2009 First International Workshop on Near Field Communication. IEEE Xplore Digital Library, Hagenberg (2009)
8. Feldhofer, M., Dominikus, S., Wolkerstorfer J.: Strong authentication for RFID systems using the AES algorithm. In: Cryptographic Hardware and Embedded Systems—CHES 2004, pp. 357–370. 6th International Workshop Cambridge, MA, USA, Proceedings. Springer, Berlin (2004)
9. Nasution, S.M., Husni, E.M., Wuryandari, A.I.: Prototype of train ticketing application using Near Field Communication (NFC) technology on Android device. In: 2012 International Conference on System Engineering and Technology (ICSET). IEEE Xplore Digital Library, Bandung (2012)
10. Komsta, H., Brumercikova, E., Bukova, B.: Application of NFC technology in passenger rail transport. Transp. Probl. **11** (3), 43–53 (2016). Wydawnictwo Politechniki Śląskiej
11. Weller, Ch.: A Swedish rail line now scans microchip implants in addition to accepting paper tickets. Business Insider. http://www.businessinsider.com/swedish-rail-company-scans-microchip-tickets-17-6?IR=T

12. Coffey, H.: Swedish commuters can use futuristic hand implant microchip as train tickets. The future is here—a Swedish rail company is trialling letting passengers use biometric chips as tickets. http://www.independent.co.uk/travel/news-and-advice/sj-rail-train-tickets-hand-implant-microchip-biometric-sweden-a7793641.html
13. Kampczyk, A., Strach, M.: Koncepcja zastosowania techniki RFID w inwentaryzacji elementów infrastruktury transportu szynowego (Idea of applying the RFID technique in inventorying of the infrastructure of the rail transport). Conference: Nowoczesne technologie i systemy zarządzania w transporcie szynowym (Modern Technologies and Management Systems for Rail Transport). Vol. Zeszyty Naukowo-Techniczne Stowarzyszenia Inżynierów i Techników Komunikacji Rzeczpospolitej Polskiej. Oddział w Krakowie (Research and Technical Papers of Association for Transportation Engineers in Cracow). Seria: Materiały konferencyjne, nr 3, pp. 157–166, Kraków (2013)
14. Rozporządzenie Ministra Infrastruktury i Rozwoju z dnia 20 października 2015 r. w sprawie warunków technicznych, jakim powinny odpowiadać skrzyżowania linii kolejowych oraz bocznic kolejowych z drogami i ich usytuowanie (Dz.U. 2015 poz. 1744) (Regulation of the Minister of Infrastructure and Development of 20 October 2015 on technical conditions that must be met by railway lines and sidings level crossings and their location, Journal of Laws Dz.U. of 2015, item 1744)
15. Rozporządzenie Ministra Infrastruktury z dnia 7 sierpnia 2008 r. w sprawie wymagań w zakresie odległości i warunków dopuszczających usytuowanie drzew i krzewów, elementów ochrony akustycznej i wykonywania robót ziemnych w sąsiedztwie linii kolejowej, a także sposobu urządzania i utrzymywania zasłon odśnieżnych oraz pasów przeciwpożarowych (Dz.U. 2008 nr 153 poz. 955) (This distance is incompliant with §1 of the Regulation of the Minister of Infrastructure of 7 August 2008 on requirements regarding distances and conditions permitting localization of trees and bushes, noise protection and ground works performance elements nearby the railway lines, and management and maintenance of snow fences and firebreaks, Journal of Laws Dz.U. 2008, No 153, item 955)
16. Kampczyk, A.: Przykładnica magnetyczno – pomiarowa i jej zastosowanie. Zgłoszenie patentowe: P. 420214. (Magnetic-measuring square and its application. Patent pending: P. 420214)
17. Standard techniczny określający zasady i dokładności pomiarów geodezyjnych dla zakładania wielofunkcyjnych znaków regulacji osi toru Ig-7 (Technical standard determining the principles and accuracies of surveying measurements for installation of multi-functional Ig-7 track axis configuration signs). PKP Polskie Linie Kolejowe S.A. Warszawa (2012)

Computation of the Synthetic Indicator of the Economic Situation of the Rail Transport Sector in Poland

Jakub Majewski, Michał Suchanek and Michał Zajfert

Abstract The purpose of this article is to present the computation results of the synthetic indicator of the economic situation of the rail transport sector in Poland, referred to as 'the Rail Barometer'. It was developed in 2017 based on the methodology presented by Zajfert, Antonowicz, Majewski, Wołek, members of the Scientific Council of the 'Pro Kolej' Foundation, and published in the article: 'Assumptions for the synthetic indicator of the economic situation of the rail transport sector in Poland' (Antonowicz et al. in Ekon. Transp. Logist. 74:467–481, 2017 [1]). The article describes the effects of work on implementation, provision of statistical content and calibrating the tool for monitoring changes in the rail sector, including the database structure and volatility of the value expressed as a number of points of 'the Rail Barometer' computed on its bases.

Keywords Rail transport · Analysis of the economic situation in transport · Economic indicators

1 Introduction

The synthetic indicator of the economic situation of the rail sector, referred to as 'the Rail Barometer', has so far been the most comprehensive tool used to analyse

J. Majewski (✉)
'Pro Kolej' Foundation, Warsaw, Poland
e-mail: j.majewski@prokolej.org

M. Suchanek
Faculty of Economics, University of Gdansk, Gdańsk, Poland
e-mail: ekomsu@ug.edu.pl

M. Zajfert
Institute of Economic Sciences of the Polish Academy of Sciences, Warsaw, Poland
e-mail: mzajfert@inepan.waw.pl

M. Suchanek (ed.), *Challenges of Urban Mobility, Transport Companies and Systems*, Springer Proceedings in Business and Economics,
https://doi.org/10.1007/978-3-030-17743-0_22

the situation on the rail market in Poland.[1] It allows for an analysis of individual segments of the sector based on the data published in generally accessible sources or established on the basis of a current offer of carriers and infrastructure managers. The elements and measurements that constitute part of the final value of 'the Rail Barometer' take into account also macroeconomic data, commercial information and stock exchange quotations pertaining to the appraisal of companies operating on the rail market. Such a wide range of data representative for the entire country is to present, in the most objective manner possible, the current situation of the rail sector. In order to preserve high dynamics, an interval of a month and computation formulas which eliminate the impact of transport seasonality and dependence on the seasons of the year were taken into account in updating the database.

The selection of data for the synthetic factor of the economic situation of the rail market takes into account the availability and frequency of publication of subsequent figures and allows to monitor the rail market in a continuous manner as well as to assess its current condition based on the figures updated monthly.[2]

It is worth mentioning that apart from the scientific application, 'the Rail Barometer' may be a valuable instrument of analytical support in management of entities operating on the rail market. The rail market is characterised by a high volatility which is due to both macroeconomic trends and transport sector conditions as well as rail as one of its branches. With a very limited availability and substantive delays in access to cross-sectional statistical data, rail sector managers are confronted with a deficit of knowledge and points of reference needed to take decisions that have long-term consequences as well as background information to assess the accuracy of their own predictions and past actions.

Lack of a similar take makes it problematic to make even simple comparisons of tendencies characteristic for the sector and indicators for other sectors of the economy and macroeconomic situation. Given the above, the practical aspects of 'the Rail Barometer' consist in:

- a unified database which is the result of collecting in one place, in an ordered form data describing the rail sector,
- easy access to a regularly updated numerical value which describes the current economic situation and comparisons with past periods,
- a tool allowing for setting trends for the entire market and its individual segments,
- a tool allowing for carrying out individualised analyses for each market segment,
- data structure that may be used in forecasting trends and future development of the situation.

---

[1] Similar indicators are rare also for other rail markets of the European Union. A German example is SCI Rail Business Index published every quarter by an advisory company SCI Verkehr. Cf.: https://www.sci.de/trends/sci-railindex/.

[2] At the same time, the assumption constitutes a significant methodological constraint resulting from availability and registration rate of data. That leads to a restriction of the number of components of the synthetic indicator solely to the elements which are available with a delay of no more than one month, and in specific situations—a quarter.

The computational model of the synthetic indicator of the economic situation on the rail market was developed on the basis of comparison of data read monthly with data for a given month in the base month.[3] By definition, the model is to be used to describe changes in time of the data and not to focus on the assessment of absolute values attained. The authors are of the opinion that the biggest advantage of this formula is regularity in which the value expressed in points is determined based on the database updated every month and the possibility of comparing it with the generally available economic and social indicators.

2 Structure Assumptions of 'the Rail Barometer'

The synthetic indicator of the economic situation on the rail market was determined in a monthly cycle, and the range of considered data allows for monitoring the rail sector all over Poland. Availability of statistical and operational archive data constituted a limitation in analysing the data for the given time frame. Finally, the data collected pertain to the period starting in January 2013.

The value of 'the Rail Barometer' does not compare different populations (e.g. different market segments, regions or countries), but describes a change in time of the criteria for one population. Thus, it does not allow for making judgments as to the final result obtained but rather it allows for a discrete analysis (increase, decrease, no changes). Only prolongation of the time perspective by sequence determination of the indicator for subsequent time sections (months) allows for a qualitative assessment of the numeric values obtained by the indicator.

In the case of factors of financial nature, their real value was taken into account—without the impact of inflation. In the case of factors of seasonal nature, the dynamics was determined based on a comparison of corresponding months in a base year. In order to improve the transparency of the presented data, in the process of model calibration it was assumed that the indications of 'the Rail Barometer' will be compared with the base year, i.e. 2014, which makes it possible to make first comparisons as to the dynamics.

At the current stage of works, the synthetic indicator of the economic situation on the rail market in Poland is computed based on five components which describe each segment of the rail sector and factors that condition its functioning [2–7]:

(1) assessment of the economic situation in passenger services,
(2) assessment of the economic situation in freight services,
(3) indicators that describe the infrastructure,[4]
(4) indicators that describe the economics of rail operations of rail companies,
(5) indicators monitoring transport branches that compete with the rail.

[3]Such an approach determines the manner of setting trends for the analysed data: growth, decline or stagnation phase.

[4]Due to the restrictions pertaining to archive data, this element has been monitored since the beginning of 2017 and the synthetic indicator has included it only since January 2018.

3 Assessment of the Economic Situation in Passenger Services

The general economic situation indicator makes it possible to make independent computation of components for the market of passenger and freight services. The situation of the segment of passenger services was determined on the basis of a number of quantitative and qualitative factors specified on the basis of data published by the Office of Rail Transport, data provided by the carriers and infrastructure manager. They include the following elements:

- the number of passengers served,[5]
- operational work completed,[6]
- passenger service punctuality,[7]
- average delays of trains,[8]
- ticket price indicator,[9]
- cost of access to the infrastructure.[10]

It should be noted that the value describing passenger services used in the computations takes into account the number of passengers and not the transport activity, as the authors are of the opinion that the first value reflects much better the change in popularity of passenger services. In the case of operational data including transport activity[11] the assessment of the dynamics of changes would favour long-distance journeys; however, journeys effected at various distances are equally important for the development of the rail market.

The graph shows a presentation of the indicator for the passenger market in the years 2014–2017 (Fig. 1).

In the years 2014–2017, the indicator for the passenger market was characterised with a high volatility. At the beginning, it systematically decreased to a record low—at the level of 50 points, which resulted from accumulated adverse phenomena in January 2015. Since that time, the general tendency changed and after nearly two years of improvement, in November 2016, it achieved the initial value and then exceeded it, pointing to a good economic situation and development of this segment of the rail market.

---

[5]The change in the number of journeys measured on the basis of the number of passengers [pass.] in the last three months compared with corresponding months of the base year.

[6]The change of the value of the operational service (product of the number of trains made operational and the distance they covered [train*km]) in the last three months compared with the corresponding months of the base year.

[7]The percentage of trains that are on time, i.e. delayed by less than 5 min on their arrival at the terminus.

[8]A mean value of delay of trains, taking into account delays up to 5 min on arrival at the terminus.

[9]An average cost of covering 1 km for ten largest carriers calculated on the basis of a standard fare adopted for an average service distance characteristic for a given carrier.

[10]An average cost of access for ten largest carriers for the journeys characteristic for those carriers of a distance that equals two average service distances.

[11]i.e. the product of the service volume and service distance [pass*km].

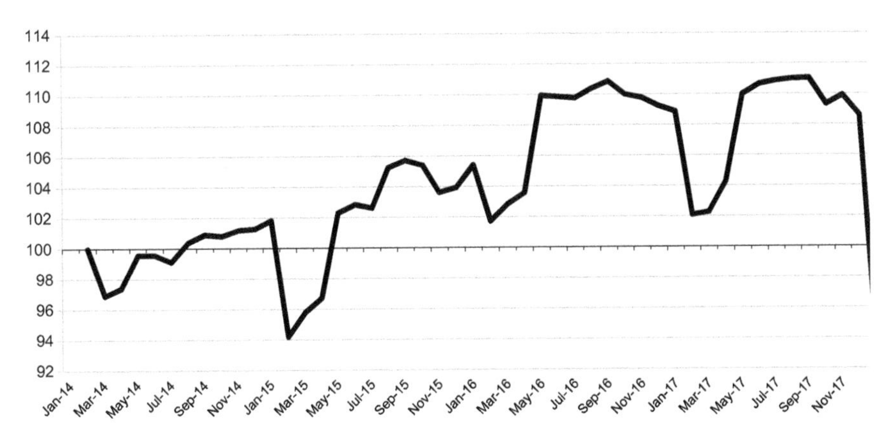

Fig. 1 Synthetic indicator for the passenger market in the years 2014–2017. *Source* Own computations

4 Assessment of the Economic Situation of Freight Services

The situation of the segment of freight services was determined on the basis of a number of quantitative and qualitative factors. They were specified based on operational data published by the Office of Rail Transport as well as the costs of operational activity of carriers. The final indicator is composed of the following elements:

- transport activity,[12]
- operational activity,[13]
- freight service punctuality,[14]
- average delays of freight trains,[15]
- cost of access to the infrastructure.[16]

The above list includes the operational value pertaining to the transport activity which in the opinion of the authors best reflects the changes of this market segment. For freight services, they are the effect of both changes in the value of freight services themselves and the distance on which such services are rendered. The value of the operational activity for the currently available data seems, however, to best reflect

[12]A change in the value of the transport service [tonnes*km] in the last 3 months compared with the corresponding months in the base year.

[13]A change in the value of the operational activity [train*km] in the last 3 months compared with the corresponding months in the base year.

[14]The percentage of trains operating on time, e.g. delayed by less than 5 min at the arrival at the terminus.

[15]The average value of train delays, taking into account delays of up to 5 min at the arrival at the terminus.

[16]The average cost of access to the infrastructure for the adopted ten lines characterising freight services typical for Poland, whose total length amounted at 4531 km, which constitutes 24% of the total length of the railway network in Poland.

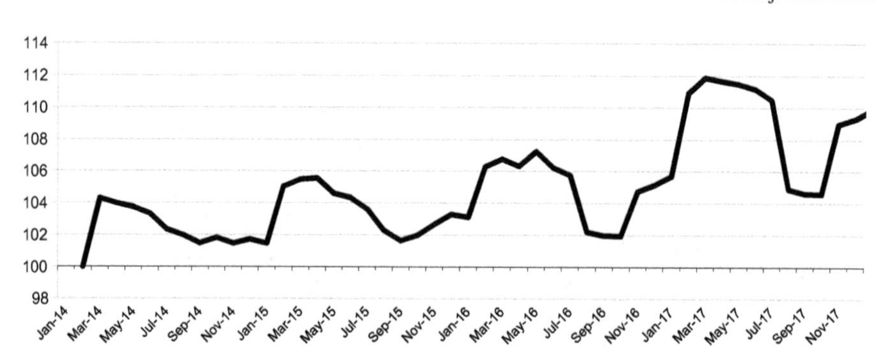

Fig. 2 Synthetic indicator for the freight market 2014–2017. *Source* Own computations

the aspect of organising services and, as a result, points to the effectiveness of the transport activity.

The graph shows a presentation of the indicator for the freight market in the years 2014–2017 (Fig. 2).

For most of the period 2014–2017, the indicator for the freight market adopted negative values—below 100 points—which illustrates the adverse situation of this market segment. Following a rapid decrease at the level of 60 points, the barometer indications oscillated at the level of 70 points. A six-month increase tendency was observable only in July 2016, but it failed to reach the initial value and ended up in another decrease. Only in September 2017 did the indicator attain the value of more than 100 points, which shows a positive economic situation and development of this segment of the rail market.

5 Indicators Describing Infrastructure

Infrastructure is one of the key elements underlying the quality of services in the rail sector and as such its competitiveness on the transport market. Hence, the main point of reference in the structure of the synthetic indicator of the situation on the rail market is its utility. As a result, the following elements of monitoring this segment of the rail market were adopted:

- average maximum velocity for passenger trains,[17]
- average maximum velocity for freight trains,
- average axis load,
- number of kilometres of lines covered with operation restrictions.[18]

[17] Calculated as a weighted average for rail network sections, taking into account the values for car sets, electric multiple units and railbuses.

[18] This value includes sections closed due to investment and maintenance works and lines where the permissible train velocity is 0 km/h.

Due to the fact that 96% of rail infrastructure in Poland is currently operated by PKP Polskie Linie Kolejowe, it was assumed that the data for this manager will be representative for the entire market.[19]

6 Indicators Describing the Economics of Rail Operation of Rail Companies

Components of the synthetic indicator of the economic situation of the rail market describing the economics of rail operation consist in two parts—financial standing of companies operating in this field and the area of key costs necessary to perform transport services.

At the current stage—due to lack of alternative data—monitoring of the financial standing of rail companies has been based on the share prices of companies connected with the rail sector listed on the Warsaw Stock Exchange. For the purposes of the computation, a special sub-index of rail companies has been developed,[20] with an assumption that its value referred to average stock exchange quotations will illustrate the confidence that investors have in the sector.

The costs of the operational activity of rail carriers have been assumed on the basis of access to the infrastructure, energy and remunerations. The following elements have been used to determine trends in the above area:

- cost of access to the infrastructure for passenger services,
- cost of access to the infrastructure for freight services,
- cost of operating system energy,[21]
- average monthly nominal gross remuneration in the corporate sector.

7 Indicators of Monitoring the Rail Competitive Environment

The indicator of the competitive environment, constituting an element of 'the Rail Barometer', was designed as a derivative of the conditions of operation competitive to road transport. In this case, the elements that best reflect the possibility of operation of road services include:

- length of public roads covered by the ViaTOLL system,

[19]It may be assumed that in future the infrastructure indicators will include also other line infrastructure managers as well as service infrastructure structures, such as intermodal terminals.

[20]Operationally referred to as WIG-Rail.

[21]Determined on the basis of the cost of MWh as per the price list of PKP Energetyka S.A. and the wholesale price of the Diesel fuel at the ratio of 80/20, ensuing from the indicator of the number of trains using electric and diesel multiple unit.

- wholesale price of diesel fuel.[22]

The parameter of coverage of the system of fees for the access to road infrastructure was adopted also due to an analogy to the system of fees for the access to rail infrastructure, monitored within 'the Rail Barometer'. Covering more public roads with the toll system levels the competition of individual land transport areas.

8 Method to Determine the Weights

Based on the screening test and the analysis of values exceeding 1.0, six factors were identified and further analysed. Varimax rotation was applied in order to enhance the differences in the aspects indicated by these factors (Fig. 3).

Afterwards, confirmatory factor analysis was performed with the assumption that there are six underlying factors describing the condition of: the passenger market, the cargo market, the infrastructure, the production costs, the competitive environment and the economic situation.

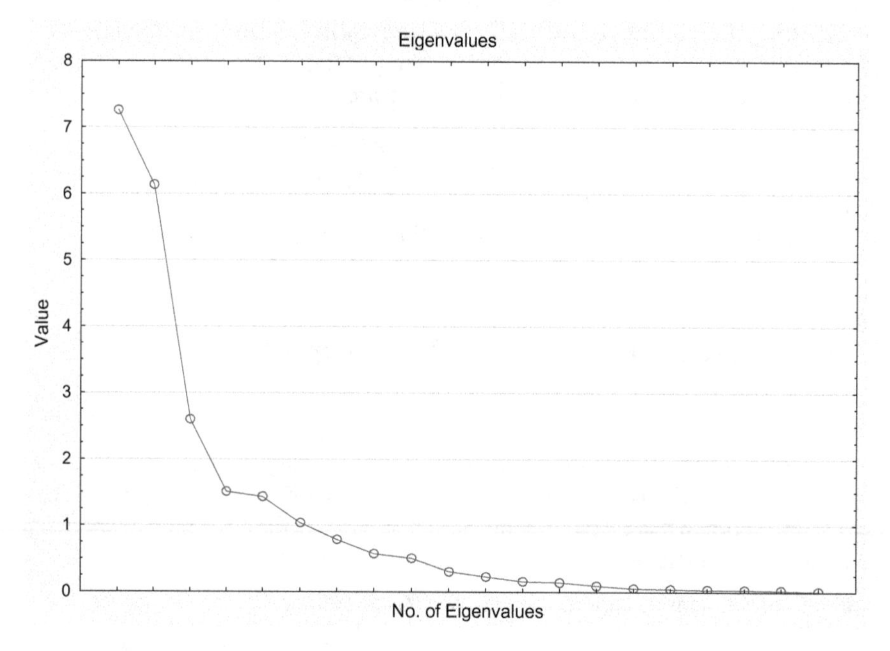

Fig. 3 Eigenvalues of the factors

[22]Diesel fuel is used both in rail and road transport. However, its price impacts more the costs of operational activity in road transport, which results from lower resistance of rolling stock and as a result a greater efficiency of rail transport than road transport, which needs ten times more fuel to carry the same unit of load. Cf.: Reference [1, p. 478].

Out of the 23 variables affecting the six factors, 20 are statistically significant when applying the 0.1 significance level and 18 are significant when applying the 0.05 significance level. Average speed of the passenger train seems not to be a factor showing the state of the passenger market. The price of ordering a cargo train does not affect the production cost in the sector, and the evaluation of the portfolio does not affect the overall state of the sector economy (Table 1).

Table 1 Indicators and values of the weights

	Model values			
	Factor	Error	*T*-stat	*p* value
(PassMrkt)-1 → [thoupass]	0.045	0.006	7.564	0.000
(PassMrkt)-2 → [Wtrainkm]	0.088	0.008	10.392	0.000
(PassMrkt)-3 → [Punctual]	−0.026	0.009	−2.813	0.005
(PassMrkt)-4 → [PriceTic]	0.007	0.002	4.156	0.000
(PassMrkt)-5 → [AVGSPP]	−0.001	0.001	−0.966	0.334
(CrgoMrkt)-6 → [Mlntonne]	0.097	0.009	10.392	0.000
(CrgoMrkt)-7 → [ThouTrKm]	0.083	0.008	9.914	0.000
(CrgoMrkt)-8 → [CargoPun]	−0.107	0.013	−8.346	0.000
(CrgoMrkt)-9 → [ShareITM]	0.006	0.002	3.561	0.000
(InfrStrc)-10 → [MAXSPDPS]	0.041	0.004	10.392	0.000
(InfrStrc)-11 → [MAXSPDCG]	0.021	0.002	10.077	0.000
(InfrStrc)-12 → [Force]	0.006	0.001	5.440	0.000
(InfrStrc)-13 → [PLKLine]	0.015	0.002	8.378	0.000
(ProdCost)-14 → [PricePas]	0.059	0.006	10.392	0.000
(ProdCost)-15 → [PriceCar]	0.003	0.003	0.976	0.329
(ProdCost)-16 → [EnergyPR]	−0.009	0.003	−3.453	0.001
(ProdCost)-17 → [Salary]	−0.015	0.009	−1.678	0.093
(CompEnvm)-18 → [PublRoad]	0.075	0.020	3.762	0.000
(CompEnvm)-19 → [NewCars]	0.256	0.069	3.733	0.000
(CompEnvm)-20 → [FuelPR]	−0.031	0.012	−2.606	0.009
(Economy)-21 → [Portfoli]	0.013	0.019	0.710	0.478
(Economy)-22 → [INTTRade]	0.051	0.028	1.868	0.062
(Economy)-23 → [PMI]	0.030	0.003	10.392	0.000

Source Own computations

9 Rail Market Synthetic Indicator

All the five components of 'the Rail Barometer', referred to at the beginning and then presented, were put together in the form of the rail market synthetic indicator. So as not to prioritize individual segments at this stage of works, it was assumed that the volatility of the final component will be the sum of volatility of its elements and at the same time no corrections in the form of weights were applied in the prepared model.

As a result of completing the methodological framework developed in 2017 with data, a cycle of monthly values of 'the Rail Barometer' was computed. The starting point and at the same time the first analysed month was January 2014 as it was the first month the data for which may have been compared with the preceding year. Those values were adopted as the starting point, which was attributed to the value of 100 points. The last value presented in this article was computed for December 2017 (Table 2).[23]

A graphic presentation of the indicator of the economic situation or the rail transport sector in the years 2014–2017 is included in Fig. 4.

The analysis of the value of 'the Rail Barometer' in the years 2014–2017 shows a significant volatility of the situation on the rail market. For nearly three subsequent years, the value of the indicator was negative—below 100 points—which reflects an adverse sector situation. At the same time, in mid-2015, one may observe a growth tendency—as a result, the base value was attained in November 2016 and the record value of the indicator stands at 148 points—at the end of the analysed period.

Table 2 Values of the synthetic indicator of the economic situation on the rail market in the years 2014–2017

1	2	3	4	5	6	7	8	9	10	11	12
2014											
100.0	100.9	100.6	100.1	99.4	97.7	97.9	97.5	97.4	97.7	98.8	97.9
2015											
99.0	98.2	99.6	98.9	98.9	98.1	98.6	98.1	98.0	98.3	99.2	101.8
2016											
101.0	101.9	101.7	101.2	100.2	98.5	99.3	100.1	99.7	99.9	100.5	98.3
2017											
100.0	100.8	102.0	102.0	102.1	101.5	100.5	100.0	98.6	97.4	97.2	95.8

Source Own computations

[23]Monitoring of the market with the use of 'the Rail Barometer' will be continued also in 2018 and in subsequent years.

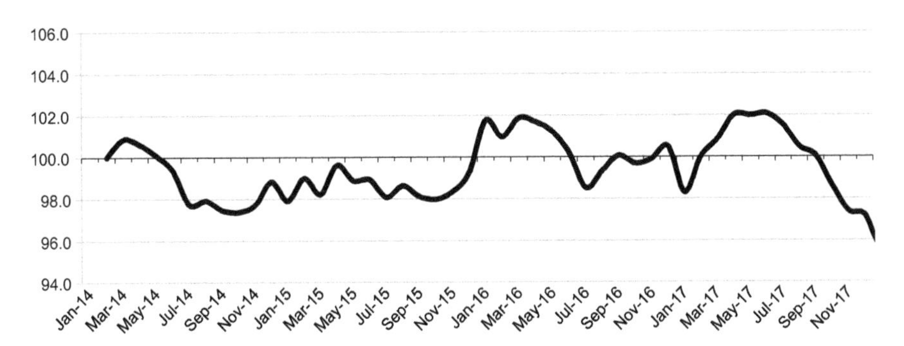

Fig. 4 Rail Barometer 2014–2017. *Source* Own computations

10 Summary

Positive experience resulting from practical application of the methodology of monitoring the rail market situation, developed in 2017, shows that a specially prepared analytical tool coupled with a database may be useful in monitoring and setting trends for individual sectors of the economy. Monitoring the rail sector, due to its specific technical and organisational nature as well as its complexity and significant diversity of individual segments, should not base solely on simple values connected with the services rendered.

The synthetic indicator of the rail market situation presented in the article has a significant development potential. Statistical data ordered on its bases and arranged in a few-year time sequences constitute a big source of analysis and a starting point for further studies.

It is worth emphasising that in consequence of further studies the model should be supplemented and modified in order to illustrate to the greatest extent possible the volatility of conditions of conducting rail activity.[24]

References

1. Antonowicz, M., Majewski, J., Wołek, M., Zajfert, M.: Założenia konstrukcji syntetycznego wskaźnika koniunktury sektora transportu kolejowego w Polsce. Ekon. Transp. Logist. **74**, 467–481 (2017)
2. Burnewicz, J.: Prognozowanie na usługi transportowe w Polsce do 2020. w: Liberadzkiego, B. i Mindura, L. (pod red.) Uwarunkowania rozwoju systemu transportowego Polski, praca zbiorowa. PIB Warszawa, Radom (2007)
3. Domański, E., Ożóg, Z.: Lokomotywy spalinowe serii ST44. WKiŁ, Warszawa (1984)

[24]Apart from the data taken into account in the computations made for the purpose of this article, other statistical data, such as average age of the rolling stock, profitability of rail carriers, travel times for specific lines, average permissible velocity on rail lines, are being collected.

4. Dorosiewicz, S.: Koniunktura w transporcie. Metodyka badań, wyniki, modele. Wyd. Instytut Transportu Samochodowego, Warszawa (2013)
5. Duranton, S., Audier, A., Hazan, J., Gauche, V.: The 2015 European Railway Performance Index. Exploring the Link Between Performance and Public Cost. The Boston Consulting Group (2015)
6. Duranton, S., Audier, A., Hazan, J., Langhorn, M.P., Gauche, V.: The 2017 European Railway Performance Index. The Boston Consulting Group (2017)
7. Kowalewski, G.: Próba konstrukcji złożonych wskaźników koniunktury. Ekonometria **3**(37) (2012)

How Cars Should Be Designed: The Importance of Form in New Technologies

Karol Janiak

Abstract The purpose of the article is to discuss style and structure of modern cars resulting from innovative types of drive systems and design. The author based his considerations on the experience gained during designing the solar car Eagle Two, which was constructed by the students from the Lodz Solar Team at the Lodz University of Technology, and took part in the Bridgestone World Solar Challenge in Australia. The article deals with the issues concerning how the car of not even "tomorrow" but of "today" should look like. In the moment of radical changes in the way we travel and technological breakthroughs, should we act conservatively or courageously? Are recent ideas, such as solar drive system, a solution to the difficult environmental and transportation situation, in which we are currently in? How do the new technologies affect the process of shaping the forms of the vehicles and their reception by the users? The article describes the progress of technology in transport from the point of view of the industrial designer.

Keywords Solar power · Race car · Ecology · Automotive design · Electric car · Solar car

1 Introduction

The purpose of this text is to bring closer the process of developing the form of the solar car that was constructed by the students from the Lodz Solar Team at the Lodz University of Technology. The task of the author of this text was to create, as part of the master's dissertation, a concept car design characterized by uniqueness, as well as high aesthetic and functional qualities. The vehicle was being built with a view to taking part in the Bridgestone World Solar Challenge race held in Australia.

All technological developments undergo constant development. This process gets intensified each year; however, it seems that car is a device progressing more reluc-

K. Janiak (✉)
Faculty of Industrial and Interior Design, Strzemiński Academy of Art Łódź, Łódź, Poland
e-mail: kjaniak@asp.lodz.pl

M. Suchanek (ed.), *Challenges of Urban Mobility, Transport Companies and Systems*, Springer Proceedings in Business and Economics,
https://doi.org/10.1007/978-3-030-17743-0_23

tantly than others. Communication technology has evolved through the last one hundred years from a telegraph to the Internet, while airplane, invented in the nineteenth century, made it possible to take human to the moon several decades later. Car, in fact, for the last one hundred years of its existence, has not changed. There is no doubt that it has undergone constant modifications and evolution; however, its essence remains unchanged—the process of its usage, the way it is powered and how it functions remain unchanged. However, a breakthrough is coming. Even automotive traditionalists have to admit that conventional cars powered by internal combustion engines will be going into non-existence in the near future. Humankind is facing such problems as global warming, or rising costs of extraction of petroleum. One of the products, which have to change in the light of the forthcoming revolution, is the car [1].

We are now in a perfect moment for creating a concept of a new car, and for practical and visual solutions exploration. For the first time in history, we have real technical possibilities to depart from the well-known model of a car and take a new route. Development of computer technologies, electric drives, launching of lithium batteries, as well as experiments with fuel cells, or finally introduction of relatively efficient solar panels opens the door for new solutions in the field of transport. We are also facing the necessity to introduce changes as further degradation of natural environment, unless we implement radical actions, will lead to serious consequences in the near future. It is not the lack of petroleum that will force us to do it, but the fact that further exploration of the one we have is going to cause climate collapse.

Nowadays, everyone knows what a car looks like. If we asked random people to symbolically draw a car, everyone would roughly sketch the same. In the modern world, it is an extremely important element of existence, used every day as a means of communication, transport of goods, and it often becomes an object of passion or a tool of manifestation of wealth or status [2]. In many cases, it becomes a flagship—it sends out specific messages with its appearance, purpose or usage. Its basic form, however, has not changed much for years. If we ignore the marginal and exotic cases, most cars have a similar shape. This is due to the fact that car is a device specialized for a specific purpose—transporting people and goods in a specific way—using an internal combustion engine. Its shape is therefore a logical consequence of function and technology. In case of modern cars, the basic function will remain unchanged—the car will still be the means of locomotion. It is the technology that will make this locomotion possible that will be changed [3]. So how should a car worthy of the modern days look like? There are two options—evolution or revolution.

On the journey to a better tomorrow, we must be accompanied by the common sense—you cannot replace the old technology with the new one from one day to the next, you need time for this. Ideas to prohibit sale of combustion engine cars, or introduce fees for their use, or other types of methods forcing people to assimilate the latest solutions may paradoxically lead to ecological and economic disaster. The new technology must live in symbiosis with the old and gradually replace it. Apart from infrastructural aspects or energy problems, in order for this type of transport to become popular in the whole society, first it must fully spread on the new car market. This will lead to the reduction in prices as well as the introduction of electric vehicles on the secondary market, what is necessary to popularize the new means of transport. The project discussed here is a form of a car that could be a prospect for a better tomorrow, but in no way eliminating the current reality. The earlier creation of a new prototype of a car will allow the society to get accustomed to it faster and smoother, which will have a positive influence on assimilation of the new technology.

2 Car of the Future

We are now facing a new revolution in the world of automotive industry. For the first time in one hundred years, significant changes in how the car works have occurred. Although the history of automotive industry knows examples of electric cars, as early as those from the first half of the nineteenth century [1, 4], only now the prospect of departing from combustion engine cars powered by fossil fuels seems to become realistic due to ecological aspects and increasingly difficult, hence, more expensive petroleum extraction. There is a turn, or even a return, towards electric drives. Former electric vehicles, from the beginnings of automotive industry, did not have an exceptional character, due to the fact that car itself did not have it yet. Today, the situation is different, and the electric drive offers many possibilities. In such a solution, we do not have large engines, gearboxes, fuel tanks and powertrains. Electric motors are relatively small, and the batteries allow great flexibility in locating them. This space saving gives the freedom to create completely new forms impossible to be achieved before. What is more, the new type of a drive is the perfect pretext for manufacturers to express the unique character of a car through style. Due to the fact that car is an object very strongly embedded in the consciousness of people—shaping its form is a big challenge. Previous attempts to depart from the well-known model were not received with enthusiasm. When for the first time mono-construction cars, which broke with the widely held pattern, appeared, they did not meet with the kind reception of users. A designer who faces the task of designing a "new" car, whose character is completely different from what it was before, has to face a dilemma—whether to create some kind of a skeuomorphic lie (imitating known patterns) and design a "combustion engine" car electrically powered or leave the well-known pattern completely and build a new form around the new technological and functional reality. In recent years, many manufacturers have dealt with this problem by taking one of the above paths. It should not be forgotten that marketing departments and car manufacturers, which do not quite know how an electric car should look like, are also involved in the design process. The radiator grille is often a testimony to these dependencies. An electric car does not need as much cooling as a combustion engine car, but "grille" is often a characteristic element of the stylistic language of the brand and must be included in the design, although it is not needed from the point of functioning [5]. This type of decisions results also from the fact that non-combustion engine cars are so different; they have to be made similar to the already known vehicles. You can also look at the problem from yet another side, whether the electric car needs a new language at all, or maybe it is just an artificial invention of marketing departments, which in the pursuit of sales constantly creates new classes and niches in the automotive market. An electric car must meet the norms set by new technologies, but they are mostly the same as for its combustion relatives [6]. Maybe the approach to ecology should be also changed? Thinking of it as about a process of increasing efficiency via different methods, and not by excluding the old ones. Owing to this, as in many cases of car design improvements from the past, today it is possible to create efficient, interesting and exciting vehicles.

There is a fundamental problem in the perception of ecological cars. Designers and producers create cars, which with their character so strongly and unfortunately express their dissimilarity, and instead of gaining, they lose aesthetically, and their reception is negative. Most cars that compete in the Bridgestone World Solar Challenge race may be taken as examples, which by their extreme focus on function and efficiency have lost their character, clarity, usability or aesthetics. Many companies producing ecological cars also face similar problems, including Mitsubishi E-miev, Toyota Prius or Nissan Leaf. Another example of how not to design cars of "tomorrow" is Waymo-autonomous Google car, characterized by an almost infantile style [7]. However, not everything is lost yet. Many producers show that ecology can be used to its advantage. They prove that ecology can be a means of expressing emotions. The first connotation is Tesla. The American Giant is currently the best-selling luxury sedan in the USA and has achieved this success with fully electric cars [8]. There is no better proof of the possibility of symbiosis of emotion with ecology than a four-passenger sedan accelerating to a hundred at the speed of Bugatti Chiron [9]. There are plenty more of such examples. The first ideas of electric supercars, such as Rimac Concept One [10] or Lamborghini Terzo Millennio [11], have already been drawn. Many manufacturers see the future in bright colours; it is worth to familiarize oneself with the project Fisker Emotion [12], or BMW Vision Next 100 [13]. The car described in the text was to be qualified for this group.

3 Solar

In order to verify the hypothesis that car should change its form, the author decided to collect data by carrying out an open participant observation. It took place during the project of a solar car in Poland in 2016–2017. Case study [14] was used as the research method for the analysis of data. The choice of the method was dictated by the initial stage of research and an unusual phenomenon, which is a solar car, or the desire for extensive analysis and better understanding of the phenomenon itself.

3.1 Cephei—The First Version of the Project

The purpose of the project was to create a car distinguished by its style, usability and good design. In the matter of limitations, certain parameters established at the beginning and the regulations [15] had to be taken into account. Due to the way how the points were granted, it was decided that the constructed car would be a two-seater sports car.

The key aspect for the team was the maximum reduction in air resistance, so the frontal area of the vehicle, as well as the coefficient of the CX frontal aerodynamic resistance, results from the shape of the car body. Starting from these assumptions, the initial dimensions of the car were set at 5 m length (the maximum allowed by the regulations, necessary due to the large number of panels), 1 m width as well as 1 m height (for the most possible radical lowering of the frontal area).

After establishing those initial settings, the design process and work on what form the car should take started. It was decided that the designed vehicle would not be modelled on the existing ones. The message of the form was to be honest, reflecting the implemented technology. Another aspect that was to be included in the project was the semantic layer—the created car was supposed to be exciting, like a symbol of enthusiasm and hope for the future. You can therefore express the assumed style in a few simple points: dynamics, modernity, technologicality, energy, enthusiasm, honesty of the form, legible semantics.

Car powered by solar energy does not need an engine, a drive system and fuel, and thus loses very large elements that define, by their dimensions, the shape of the vehicle. Therefore, the only right solution was to "cover" the remaining functional and technological elements with as little amount of material as possible, eliminating, semantically confusing, empty spaces and stretches of the car body. Apart from the purely reasonable aspects (e.g. saving material), it was supposed to be a solar vehicle, so its shape could not leave any doubts about it.

In order to achieve the smallest possible loss in air resistance, a very small frontal area—1 m^2—was set. However, this approach to the subject creates a problem—the car would be very unstable. In order to avoid that effect, following the example of the Deltawing [16] vehicle, the rear track was extended. To solve the problem of a large frontal area, a concept was proposed in which air would literally fly straight through the car. That was how the vehicle was created, which consisted of a drop of a body suspended under the roof wing with solar panels supported on the edges with "razor blades" finished with wheels. This solution would allow keeping a small frontal area and at the same time the increase in the rear track width and the large surface of the solar panels.

The final form was very advantageous from the point of view of aerodynamics. It had another advantage as from the outside it was visible that the car did not have a place where any engine could be found—the created form was unusually honest (Fig. 1).

3.2 *Albireo—The Second Version of the Project*

The regulations of the forthcoming competition Bridgestone Solar Challenge 2017 [17] were published upon finishing the first stage of work. In connection with the

Fig. 1 Cephei, visualization. *Private archive* Karol Janiak

Fig. 2 Albireo, visualization. *Private archive* Karol Janiak

changes, a decision was made to start work from the beginning, this time on a four-seater car what significantly influenced the guidelines. Width was enlarged to 1.5 m, and it was decided to make a mono-body-shaped car. Length and height remained unchanged. Despite increasing the width of the car by 50%, its proportions were still unfavourable. Long tail was very visible, and the car looked unattractive and unnatural. It was proposed to use a module construction based on a detached tail. That solution would allow to diametrically improve the appearance, proportions and practical aspects of the everyday usage of the car, parallel to keeping high technical values, required during the race or a longer journey.

An important element strongly defining the car body was the windshield; the intensely sloping glass pane gave the car body an expression of momentum and dynamics. Curling sideways and moving backwards "A" pillars would improve visibility. The front of the car was supposed to send a clear message that there was no engine in it that would normally be expected to be there. Therefore, the windshield reached the very top of the car body, and thus, the space of the cab also. It was supposed to create an impression of openness, being in a glass bubble.

Strong elements of the front of the vehicle were the bulging wheel arch panels that gave the impression of manoeuvrability and stability of the vehicle. In the front projection, the cab is narrower than the car body underneath, which places the car on the ground and creates the impression of stability. Due to the potential difficulties in production, excessive material usage and problems related to the necessity to find a solution hot to fix the tail, the concept was rejected (Fig. 2).

3.3 Helios Aka Eagle Two—The Final Version of the Project

Finally, it was decided that the car would have a four-door, fully mono-shaped body. Accepting that type of a solution forced generating an aesthetical and attractive form of the car with extremely unfavourable proportions. The car was supposed to be 1.5 m wide, 1 m high and 5 m long. It gave the length of the car bigger than in a multitask military vehicle type HMMWV "Hummer", with frontal area of the category of a supermini car of the 1970s, i.e. original Austin Mini, which was extremely unfavourable from the design perspective.

Work began from pulling the back of the car from the previous concept to the desired length of 5 m. The form was being changed, trying to hide the extremely unfortunate proportions. The wheels maximally shifted to the front together with the long rear were supposed to give the impression of momentum and dynamics. That effect was reinforced by an intensely sloping, panoramic windshield, which also provided excellent visibility (also of the traffic lights) and allowed a lot of light into the interior. The front of the car was connected to the rear by a body side moulding running from the front wheel arch panels to the edge of the roof run-off. It added a lightness to the car and diversified its profile. The biggest stylistic challenge in the new concept was the rear of the car. Due to its considerable lengthening, the car gained a large, awkward "abdomen". In order to improve its appearance and optical slimness, it was lifted up, levelling more with the roofline and cut strongly from the bottom, creating a convergent shape ended with a thin, rear edge of the shape of the car. It was also decided to lower the side roofline, which further hid the thickness of the vehicle at the height of the rear wheels. That detail together with highlighted wheel arch panels created attractive, wide "hips" of the car, following the pattern of sports cars like Porsche 911. Then, the form was optimized in terms of production capabilities and technical parameters of photovoltaic panels (Fig. 3).

Fig. 3 Eagle Two. *Archive* Łódź Solar Team, Sebastian Górecki

3.4 Scuti—Concept Design

Upon completing the work on the production project, a decision was made to develop the project in the form of a stylistic vision. The form was subjected to strong radicalization and exaggeration of key stylistic aspects. The object created was more like a "carving" of the car, a spatial concept drawing, rather than the vision of the actual vehicle. Changes began with an improvement in the overall proportions. The car body was widened and the profile lowered. Extremely large wheels were used in the concept version, which worked in favour of proportions and placed the car firmly on the ground. There was a return to the key element of the stylistics, which was the intensely sloping windshield, which gave its shape the momentum and made it strongly dynamic. The cab was made much more rounded and without edges. The front of the car was much lower. It was characterized by two main features: the moulding running through the entire width of the front, merging with the headlights, and a much highlighted splitter, i.e. the front spoiler, bringing associations with motorsports and giving aggressive expression. The concept also solved the problem of a little aesthetic side surface at the height of the rear wheel. There was a return to the rounded shapes; the side line of the roof was pulled down again. Such a solution allowed to visually lower the car and restored the attractive appearance of the rear wheel arch panels. The rear of the car itself was defined by sharp as a razor edge of the roof run-off. The rear lights, similar to the front ones, merge with the body of the car. Due to the fact that the car had to accommodate four people and have 5 m^2 panels, its shape was quite bulky, even in the concept. That may give the wrong impression that there was a combustion engine, placed centrally in the car, as in super-sports cars. In order to reduce that impression, air ducts beginning in front of the rear wheels and finishing in the diffuser were used. A thin and strongly converging back also deepened that impression.

That concept design was very impressive and that was its only goal. The implementation of a similar vehicle into production would be very problematic, at least in an unchanged form. That was an extremely idealized vision of the car that was created. When we compare those two projects with each other, it may be difficult to find similarities, as they were such different proposals. That was due to limitations and adversities that had to be addressed when creating the actual physical car. That vision, however, showed the direction which, according to the author, should be taken into account in the design of the cars of the future (Fig. 4).

4 Conclusion

Plenty of factors influence what form should the project take. This type of a situation takes place almost always when the task is to create a production solution, or a more advanced one than a purely visionary–conceptual creation. It was no different with that project. The task to build a solar car from scratch is a titanic challenge. It requires

Fig. 4 Scuti, visualization. *Private archive* Karol Janiak

a creation of a design project, meeting all of the technical requirements, designing all of the necessary systems, constructing the whole car and then assembling it. And lastly, the final project must be efficient and useful. Out of the factors that influenced mostly the final form of the car, excluding the vision, were not technicalities or not even strictly technical or technological possibilities but the main force, which made the final creation of the car, was the team, or rather its limitations. Most of the interesting or innovative ideas were rejected. It is a conflict that often occurs in this type of activities. "Artist-designer" creates a vision that will be a kind of a utopia, and the goal of engineers is to adapt this concept to the realities of production and technical and technological aspects. They must minimize costs and simplify the construction as much as possible. None of the innovative, visionary ideas were used, but despite that, the created project was interesting, original and distinctive, through aesthetics and idea, against the competition from the BWSC race.

Designing and building race cars seems a vain activity due to the fact that apparently its achievements do not bring anything to the lives of the majority of people and serve as an entertainment for only a tiny group. Nothing could be further from the truth. Many technologies that we use today had it beginnings in the hands of small groups of people, be it military, or precisely in racing cars. Such experiments and research are a testing ground for new ideas, searching for solutions that will be used in road cars after years. The construction of Eagle Two is not a breakthrough in the development of the automotive industry, but it is a piece added to the work of evolution, an object that provokes a discussion on how we should travel in the near future. It is also a manifestation of enthusiasm and possibilities, how vehicles in the forthcoming times should be designed in an interesting, unusual and innovative way.

References

1. Rychter, W.: Dzieje Samochodu. Wydawnictwa Komunikacji i Łączności, Warszawa (1979)
2. Barth, L.: Cars as Status Symbols. www.consumerreports.org, 15 May 2017
3. Newcomb, D.: At CES and Beyond, it's the End of the Car as we Know it. www.forbes.com, 15 May 2017
4. Rostocki, A.M.: Historia Starych Samochodów. Wydawnictwa Komunikacji i Łączności, Warszawa (1981)
5. Holzhausen, F.: Creating a Car from the Ground Up. www.carbodydesign.com, 25 May 2017
6. Stevens, P.: Do Electric Cars Need a New Design Language. http://cardesignnews.com/articles/design-essey/2014/12/do-electric-cars-need-a-new-design-language-by-peter-stevens, 23 May 2017
7. Philips, T.: Chris Bangle, Peter Stevens and Tom Matano Speak Out About Google's Driverless Car. http://cardesignnews.com/new_cars/display/store4/ote,301248, 22 May 2017
8. http://www.autoblog.com/2016/10/14/tesla-model-s-best-selling-us-luxury-sedan/, 25 May 2017
9. https://www.bugatti.com/chiron/
10. http://www.rimac-automobili.com/en/supercars/concept_one/
11. https://www.lamborghini.com/en-en/brand/innovation-excellence/terzo-millennio
12. https://www.fiskerinc.com/
13. https://www.bmwgroup.com/en/next100.html
14. Yin, R.K.: Case Study Research and Applications: Design and Methods, 6th edn. Sage, Thousand Oaks, CA (2018)
15. https://www.worldsolarchallenge.org/files/522_2015_world_solar_challenge_event_regulations.pdf
16. http://www.deltawingracing.com/
17. https://www.worldsolarchallenge.org/event-information/2017_regulations

S-mile Visualizer Tool as an ITS Component

Marcin Staniek

Abstract The article describes individual assumptions underlying ITS, the scope of its application as well as advantages resulting from its deployment in the city space. It also provides characterisation of S-mile Visualizer Tool, its structure as well as data collection and storage standards. Data processing mechanisms and data presentation technologies embedded in the tool have been defined with reference to interactive thematic maps. The author has indicated individual options for supporting ITS with a data set obtained from the S-mileSys platform being currently under development, under the ERA-NET Transport III programme.

Keywords ITS · Smart cities · Freight transport · Local authorities

1 Introduction

In general terms, intelligent transport systems (ITS) form a collection of highly advanced measurement, communication and information technologies as well as management techniques which enable efficient use of a transport system, improvement in traffic safety and protection of natural environment [1]. They are also relevant and effective tools for implementation of transport policies in many cities. The rationale behind these solutions is to enable developing the city as a space which proves to be both inhabitant- and tourist-friendly [2]. Owing to the available ITS features and components, city authorities can actually manage private as well as public transport, both passenger and freight transport. They encompass control of the flow of means of transport for the sake of increased comfort and reduced travelling time, including solutions intended for vehicle weight monitoring aimed to prevent the permissible axle weight from being exceeded, thus precluding damage to pavement of roads not suitable for heavy vehicle traffic [3]. They also comprise solutions which

M. Staniek (✉)
Faculty of Transport, Silesian University of Technology, Katowice, Poland
e-mail: marcin.staniek@polsl.pl

M. Suchanek (ed.), *Challenges of Urban Mobility, Transport Companies and Systems*, Springer Proceedings in Business and Economics,
https://doi.org/10.1007/978-3-030-17743-0_24

enable imposing specific constraints on the access to individual city zones, these being designed in such a manner that private vehicle traffic as well as emission of pollutants into the environment can be reduced, thus improving the living comfort of people dwelling in these areas [4, 5].

2 ITS Architecture

The ITS architecture is a set of statements of general nature that enable planning of development of mutually cooperating applications and services. It creates a common milieu for planning, defining and integrating individual components. It is defined with reference to specific needs identified among the transport system users as well as their expectations and preferences, and it consists of the general, the logical, the physical and the communication structure. The general ITS structure is the system's sphere where the general concept as well as individual operating principles are taken into account. The logical structure covers the features that the system being developed is assumed to deliver by taking needs and expectations of its users into consideration. It defines the interactions between the system and its users as well as between the system itself and its closest environment. It covers the characteristics which define data sets required in the processes of control and management. The physical ITS structure is a set of specifications explicitly defining the technical solutions underpinning the system components being implemented. It stands for the characteristics of the software deployed, required to deliver the pre-defined system features. The communication layer defines the solutions that enable information exchange within the system between its components [6, 7].

The ITS architecture can be described in different ways; however, the very substance of how ITS is intended to function must be defined explicitly. In accordance with Directive 2010/40/EU [8], the notion of architecture is understood as a conceptual design that defines the structure, the operating principle and the integration of the system being developed in the space which surrounds it. In terms of standardisation of the ITS architecture, key roles are performed by domains and groups of services. Paper [9] describes the structure of ITS domains/services defined by the ISO 204 technical committee. The output document was prepared on the basis of national and international specifications originally created in the USA, the European Union or Japan. It represents the highest level of the ITS architecture breakdown, specifying the structure of the functional goals it pursues. The status of this document clearly indicates the non-obligatory nature of the ITS architecture description according to [10] as well as the potential for utilising individual technical and technological capabilities at hand according to the actual needs.

Bearing the subject of this article in mind, the ITS services that function in the freight transport domain have been listed below:

- commercial vehicle pre-clearance,
- commercial vehicle administrative processes pre-clearance,

- automated roadside safety inspection,
- commercial vehicle on-board safety monitoring,
- freight transport fleet management,
- intermodal information management,
- management and control of intermodal centres,
- management of dangerous freight.

3 S-mileSys Platform as a Means to Support the Freight Transport Area

The priorities underpinning creation and development of sustainable logistics are fostered by numerous institutions, including the European Commission [11–13], and have become the germ of design and implementation of an integrated IT system based on diverse ICT solutions, making it possible to support freight transport over the first/last mile [14]. The most fundamental requirements assumed for the S-mileSys platform include efficient route planning for distribution and collection of goods by taking innovative optimisation criteria into consideration. The platform in question, currently under development, is generally intended as a means of support for three groups of stakeholders:

- clients searching for first/last mile freight companies,
- first/last mile freight companies,
- local authorities, including road infrastructure administration bodies.

The key system component is S-mile Transport Planner Tool which enables planning of goods collection and distribution over a single optimised route based on the assumed transport service optimisation criteria. Besides typical route optimisation criteria based on time, distance or cost, what this tool also takes into account in the planning process is such a set of innovative criteria as society- or environment-oriented factors or optimisation with regard to quality of transport routes. S-mile Transport Planner Tool makes use of information contained in the system's databases and concerning transport companies, their capabilities and limitations, as well as generally available data, such as topology of the road network area where the given transport service is to be performed.

S-mile Fleet Management Tool and S-mile Freighter Tool are both tools which directly affect the manner in which transport companies operate. The solution that supports the dispatcher's operations, S-mile Fleet Management Tool, sends information about the transport services to be performed to a driver and updates the transport routes previously planned. S-mile Freighter Tool is a solution which covers mobile applications installed on mobile devices on board of the fleet vehicles, supporting drivers by navigating them over the road network and helping them organise their job in accordance with pre-established schedules of transport services.

The system component which initiates the process of processing freight transport-related requests is S-mile Market Tool, ensuring integration of freight transport and enabling first/last mile carriers to be searched for. A tool that has been tailored for local authorities is S-mile Simulation Tool which enables them to simulate motivational and organisational actions undertaken in the transport system's space. It allows for assessment and verification of the simulation results from the economic, environmental and social perspective. S-mile Visualizer Tool has been designed for detailed presentation of data concerning the freight transport services performed in the system, road traffic parameters, road infrastructure condition or pollutants emitted by heavy goods vehicles. It has been described in detail in the next section of the paper.

The S-mileSys platform is developed as a means to support the first/last mile freight transport under the international project entitled "Smart platform to integrate different freight transport means, manage and foster first and last mile in supply chains (S-mile)", co-financed under the "Sustainable Logistics and Supply Chains" call within the framework of the ERA-NET Transport III programme. The project is implemented by an international consortium of businesses and institutions from Spain, Turkey and Poland. They represent the sectors of civil engineering, environmental protection and IT as well as the higher education milieu.

4 S-mile Visualizer Tool as a Data Visualisation Solution

S-mile Visualizer Tool is one of the most crucial components of the S-mileSys platform, and its main functions are identification and visualisation of the transport services being rendered. By means of the solutions developed under the S-mile project, real-life data of routes actually covered by vehicles of transport fleets can be obtained. S-mile Freighter Tool integrated with Main Mobile App for smartphones has been designed for various purposes, including to acquire information on attributes of description of transport routes, while the Road Condition Tool mobile application has been developed to record dynamics of vehicle traffic using GPS and MEMS modules.

4.1 Data Processing for Visualisation

What is required for the visualisation process is data which describe the transport route in the road network, including the vehicle traffic dynamics, as well as attributes that identify the given transport service in the S-mileSys system database. The attributes which explicitly identify the transport service are the transport company identifier, the route identifier and the vehicle identifier. Furthermore, in order to estimate and visualise indicators of harmful emissions and impact on the society, one must first acquire data concerning the vehicle type, cargo weight and some additional parameters describing the route, including its length, elevation profile and traffic volume load (traffic intensity). The data required for assessment of the road infrastructure condition are instantaneous values of vehicle speed, GPS location and linear accelerations in a three-dimensional system.

From the operating point of view, S-mile Visualizer Tool is a web-based solution linked with the RCT server database and the RCT processing application, where the latter processes input data and calculates pre-defined indicators of description of freight transport routes. The data processing is performed in consecutive steps, following which it is possible to run full visualisation of results according to user-defined criteria. Transport route description data are recorded by transport fleet vehicles equipped with mobile devices with the RCT mobile application installed and running, after which they are sent to the RCT server database. Following the data receipt confirmation, the first processing step, i.e. data filtering, begins. Subject to analysis is the correctness of data stored in the route file, established by determining continuity of a data series as well as correctness of the values obtained as per the assignment of the data to a pre-defined range. Data discontinuity and high deviations from the mean initiate the data filtering function against a part of the data series, and it is also possible that an entire data set is rejected. The second step consists in conversion of the route recorded as a series of GPS coordinates over a set of the OSM map sections which reflect the actual route. It is performed according to a complex data conversion algorithm which applies a number of criteria in the assignment process: the criterion of the GPS track coverage by OSM sections, the criterion of conformity between the directionality of the resultant GPS track vector and the OSM section as well as the criterion of continuity of mapping of the GPS track assigned to successive OSM sections. It should be noted that, in order to characterise the assignment algorithm, an OSM section may be understood as per two definitions. According to one of them, it is a segment, i.e. a section between consecutive GIS coordinates on a map, where the section length is considerably diversified and results from the road network specificity. According to the second one, it is referred to as a part, i.e. a section of fixed length defined by the S-mileSys system administrator.

The step following the conversion of the GPS track into a set of OSM sections is determining parameters of description of the transport service route. It consists in determination of environmental pollution indicators (CC, DALY, NOISE) [15, 16] and estimation of the road pavement condition assessment index (RC) [17]. What is created as an outcome of these operations is a file with route mapping in the OSM description containing the following header: ID company; ID route; ID vehicle; vehicle Type; freight Weight; max Freight Weight; vehicle Weight; step; trip Start; and a series of data broken down into parts with the following parameters: RC, SPEED, CC, DALY, NOISE.

In the next step, the indicator values established for parts are subject to calculations aimed at determination of their equivalents applicable to segment description. Mean values are calculated for all indicators, except for noise, for which the maximum value is assumed and extracted from the set of values calculated for parts.

4.2 Data Visualisation Tool's Capabilities

The route description parameters established for the transport services rendered are entered into database tables on which the visualisation process performed in S-mile Visualizer Tool is based. Figure 1 shows the visualisation outcome along with the user interface.

According to the assumptions underlying the S-mile project, the data visualisation tool has been designed and implemented in a manner that enables identification and presentation of the following data:

- Road condition,
- Traffic congestion,
- Vehicles distribution,
- Distribution of load masses,
- Emission impact CC,
- Emission impact DALY,
- Emission impact NOISE.

Users defined in the S-mileSys system can use S-mile Visualizer Tool for presentation of the foregoing data by first defining their visualisation criteria, i.e. the transport service performance time and the area of its impact. They can also select data which

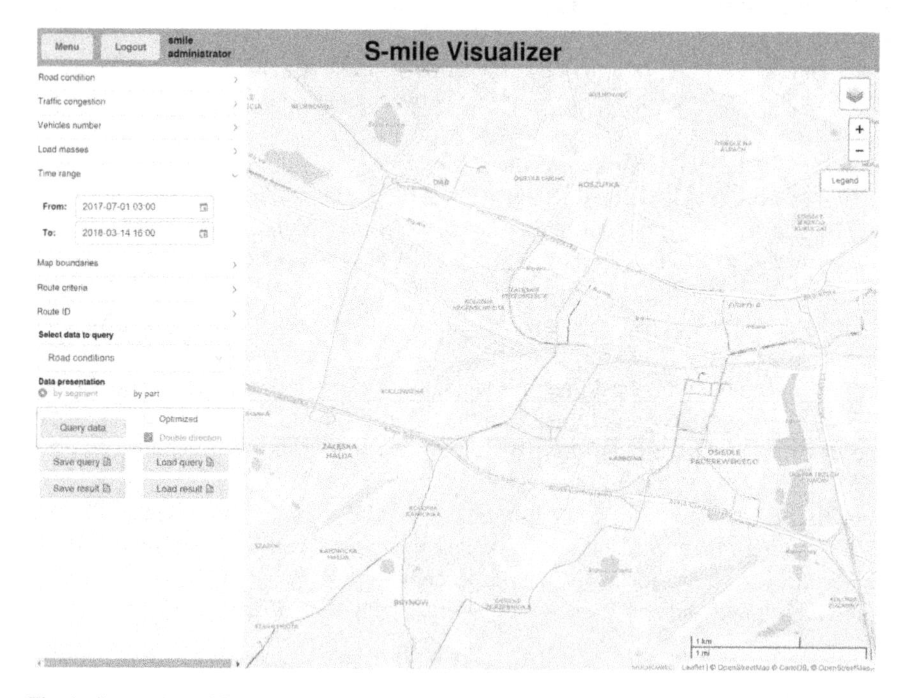

Fig. 1 Screenshot of S-mile Visualizer Tool web application

Fig. 2 Examples of interactive maps describing traffic situations

describe the routes covered while rendering transport services by identifying the company, the vehicle as well as the vehicle type. All data stored in the database and meeting the above criteria are visualised. The solution features an additional function which allows for visualisation of a single transport service delivery provided that an identifier of a route previously covered (ID route) has been defined.

Figure 2 shows visualisation of data acquired while vehicles traversed the Upper Silesian conurbation as a part of functional tests of selected S-mileSys tools. Traffic congestion values have been provided on the left-hand side of Fig. 2, while the right-hand side contains noise emission values.

4.3 Application of S-mile Visualizer Tool in ITS

The main purpose assumed for S-mile Visualizer Tool is graphical representation of data concerning routes by which transport services have been delivered. It is for the graphical data presentation that the system user can make adequate decisions in the scope of planning and organisation of freight transport on both the operating and the strategic level. The user may be set by two groups of the S-mileSys system's beneficiaries, namely transport companies and local authorities, the latter being responsible for ITS management. They differ completely in terms of the priorities attached to handling of transport services. The former pursue profit maximisation by minimisation of transport service costs, while the latter, i.e. municipal decision makers, also take into account the environmental and social aspects of provision of transport services in the territory subject to their administration as they design and shape urban space.

The diverse ITS solutions currently implemented in the sphere of freight transport should ensure automated road safety oversight as well as safety monitoring features on board of each vehicle. ITS should enable efficient freight transport fleet management, with particular regard to supporting transport of hazardous cargo within urban areas. The system developed under the S-mile project, i.e. S-mileSys, is capable of supporting the aforementioned services of the ITS architecture. It should also be emphasised that the solution described above, i.e. S-mile Visualizer Tool, not only offers the graphical data presentation feature, but also makes it possible to submit queries to the database of completed transport services which returns JSON output files. The queries can be defined via an internal system—the ITS module. Depending on pre-assumed solutions, one can develop a protocol in the existing ITS dedicated to exchange of information on the freight transport performed in the city, covered by the S-mileSys solutions. It also makes the integration process easier when an ITS is developed. Furthermore, the pressure imposed by local authorities on the businesses which render freight transport services in urban areas, pushing them towards application of S-mile Freighter Tool, will allow for comprehensive impact to be exerted on transport for the sake of improved safety as well as implementation of socially and environmentally oriented solutions.

5 Conclusions

The paper provides characterisation of S-mile Visualizer Tool's features and discusses methods used to acquire and process data required for graphical presentation of routes by which transport services have been rendered. The author has highlighted the wide spectrum of capabilities of transport description available under the implementation of the S-mileSys platform within the chosen city or urban area. This article addresses the opportunities arising from utilisation of the data first acquired and then identified, stored in the S-mileSys system, for building an advanced intelligent transport system assumed to inherently support transport of goods.

Acknowledgements The present research has been financed from the means of the National Centre for Research and Development as a part of the international project within the scope of ERA-NET Transport III Sustainable Logistics and Supply Chains Programme "Smart platform to integrate different freight transport means, manage and foster first and last mile in supply chains (S-mile)".

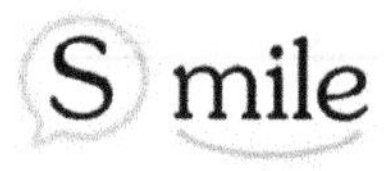

References

1. Williams, B.: Intelligent Transport Systems Standards. Artech House, London (2008)
2. Bamberg, S., Ajzen, I., Schmidt, P.: Choice of travel mode in the theory of planned behavior: the roles of past behavior, habit, and reasoned action. Basic Appl. Soc. Psychol. **25**(3), 175–187 (2003). Taylor & Francis
3. Anand, N., Quak, H., van Duin, R., Tavasszy, L.: City logistics modeling efforts: trends and gaps—a review. Procedia Soc. Behav. Sci. **39**, 101–115 (2012). Elsevier
4. Koźlak, A.: The role of the transport system in stimulating economic and social development. Res. J. Univ. Gdansk Transp. Econ. Logistics **72**, 19–33 (2017). Gdańsk
5. Sussman, J.S.: Perspectives on Intelligent Transportation Systems (ITS). Springer Science & Business Media, New York (2008)
6. Chowdhury, M.A., Sadek, A.W.: Fundamentals of Intelligent Transportation Systems Planning. Artech House, London (2003)
7. Kosch, T., Kulp, I., Bechler, M., Strassberger, M., Weyl, B., Lasowski, R.: Communication architecture for cooperative systems in Europe. IEEE Commun. Mag. **47**(5) (2009). IEEE
8. Directive 2010/40/EU of the European Parliament and of the Council of 7 July 2010 on the framework for the deployment of Intelligent Transport Systems in the field of road transport and for interfaces with other modes of transport
9. ISO 14813-1:2007 Intelligent transport systems—reference model architecture(s) for the ITS sector. Part 1: ITS service domains, service groups and services
10. Bělinová, Z., Bureš, P., Jesty, P.: Intelligent transport system architecture different approaches and future trends. In: Data and Mobility, pp. 115–125. Springer, Berlin, Heidelberg (2010)
11. Żak, J., Galińska, B.: Multiple criteria evaluation of suppliers in different industries-comparative analysis of three case studies. In: Advanced Concepts, Methodologies and Technologies for Transportation and Logistics, pp. 121–155. Springer, Cham (2016)
12. Galińska, B.: Multiple criteria evaluation of global transportation systems—analysis of case study. In: Sierpiński, G. (ed.) Advanced Solutions of Transport Systems for Growing Mobility. TSTP 2017. Advances in Intelligent Systems and Computing, vol 631. Springer, Switzerland (2018)
13. Carter, C.R., Liane Easton, P.: Sustainable supply chain management: evolution and future directions. Int. J. Phys. Distrib. Logistics Manage. **41**(1), 46–62 (2011). Emerald
14. Staniek, M., Sierpiński, G.: Smart platform for support issues at the first and last mile in the supply chain-the concept of the S-mile project. Sci. J. Silesian Univ. Technol. Ser. Transp. **92**, 141–148 (2016). Publishing House of Silesian University of Technology, Gliwice
15. McKinnon, A., Browne, M., Whiteing, A., Piecyk, M.: Green logistics: improving the environmental sustainability of logistics. Kogan Page Publishers, London (2015)
16. McKinnon, A.C.: Logistics and the environment. In: Handbook of Transport and the Environment, pp. 665–685. Emerald (2003)
17. Staniek M.: Road infrastructure condition assessment as element of road traffic safety—concept of the RCT solution in the S-mileSys platform. In: Macioszek, E., Sierpiński, G. (eds.) Recent Advances in Traffic Engineering for Transport Networks and Systems. TSTP 2017. Lecture Notes in Networks and Systems, vol. 21. Springer, Cham (2018)

Multi-criteria Evaluation of Global Transportation Corridors Effectiveness. Case Study Analysis

Barbara Galińska and Renata Pisarek Bartoszewska

Abstract Transportation corridors are basic elements of whole transportation system. The selection of particular solution (transportation corridor) directly determines the effectiveness of whole transport. Thus, such selection should be based on detailed and thorough analysis, and evaluation factors should include all aspects contributing to transport effectiveness, i.e., time of transport, costs of transport, timeliness of transport or finally, transport reliability or flexibility. The overall research goal of this paper is to evaluate the global transportation corridors, considering effectiveness factor as the most significant. The authors claim that this aspect has a multiple-criteria character, and thus, they develop the proposed approach based on the principles of multiple-criteria decision making/aiding. The challenge and the novelty of this work are to distinguish all factors contributing to transport effectiveness and apply them to the proposed methodology, in particular, to the indicated family of evaluation criteria of the alternative options—global transportation corridors. To the best of the authors' knowledge, such a contribution has not been reported in the literature, so far.

Keywords Global supply chains · Freight transportation systems/corridors · Efficiency of transport · Multiple-criteria decision making/aiding

1 Introduction

Freight transportation, as a necessary element of each supply chain, is the set of activities connected with relocation of shipments in time and space with proper

B. Galińska (✉)
Faculty of Management and Production Engineering, Lodz University of Technology, Łódź, Poland
e-mail: barbara.galinska@p.lodz.pl

R. Pisarek Bartoszewska
Department of Business Analysis and Strategy, Faculty of Economics and Sociology, University of Lodz, Łódź, Poland
e-mail: renata.pisarek@uni.lodz.pl

M. Suchanek (ed.), *Challenges of Urban Mobility, Transport Companies and Systems*, Springer Proceedings in Business and Economics,
https://doi.org/10.1007/978-3-030-17743-0_25

means [1, 2]. It is covering a distance or a change of place of goods using the transport facilities [3]. The general definition of the transportation defines it as a process which is the finite sequence of activities necessary to relocate shipments [4]. The set of transportation processes creates the transportation system. It is defined as a set of components such as transportation infrastructure, fleet of vehicles, human resources and governing rules that ensure a coordinated and efficient transfer of goods from their origins to destinations in a certain area [5]. One of the types of transportation system is the global transportation system which covers the whole world, continent, a specific group of countries or an economic group [6]. It has a huge influence on functioning of the international trade exchange where costs and time of transportation are very important [7].

What is more, effectiveness mainly determines transportation process as it involves transportation of the goods to the indicated destination and its efficiency, meaning the ability to optimize all resources in order to conduct the whole process effectively. There are, however, several methods which may increase the transport efficiency. One of such measures is decision making, based on the careful analysis which takes into account all gauges for transport evaluation. They are time of transport, costs of transport, reliability of transport, its timeliness and flexibility.

Next important part of the global transportation system is the global transportation corridor/solution. It ensures the transfer of significant passenger and freight traffic flows between separate geographic regions. It also contains infrastructural objects (mobile means of transportation and stationary equipment) of all transport modes occurring in a given corridor, as well as all technological, organizational and legal conditions for carrying out these transports [8]. This is a concept of moving goods between supply chain links on a described scale with the application of a single-mode, multimodal, intercontinental and worldwide transportation. The transportation corridors must fit the local or global configuration of manufacturing and distribution systems and strategies to provide the desired customer service at the lowest possible cost and to maximize the supply chain profit [9].

Effectiveness factors mainly determine the whole transportation process, which should be taken into account while creation of freight corridors. Also, selection of necessary corridors should be based on the complex analysis, including all factors which may have an impact on transport effectiveness.

The overall research goal of this paper is to evaluate the global transportation corridors, taking into account their effectiveness. The authors claim that this aspect has a multiple-criteria character, and thus, they develop the proposed approach based on the principles of multiple-criteria decision making/aiding (MCDM/A). The challenge and the novelty of this work are to present a coherent set of evaluation criteria for the transportation corridors, including all gauges considered in evaluation of transportation effectiveness.

The hypothesis of this work indicates the way of evaluating and selecting the most efficient global transportation solution, generating various possibilities based on the multiple-criteria analysis.

In the practical part of this paper, the authors describe alternative global transportation corridors between China and Central Europe (Poland) based on a multimodal transportation process used by the enterprise operating in energy and automation industry. Next, the authors evaluate them with a consistent family of criteria, model the decision maker's (DM's) preferences and carry out a series of computational experiments with the application of selected MCDM/A ranking methods. As a result, the final rankings of transportation options are generated that give the DM the most efficient transportation solution.

2 Methodological Background of the Research

2.1 Factors Contributing to Transport Effectiveness

The term "effectiveness" plays a crucial role in modern logistics although it leads to many problems on operational, tactical or strategy level. Thus, the evaluation of transportation corridors effectiveness should be subject to separate research which would involve all factors determining the corridors effectiveness, e.g., transportation process. Various researchers indicate logistics gauges as relevant to transport evaluation, which are all presented in Table 1.

Usage of logistics gauges may provide information about transportation process, detect all deviations from the transportation plan and also improve whole transport and its particular elements (including transportation corridors) making it more competitive. Based on the analyzed literature, evaluation of the global transportation corridors effectiveness (referred to as variants in the next section of this paper) included various logistics gauges (variants evaluation criteria), as described in Table 2.

Table 2 forms a relevant family of evaluation criteria which is one of the most important elements of applied methodology of multiple-criteria decision making—further presented in the next section of this paper.

2.2 Principles of MCDM/A and Description of the Applied Methods

Multiple-criteria decision making (MCDM) also known as a multi-criteria decision aid or multi-criteria decision support derives from operational research [17, 18]. Such method supports DM (person who defines decision problem) with rules, tools and methods in solving complex decision problems, considering several—often contradictory—points of view [19, 20]. Evaluation process involves different aspects of the considerate variants of multi-dimensional nature (which are also hardly comparable) in order to select the best alternative [21, 22].

The main components of multiple-criteria decision problems are:

- a set of solutions (variants) A which are analyzed and evaluated in decision process (global transportation corridors in this case study);

Table 1 Logistics gauges in transport evaluation

Gauges	Twaróg [10]	Nowicka-Skowron [11]	Pfohl [3]	Kisperska-Moroń [12]	Litman [13]	Los [14]	Rodrigue [15]	Waściński and Zieliński [16]
Quantitative indicators								
Costs of transport: – For km – For carriage	X	X	X			X	X	X
Completed tonne-kilometers	X		X					
Number of kilometers driven		X						X
Time for one transport order completion	X				X	X	X	
Number of vehicles used for transport		X						
Number of employees in transport sector		X		X				
Number of completed consign-ments (number of transported shipment)	X		X				X	
Number of operational disturbances (including number of accidents)	X	X						X
Actual length of means of transport performance	X		X		X			
Means of transport utilization rate	X		X		X	X	X	X

(continued)

Table 1 (continued)

Gauges	Twaróg [10]	Nowicka-Skowron [11]	Pfohl [3]	Kisperska-Moroń [12]	Litman [13]	Los [14]	Rodrigue [15]	Waściński and Zieliński [16]
Working time utilization rate	X							X
Fuel consumption				X				
Value of the means of transport owned		X						
Means of transport maintenance costs		X						
Qualitative indicators (productivity)								
Timeliness of transport—within agreed time	X		X	X	X	X	X	X
Reliability of transport	X		X			X	X	X
Flexibility of transport (transport readiness)	X		X				X	X
Security of transport (number of goods damaged during transport)	X		X	X				X
Transport documentation accuracy				X				
Accessibility of transport					X	X		
Comfort for customer (quality of customer service)	X			X		X	X	

Table 2 Global transportation corridors effectiveness evaluation criteria

No.	Logistics indicator (criterion)	Description
K1	Costs of cargo transportation	The criterion specifies an overall unit cost of load unit transport (40 ft. container) from China (supplier's warehouse) into company's warehouse (Poland, Łódź). The criterion was formulated on the basis of the data provided by the company and offers from the freight forwarders. It is a minimized criterion. Formula: $\frac{\text{costs of cargo transportation}}{\text{number of cargo transportation}}$ (PLN/cargo transportation)
K2	Time of a single cargo transportation	The criterion specifies an overall time for unit load transportation (40 ft. container) including the duration time of handing over of consignment by Chinese supplier until delivery to the company's warehouse. The criterion is minimized. Formula: transport time duration + customs clearance duration + cargo loads duration + unloading time (number of days)
K3	Means of transport utilization rate	The criterion defines transport adaptation to the specificity of the transported material and expresses the percentage of available capacity utilization of transport unit (40 ft. container). It is a maximized criterion. Formula: $\frac{\text{actual goods carriage load }(\text{m}^3)}{\text{possible goods carraige load }(\text{m}^3)} \times 100\%$ (%)
K4	Timeliness of transport	The criterion specifies the percentage of consignments delivered on time. The results are based on the statistical data from the company and carriers. It is a maximized criterion. Formula: $\frac{\text{timely delivery of orders}}{\text{total amount of deliveries}} \times 100\%$ (%)
K5	Reliability of transport	The criterion specifies the percentage of transport requirements realization in relation to all transportations carried in a month. It was assumed that reliability of the transport decreases with the growth of trans-shipments during transport (number of indirect operations). It is a maximized criterion. Formula: $\frac{\text{number of transport requirements realizations}}{\text{total amount of transport requirements}} \times 100\%$ (%)

(continued)

Table 2 (continued)

No.	Logistics indicator (criterion)	Description
K6	Flexibility of transport	The criterion measures the time of variant's response (including its contractors–carriers) toward unexpected road events. It is a percentage of immediate deliveries of orders in relation to all performed deliveries. It was assumed that flexibility of the transport increases with the growth of transport capacity during a month period. The criterion is maximized. Formula: $\frac{\text{immediate deliveries of orders}}{\text{total number of deliveries}} \times 100\%$ (%)
K7	Security of transport	The criterion specifies the percentage of damaged goods in transport unit load (40 ft. container), in a month period. It is a minimized criterion. Formula: $\frac{\text{number of damaged goods}}{\text{total number of goods}} \times 100\%$ (%)
K8	Customer's comfort	The criterion defines various communication aspects such as option to track the package during transportation. It is expressed in 1–3-point scale. It is a maximized criterion. Formula: 1 point for tracking option during transport, door-to-door transport, option for free storage of goods in the port (points)

- a consistent family of criteria F (global transportation corridors evaluation criteria in the case study).

Each criterion in the family of criteria F is used to evaluate the A set and represents the DM's preferences in relation to a proper aspect of a decision problem.

To solve multiple decision problems, various tools, procedures or methods can be used. They can be generally divided into two groups [20, 23, 24]:

- the methods of American inspiration based on the utility function, referred to as the unique criterion of synthesis methods, e.g., UTA, AHP;
- the methods of the European/French origin, based on the outranking relation, considering incomparability relation, e.g., Electre, Promethee and Oreste.

Electre III/IV and Promethee II, two ranking methods, have been applied in order to evaluate transportation corridors effectiveness in this paper. They are indicated by the researchers as the most popular methods implementing the outranking relations framework [25].

2.3 Description of Electre III/IV Method

Electre III/IV method belongs to a family of Electre methods, proposed by Roy [26]. It is a universal, multi-dimensional ranking method, based on the outranking relation [19, 20, 22, 26, 27]. In this method, the basic set of data is composed of the following elements: a finite set of variants, a family of criteria and the preferential information submitted by the DM. The preferential information is defined in the form of criteria weights and the indifference, preference and veto thresholds [5]. (The thresholds define the sensitivity of the DM to the changes of the criteria values and the weight expresses the importance of each criterion.) Computational algorithm of Electre III/IV comprises of three stages [28]:

I. matrix evaluation construction and definition of the DM's preference model;
II. outranking relation construction;
III. outranking relation implementation.

Electre III/IV algorithm generates the final ranking of variants and orders them from the best to the worst. The following relations may occur between variants: equivalence, outranking, reverse outranking and incomparability.

2.4 Description of Promethee II Method

Promethee method was introduced by Brans et al. [29, 30] to preference rank a set of decision alternatives, based on their values over a number of different criteria. Put simply, a ranking of alternatives is based on the accumulative preference comparisons of pairs of alternatives' values over the different criteria (using generalized preference functions) [31]. The Promethee method will provide the DM with a ranking of actions (choices or alternatives) based on preference degrees. The method falls into three main steps:

I. the computation of preference degrees for every ordered pair of actions on each criterion;
II. the computation of unicriterion flows;
III. the computation of global flows.

Based on the global flows, a ranking of the actions will be obtained as well as a graphical representation of the decision problem [32].

The difference between Electre III/IV and Promethee II methods is key parameters. Electre requires indifference, preference and veto thresholds, while Promethee requires only indifference and preference thresholds.

3 Description of Decision Situation

3.1 Verbal Description

This paper presents the issue of evaluation and selection of the most effective global transportation corridor (China–Poland) for the international daughter company operating in energy and automation industry. The company is located in Łódź, Central Poland. It manufactures distribution transformers and power transformers as well as insulation components used in power transformers. Most of the suppliers' warehouses are located in Europe; nonetheless, some of the components need to be delivered from the Far East, China.

The delivery process from China is affected by a number of various difficulties, mainly caused by huge delivery distance and length of transportation process. The company uses several transportation variants offered by a forwarding company, responsible for the whole transportation process under the general contractor agreement.

It was the reason why the transportation director—DM in the decision situation—-did not undertake any analysis of different transportation variants (global transportation corridors) against the criteria of their effectiveness (including the following criteria: costs, time, degree of utilization of means of transport, timeliness, reliability, flexibility and safety or customer's comfort).

What is more, due to the last year increase in delivery costs of materials transported from China, DM finally decided to carry out a detailed analysis and evaluation of global transportation corridors. The aim was to improve their effectiveness, in particular, costs and time reduction and improvement of security and timeliness. Thus, DM would examine all possible transportation options.

3.2 Characteristics of Variants—Global Transportation Corridors

Selection of the most effective global transportation corridor is defined as a multiple-criteria problem of variants ranking. The considered variants correspond to the modes of transport (China/supplier's warehouse—Poland/Łódź) V1–V4 (Table 3). Analyzed transportation corridors include four transport modes: transport by sea, rail, air or road.

As DM has not yet undertake an evaluation of current transportation corridors, the selection of the most effective solution must be based on the comprehensive analysis. Thus, a number of criteria will be applied, determining the effectiveness of each transportation corridor.

Based on the variants evaluation criteria and the original raw data, the evaluation matrix has been constructed. The importance wages of the criteria were formulated on the basis of the interview with the DM, his preferences and aspirations (Table 4).

Table 3 Variants—global transportation corridors—in decision situation

Variant	Type and verbal description
V1	Road transport + sea transport + road transport; Stages: • road transport from supplier's warehouse to sea port in Shanghai • sea transport from Shanghai port to Gdynia port • road transport from Gdynia to Łódź warehouse
V2	Road transport + rail transport + road transport; Stages: • road transport from supplier's warehouse to railway station in Hefei • rail transport from Hefei to railway station in Małaszewicze • road transport from Małaszewicze to company's warehouse in Łódź
V3	Road transport + rail transport + rail transport; Stages: • road transport from supplier's warehouse to railway station in Hefei • rail transport from Hefei to railway station in Małaszewicze • rail transport from Małaszewicze to company's warehouse in Łódź (the company owns its siding)
V4	Road transport + air transport + road transport; Stages: • road transport from company's warehouse to Shanghai Airport • air transport from Shanghai Airport to Frankfurt Airport (Germany) • road transport from Frankfurt to company's warehouse in Łódź

Table 4 The evaluation matrix in described case study

Criterion	Weight of criterion—Electre method	Weight of criterion—Promethee method	Variants			
			V1	V2	V3	V4
K1 (PLN)	8	0.20	12,791	20,252	21,252	111,329
K2 (Days)	9	0.25	35	21	28	7
K3 (%)	5	0.08	0.90	0.80	0.80	0.95
K4 (%)	7	0.15	0.80	0.90	0.90	0.98
K5 (%)	7	0.15	0.90	0.95	0.95	1
K6 (%)	6	0.10	0.75	0.80	0.85	0.98
K7 (%)	4	0.05	0.10	0.05	0.05	0.02
K8 (Points)	3	0.02	2	2	3	2

Main global transportation corridors evaluation criteria in terms of their effectiveness are time of transport (K2), costs of transport (K1), timeliness of transport (K4) and transport reliability (K5). The results of computational experiments, based on Electre III/IV and Promethee II methods, are described in the next section of this paper.

4 Computational Experiments

Due to formal limitations of the study, the results of computational procedures were reduced to presentation of the final rankings, identifying the position of the variants in relation to global transportation corridors.

In accordance with the Electre III/IV method algorithm, the evaluation matrix has been constructed and the DM's preference model has been defined. In the second stage of the algorithm, the outranking relation has been constructed. In the third stage of the algorithm, the outranking relation has been applied and on the basis on the indexes of the variants (global transportation corridors V1–V4), the ascending and descending distillations have been performed, formulating the structure of complete preorders. Then, they have been averaged into the median ranking, and the intersection of preorders resulted in the final ranking. The results of these transportation corridors selection calculations are presented in Fig. 1a.

In accordance with the Promethee II method algorithm, the evaluation matrix has been constructed and the DM's preference model has been defined in the process of naming the wages of criteria and thresholds: indifference and preference. Due to experimental procedures, the final ranking has been obtained, as shown in Fig. 1b.

According to the final ranking, based on Electre III/IV method, the most effective global transportation corridor is variant V4, implementing air transport for the

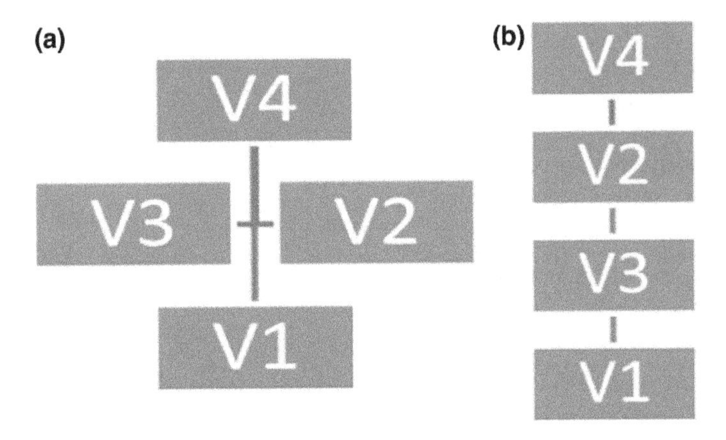

Fig. 1 The final ranking in the case study. **a** Electre III/IV method, **b** Promethee II method

longest transport route. It clearly outranks the other variants, against six out of eight evaluation criteria. Its strongest values are very short period of transport (K2), high degree of means of transport utilization (K3), excellent timeliness (K4), reliability (K5) and flexibility (K6) of transport and finally, very low rate of goods' damages during transportation process (K7). The interesting fact is that variant V4 represents the most expensive solution what is compensated by other values of the variant. The least effective global transportation corridor is variant V1, using transport by sea. Although it offers the cheapest price (K1), the other criteria are ranked very poorly.

The results generated with the application of Promethee II method are fast identical to those produced by the application of Electre III/IV method. Promethee II method indicated a slight difference between transportation corridors V2 and V3, whereas in the ranking of Electre III/IV method, they were ranked on the equivalent position. The authors of this paper recommend selection of global transportation corridor V4 as it is the most effective solution according to the conducted computational procedures. If DM would define cost criterion as the most important in the final selection, then variant V2 is the best solution, ranking on the second position in both cases.

5 Final Conclusions

The paper presents evaluation and ranking of global transportation corridors used by company operating in energy and automation industry, based on the multiple-criteria decision making/aiding (MCDM/A). The decision problem was formed as a multiple-criteria problem of ranking variants. Two methods were applied in order to generate the final ranking of global transportation corridors, namely Electre III/IV and Promethee II. The novelty of this paper was output of logistics gauges implementation into the final evaluation of global transportation corridors.

The paper contains not only methodological values but also utility functions. The methodological approach is based on the identification of the criteria which may determine the effectiveness of global transportation corridors. In practical terms, the authors demonstrate that the most effective solution is variant V4 which, although not the cheapest, is characterized by many other advantages. Thus, the authors of this paper recommend the selection of global transportation corridor V4 as it is the most effective and desired. At the same time, for the customer of lower sensitivity on the price, variant V2 is considerable.

References

1. Cavinato, J.L.: Transportation-Logistics Dictionary. Springer Verlag (1989)
2. Galińska, B.: Koncepcja Global Sourcing. Teoria i Praktyka. Difin S.A., Warszawa (2015)
3. Pfohl, H.C.: Logistiksysteme. Betriebswirtschaftliche Grundlagen. Springer Verlag (2010)
4. Borkowski, P., Koźlak, A.: Global competitiveness of Baltic Sea region in view of transport infrastructure development challenges. In: Globalizácia a jejsociálno-ekonomickédôsledky'09:

elektronickýzborníkpríspevkov z medzinárodnejvedeckejkonferencie, pp. 54–65. Žilinská Univerzita, Rajecké Teplice, Žiline (2009)
5. Żak, J.: Design and evaluation of transportation systems. In: Sierpiński, G. (ed.) Advances in Intelligent Systems and Computing, vol. 631: Advanced Solutions of Transport Systems for Growing Mobility, pp. 3–29. Springer Verlag (2018)
6. Galińska, B., Grądzki, R.: Factors which determine the achievement of benefits deriving from the concept of global sourcing. In: Conference Proceedings of CLC 2013, Tanger, Ostrava (2014)
7. Gołembska, E.: Conditions of global logistics development. In: Żołądkiewicz, K., Michałowski, T. (eds.) Meeting Global Challenges. Working Papers Institute of International Business, University of Gdańsk, Gdańsk (2008)
8. Mindur, L. (ed.): Technologie transportowe. PIB, Radom (2014)
9. Yücenur, G.N., Vayvay, Ö., Demirel, N.Ç.: Supplier selection problem in global supply chains by AHP and ANP approaches under fuzzy environment. Int. J. Adv. Manuf. Technol. **56**, 823–833 (2011)
10. Twaróg, J.: Mierniki i wskaźniki logistyczne. Biblioteka Logistyka, Poznań (2003)
11. Nowicka-Skowron, M.: Efektywność systemów logistycznych. Polskie Wydawnictwo Ekonomiczne, Warszawa (2000)
12. Kisperska-Moroń, D.: Pomiar funkcjonowania łańcuchów dostaw. Wydawnictwo Akademii Ekonomicznej im. Karola Adamieckiego, Katowice (2006)
13. Litman, T.: Sustainable Transportation Indicators. Victoria Transport Policy Institute, Canada (2003)
14. Los, P.: Mobility indicators and accessibility of transport. Slovak J. Civ. Eng. **1**, 24–32 (2007)
15. Rodrigue, J.P.: The Geography of Transport Systems. Routledge, New York (2017)
16. Waściński, T., Zieliński, P.: Efektywność procesu transportowego. Systemy Logistyczne Wojsk **42**, 221–236 (2015)
17. Hillier, F., Lieberman, G.: Introduction to Operations Research. McGraw-Hill, New York (1990)
18. Żak, J.: Application of operations research techniques to the redesign of the distribution systems. In: Dangelmaier, W., Blecken, A., Delius, R., Klöpfer, S. (eds.) Advanced Manufacturing and Sustainable Logistics. Conference Proceedings of 8th International Heinz Nixdorf Symposium, IHNS 2010, Paderborn, Germany (2010)
19. Figueira, J., Greco, S., Ehrgott, M.: Multiple Criteria Decision Analysis. State of the Art Surveys. Springer, New York (2005)
20. Vincke, P.: Multicriteria Decision-Aid. Wiley, Chichester (1992)
21. Roy, B.: Decision-aid and decision making. Eur. J. Oper. Res. **45** (1990)
22. Roy, B.: Wielokryterialne wspomaganie decyzji. Wydawnictwo Naukowo Techniczne, Warszawa (1990)
23. Żak, J.: The methodology of multiple criteria decision making/aiding in public transportation. J. Adv. Transp. **45** (2011)
24. Żak, J., Galińska, B.: Multiple criteria evaluation of suppliers in different industries—comparative analysis of three case studies. In: Żak, J., Hadas, Y., Rossi, R. (eds.) Advances in Intelligent Systems and Computing, vol. 572: Advanced Concepts, Methodologies and Technologies for Transportation and Logistics, pp. 121–155. Springer Verlag (2017)
25. Doumpos, M., Evangelos, G.: Multicriteria Decision Aid and Artificial Intelligence: Links, Theory and Applications. Wiley (2013)
26. Roy, B.: The outranking approach and the foundations of ELECTRE methods. In: Bana e Costa, C. (ed.) Readings in Multiple Criteria Decision Aid. Springer Verlag (1990)
27. Żak, J.: Wielokryterialne wspomaganie decyzji w transporcie drogowym. Wydawnictwo Politechniki Poznańskiej, Poznań (2005)
28. Galińska, B.: Multiple criteria evaluation of global transportation systems—analysis of case study. In: Sierpiński, G. (ed.) Advances in Intelligent Systems and Computing, vol. 631: Advanced Solutions of Transport Systems for Growing Mobility, pp. 155–171. Springer Verlag (2018)

29. Brans, J., Mareschal, B., Vincke, P.: PROMETHEE. A new family of outranking methods in MCDM. In: International Federation of Operational Research Studies (IFORS 84), pp. 470–490. North Holland (1984)
30. Brans, J., Vincke, P., Mareschal, B.: How to select and how to rankprojects: the PROMETHEE method. Eur. J. Oper. Res. **24**, 228–238 (1986)
31. Kun-Huang, H. (ed.): Quantitative Modelling in Marketing and Management. World Scientific Publishing Co Pte Ltd., Moutinho (2014)
32. Ishizaka, A., Nemery, P.: Multi-Criteria Decision Analysis. Methods and Software. Wiley, Chichester (2013)

Challenges for the Poland's Intermodal Corridors in the Light of Belt Road Initiative

Tomasz Radzikowski

Abstract There are four intermodal corridors running through Poland: two vertical ones and two horizontal ones. Only one of the horizontal ones offers services for maritime transport and railway from DCT and Port Gdańsk for China's foreign trade to the countries of Central and Eastern Europe. There are many competitors in rail support of Port Gdańsk—freight operators without market power. The horizontal corridors are currently being used only through the border crossing Brest–Małaszewicze for the rail transport of goods from China to Western Europe. The EU has sanctioned the trade with Russia including transit through its territory too. This is a reason for one-way freight railroad from China to Western Europe. The opposite freight direction reaches China by sea. That produces an effect of congestion on the above-mentioned border crossing. That might be suitable for southern horizontal corridor that is connected with Poland–Ukraine border crossings. The potential corridor, called the One Belt One Road project, has an opportunity to develop while China has decided to omit the territory of Russia. Belt Road Initiative would connect Central Asia and Caucasian States with Western Europe by rail. The most interesting aspect in that project is Ukraine as a transit state. It's only the convenient rail freight corridor for trade with the EU. The horizontal corridors have potential opportunities for rail freight operators' competition (especially for Polish and German freight operators). German rail freight operators stand a chance to gain market power, which will be proven in this study.

Keywords Belt Road Initiative · China · Europe · Poland · New Silk Road · Intermodal railway transport · TEN-T network · TRACECA

T. Radzikowski (✉)
Faculty of Economics, University of Gdańsk, Gdańsk, Poland
e-mail: tom.radzikowski@gmail.com

Strategy Department, PKP Polish Railway Lines, Gdańsk, Poland

M. Suchanek (ed.), *Challenges of Urban Mobility, Transport Companies and Systems*, Springer Proceedings in Business and Economics,
https://doi.org/10.1007/978-3-030-17743-0_26

1 Introduction

Nowadays, there is a trade war between China and USA. There is a list of Chinese products which will be hit with a 25% tariff in August 2018. That means that these products will be 25% more expensive for US consumers. The goal for the USA is to reduce the competitiveness of manufactured goods under China's Made in China 2025 policy.

In retaliation, China has hit US sectors such as agriculture, car industry, medical products, coal and petroleum.

The total result of the trade war may be a reduction in the GDP growth of both economies by 0.25% in 2018. In 2019, the slowdown in GDP growth may amount to 0.5% or more. The slowdown in the economic development of the USA and China may also affect countries such as South Korea, Singapore and Taiwan, which are facilities in the supply chains [1].

The European Union is going to be the most important trade partner for China. It demands changes in transport network in Eurasia including Central and West Asia and Europe.

The current network of TEN-T corridors does not take into account geopolitical and economic changes in Eurasia. When defining the corridors, attention was first drawn to the longitudinal corridors running from Polish seaports to their backup facilities—Southern Poland, Slovakia, the Czech Republic and Austria. On the other hand, less attention was paid to the latitudinal corridors, which then had no potential due to the collapse of the USSR.

Geopolitical changes in the global economy have made China become the second global economic power in the world. China and the rise of Ukraine, independent of Russia in the first decade of the twenty-first century, changed the directions of trade in Eurasia. These two continents, potentially diversified in terms of size and economic potential, set new challenges for trade based on rail transport.

How is the Polish rail transport market ready for changes in the direction of trade? Is the current network of TEN-T network corridors prepared for these changes? Is the condition of the Polish rail transport market competitive?

2 The Situation and the Condition of the Intermodal Transport Market in Germany and Poland

The structure of railway companies in Germany and Poland focuses on capital groups: DB AG (Germany) and PKP S.A. (Poland). Within these groups, there are companies with a different business profile. Mainly there are infrastructure managers, passenger carriers, freight carriers, station managers, energy suppliers and other services necessary for the operation of rail transport (Fig. 1).

The basic difference between the German and the Polish capital group is organization. DB AG is a group, and PKP S.A. is only a group of capital-related enterprises

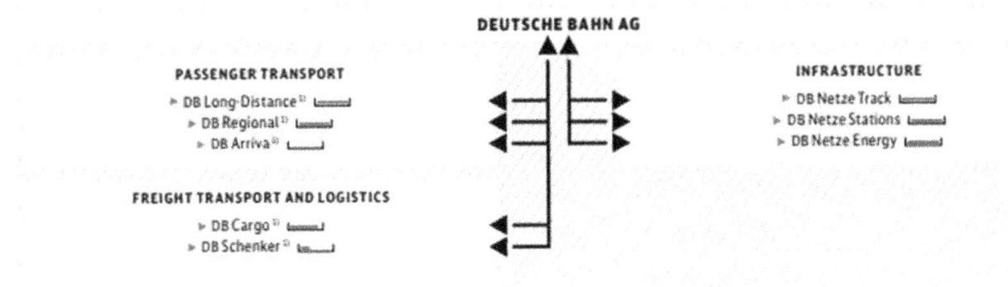

Fig. 1 DB Group structure. *Source* [2]

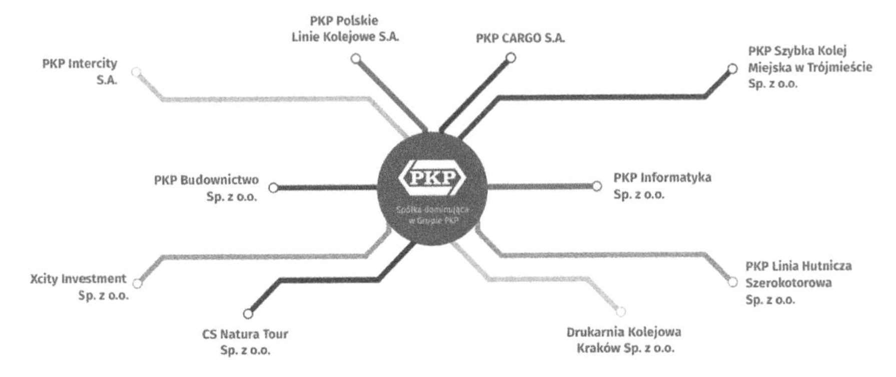

Fig. 2 PKP Group structure. *Source* [4]

that strive to establish a holding company [3]. What is important, the PKP Group has been partially privatized in the recent years. The process concerned mainly PKP Cargo and PKP Energetyka (Fig. 2).

In the case of PKP Cargo, PKP S.A. and the state treasury are still the majority shareholder. In March 2018, the carrier's shares definitely stood out on the stock market. While the main indices recorded slight drops, the value of PKP Cargo dropped by as much as 20%. At the same time, the company screwed its annual minimums [5]. From October 2017, PKP Cargo's acting CEO was Krzysztof Mamiński—CEO of PKP S.A. As a president of PKP S.A., he sought to consolidate the companies of the PKP Group. On 25 March 2018, he resigned from the office of acting CEO of PKP Cargo [6]. On 26 March 2018, the supervisory board chose Czesław Warsewicz as the new CEO of PKP Cargo [7].

PKP Energetyka remains outside the structures of the PKP Group. This causes fundamental problems when planning further electrification of the railway network, because the economic calculation is counted twice: firstly by the railway infrastructure manager and secondly by the market of electricity suppliers. As a result, in some cases electrification of railway lines may not take place without intervention at the level of ministries responsible for transport and energy.

Differences between rail market entities in Germany and Poland also apply to the passenger regional transport segment. DB Regio still operates in DB Group, while in Poland, Przewozy Regionalne was transferred to regional self-governments.

An important difference in the context of intermodal transport is the lack of a logistics segment in the structures of the PKP Group. PKP has a very poor logistics base, which is used by competition. As a consequence, intermodal terminals only serve a narrow group of operators. Thus, the competitiveness of rail transport in Poland in intermodal transport grows more slowly than in Germany.

German example shows the undeniable potential of the DB Group in pursuit of competitive entry into the Polish market. EU policy aims to liberalize and open domestic markets. Liberalization of the freight transport market is further advanced than in the case of passenger transport within the EU single market. This has a fundamental impact on increasing competitiveness on individual transport corridors. An example of this is the entry of the DB Group into transit intermodal transport through Poland [8].

3 Evaluation of Currently Planned Investments in Railway Infrastructure in Poland

Currently, the investment process in Poland is defined in the National Railway Program until 2023 (Krajowy Program Kolejowy—hereinafter called KPK). This document contains a list of railway investments planned for implementation in the current EU financial perspective, together with the initial planned investment costs.

The KPK full named "National Railway Program until 2023. Railway infrastructure managed by PKP Polskie Linie Kolejowe S.A." is a document establishing the financial framework and conditions for the implementation of the state's intentions for railway investments envisaged for completion by 2023. The program is a continuation of the long-term program called Multiannual Railway Investment Program until 2015, with a year-to-year perspective 2020. The KPK is a multiannual program in the meaning of the Act of August 27, 2009 on public finance. The program's adoption results from art. 38c of the Act of 28 March 2003 on railway transport and covers all investments of PKP Polskie Linie Kolejowe S.A. implemented with the use of funds owned by the minister responsible for transport.

The KPK complies with the requirements of the Act of December 6, 2006 on the principles of conducting development policy in the scope of development programs established in order to implement the medium-term national development strategy, the Act of July 11, 2014 on the principles of implementing programs in the field of cohesion policy financed in the 2014–2020 financial perspective and the development strategy, which in this case remains the Transport Development Strategy until 2020 (with a perspective until 2030).

The KPK is a response to the challenges of the adoption by the European Union and Poland's ambitious goals in the field of railway infrastructure and thus must ensure the sustainable economic development [9].

The KPK also specifies the sources of financing of these investments, which include:

- National funds,
- EU funds.

As part of the EU co-financing for the following programs are provided:

- "Connecting Europe Facility" (CEF),
- Operational Program "Infrastructure and Environment" (POIiŚ),
- Operational Program "East Poland" (PO PW),
- Regional Operational Programs (RPO).

The share of individual sources of financing in the KPK is presented in Table 1.

A detailed list of investments planned for completion by 2023 is presented in Annexes no. 1–4 of the KPK. The list includes a total of 225 investment projects with a total value of over 76 billion PLN.

The scope of the investment process, which will be covered by the railway network after 2023, is presented in Fig. 3.

Most of the investment covers the main railway lines, mostly belonging to the core and complementary TEN-T network. In the case of the railway lines belonging to the main freight corridors, the primary objective of the investment is to eliminate

Table 1 The KPK's projects

Financing program	Project's number	EU contribution [bn EUR]	Project's value [bn PLN]
CEF	17	3.46	18.4
POIiŚ 2014–2020	39	4.44	28.2
PO PW	8	0.33	2.1
RPO 2014–2020	51	0.93	4.6
Projects financing from national funds	34	0.00	10.0
Refundations from POIiŚ 2007–2013	60	0.52	3.0
Refundations from RPO 2007–2013	11	0.02	0.1
Total	220	9.70	66.4
CEF—reserve list. projects provided to the third call	5	2.5 (estimated)	9.7

Source Own elaboration based on the KPK

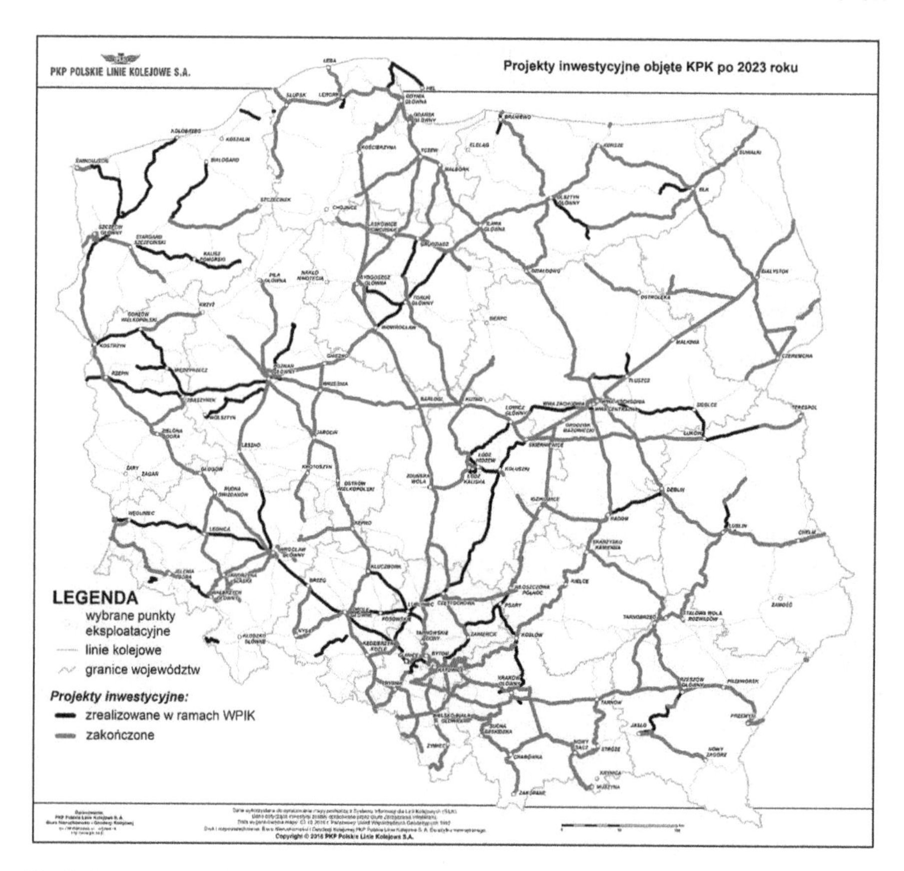

Fig. 3 Investment projects covered by the KPK after 2023. *Source* Materials made by PKP PLK S.A., www.plk-sa.pl, black and blue—investments realized since 2023

bottlenecks and to provide facilities for trains up to 750 m long. As shown in Fig. 3, the investments will cover all major longitudinal and latitudinal corridors.

In the field of cargo transport, the investment process is to ensure first of all improvement of access and increase of competitiveness of Polish seaports. Planned and implemented investments include port stations, access lines and routes servicing these ports. These investments will include Gdańsk Port Północny, Gdynia Port, Szczecin Port Centralny and Świnoujście stations, as well as corridors C-E65 and C-E59 connecting Polish seaports with Adriatic ports. The investment activities are provided on the complementary rail network, increasing reliability and traffic flow. The most important investment in this area is the creation of the so-called alternative transport route Bydgoszcz–Tri-City, which will allow efficient transport of cargo to the port of Gdynia, bypassing the crowded passenger trains running through Tczew, Gdańsk and Sopot. It will significantly improve the transport of cargo from and to Polish seaports, increasing their competitiveness on the European scale.

Investments will also include key nodes on the railway network and latitudinal corridors. However, the scope of the investment will not cover the whole infrastructure and there will still be bottlenecks and restrictions in the transport of cargo in the east–west direction. The latitudinal corridors are mixed with a large share of passenger transport. This is particularly true for the E20 and not just the corridor itself. This also applies to the contact of the standard-gauge and broad-gauge networks on the Polish–Belarusian border. The border crossing Brześć–Terespol is at the external border of the European Union. Due to customs control, congestion is created at this transition. In this connection, Russia and Belarus are looking for other routes bypassing Poland, e.g. running to the ports of the Baltic states [10].

4 Changing the Impact of External Policy on Intermodal Corridors in Poland

In connection with the debt of China's economy caused by inappropriately granted (unpaid) China bank loans which enter the capital market and the state begins to use its tools, which include stock exchange and their papers, especially bonds (sovereign debt funds), which are invested in infrastructure. Debt papers are safer than equities because they are guaranteed by national budgets, which are noted by the rating agencies (Fig. 4).

China's investments are beginning to be financed from other sources than bank loans. The first new source of financing for investments is stock exchange shares [11].

Another source is bank debt securities, with which China primarily finances investments related to environmental protection (green bond issues). In the recent years,

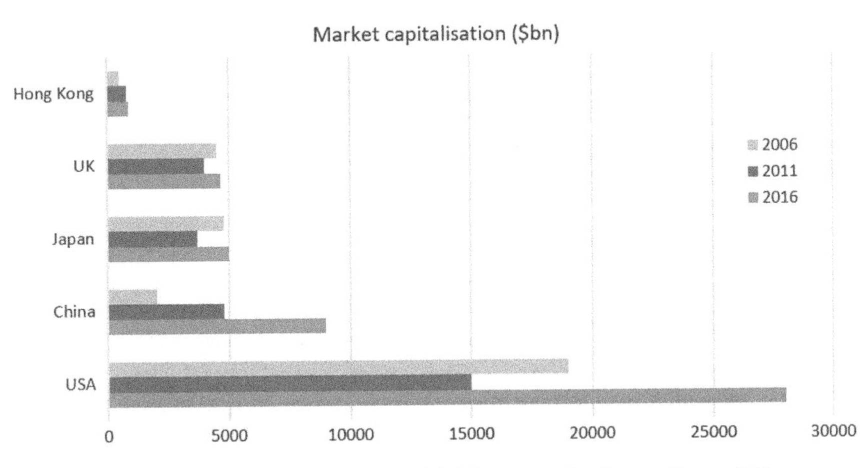

Fig. 4 The growth of Chinese equities in the global investment landscape. *Source* [11]

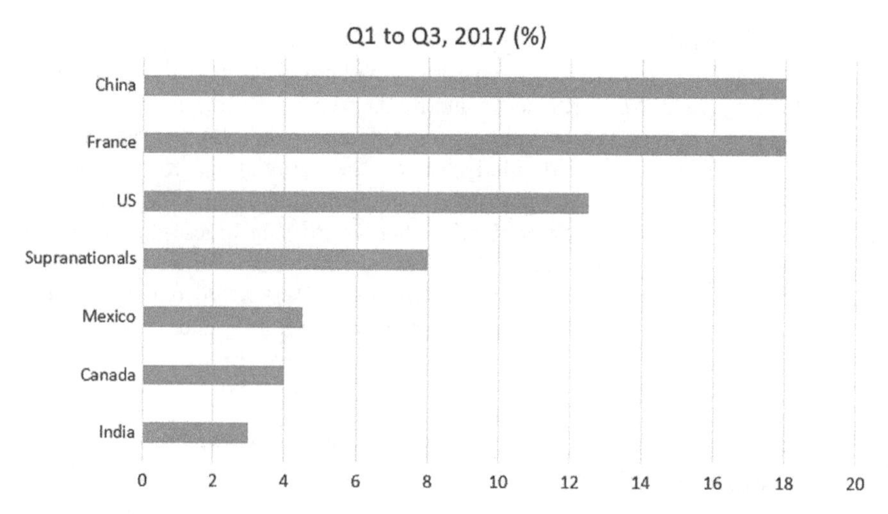

Fig. 5 Biggest green bond issuers. *Source* [12]

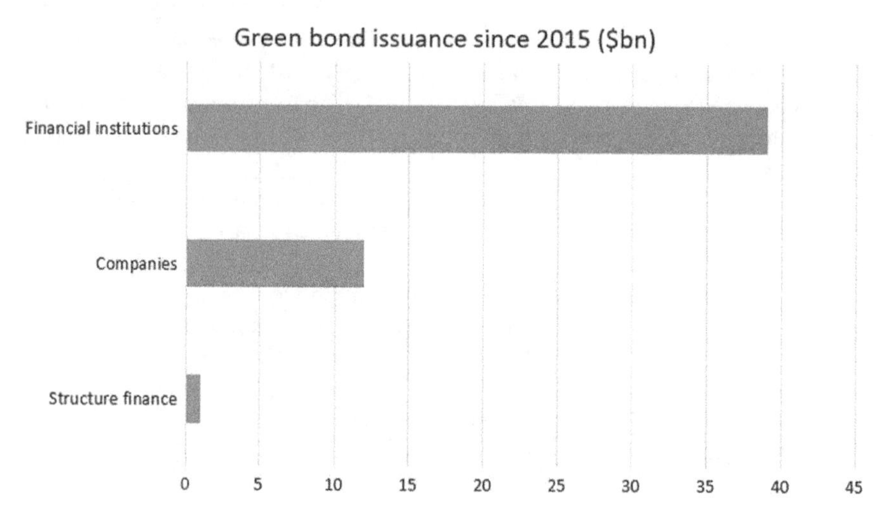

Fig. 6 China's banks lead the field. *Source* [12]

the level of green bond issues in China has reached a similar level to France. Figure 5 presents the comparison of the biggest green bond issuers.

Since 2015, China has been securing green bond issues from three sources: financial institutions, companies and structure finance. The structure of green bond issues financing is shown in Fig. 6.

China is becoming an important player in the global investment market. For European countries, it has become an important partner with high financial potential. Developing countries of Central Asia and Eastern Europe can use those opportunities to implement many infrastructure investments. Also for European Union countries

China's capital is becoming an important alternative to European funds. European funds face the challenge of Brexit, beneficial for both sides. It is doubtful whether the issue of agreement on the exit of Great Britain from the EU will be successful before the deadline of March 15, 2019 [13].

According to Joseph Stiglitz (a Nobel Prize winner and former chief economist at the World Bank), if a full-scale commercial war breaks out, China is in a better situation. They have instruments and means to help victims of a trade conflict. China has 3 trillion USD, which they can use to help the victims. In the USA, there is no economic base that would allow to react in places affected by the trade war. US fiscal measures are limited [14].

5 Stimulation of Investment Activities on the Railway Network in Poland, Boosting the Directions of East–West Transit Direction Through Poland

The currently functioning railway corridor linking China with Europe is becoming insufficient as it is based on single-track railway lines in Kazakhstan [15]. BRI (Belt Road Initiative) is an initiative that assumes the creation of a railway corridor running across various countries of Central Asia, due to the growing intermodal transport and growth of the economic potential of Central Asian countries. In addition to the existing route, called New Silk Road, new sections of the railway network, running towards Europe are created. One of the routes, also known as the Southwest Corridor, is to connect Iran with Germany. It will run through Azerbaijan, Georgia, the Black Sea, Ukraine and Poland [16].

The Southwest Corridor coincides with another very important TRACECA transport corridor—Transport Corridor Europe Caucasia Asia [17]. The corridor runs through Kazakhstan, Tajikistan, Uzbekistan, Turkmenistan, Iran, Azerbaijan, Armenia, Georgia, Ukraine, Moldova, Turkey, Bulgaria and Romania [18] (Fig. 7).

At present, Iran and Azerbaijan are preparing an investment to build a new railway connection between these countries, bypassing the conflict area of the Nachichin Autonomous Republic. The first stage finished in January 2018 involves the extension of the Azerbaijan's railway line to the Iranian part of Astara [20]. In the next stages, the construction of the Astara–Resht–Qazvin line is planned, which will connect the Iranian and Azerbaijan's railway network [21]. Azerbaijan and Georgia are modernizing the railway line from Baku to Tbilisi. In 2017, a new railway line Tbilisi–Qars (Turkey), connecting the Georgian railway network with the Turkish one, was also launched.

One of the partners of the TRACECA project is Ukraine. The corridor's course through this country also determines the potential further directions of transport across Poland. Border crossings at the end of the corridor are Mostiska–Medyka (corridor E30) and Jagodin–Chełm. The corridor through Ukraine is shown in Fig. 8.

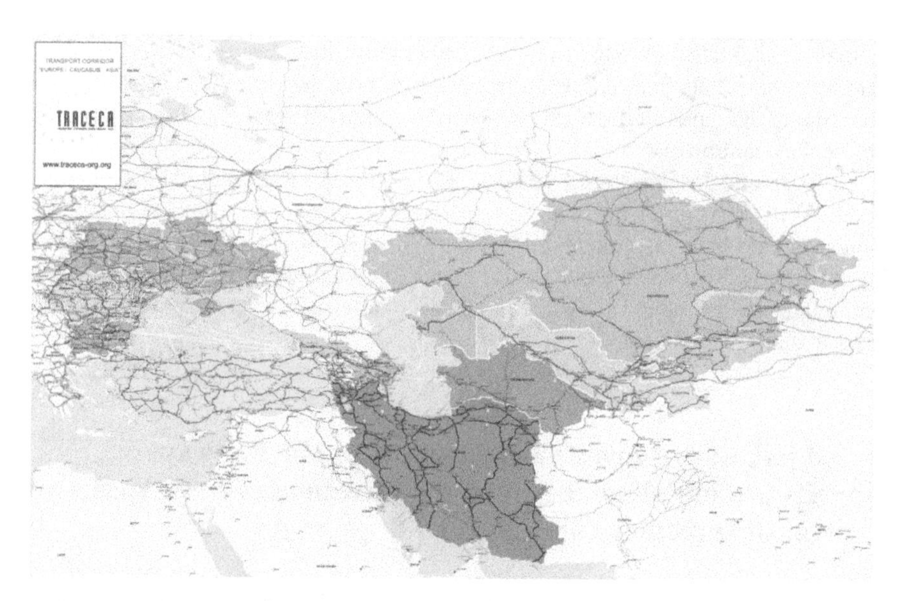

Fig. 7 TRACECA members. *Source* [19]

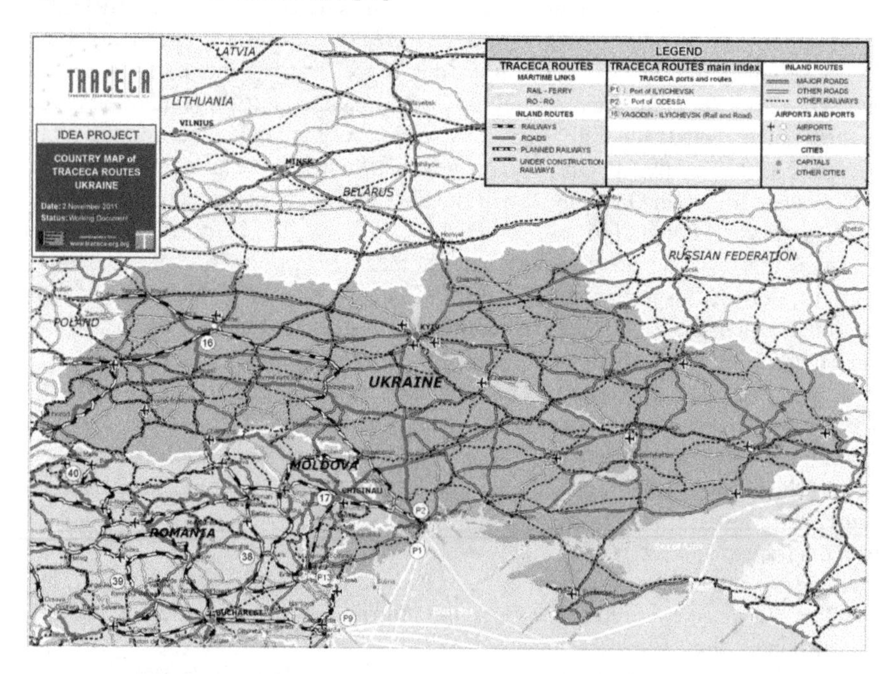

Fig. 8 TRACECA corridor in Ukraine. *Source* [22]

In Poland, the TRACECA corridor does not formally exist; however, the traffic that will be moved along this corridor will appear on the Polish rail network. Due to the direction of the corridor and the plans to create the Southwest Corridor, it can be stated that this traffic will take place in the direction of Germany. In principle, the corridor E30 (Medyka–Kraków–Śląsk–Wrocław–Zgorzelec) and the railway line No. 7 and the corridor E20 (Jagodin–Lublin–Pilawa–Skierniewice–Łowicz–Poznań–Słubice) will serve this purpose.

Investment planning on railway lines included in the above-mentioned corridors should take into account transit traffic. This applies first of all to railway line No. 7, but also to other elements of the railway network, which may become future bottlenecks in transport between Asia and Western Europe.

In this case the role of the broad-gauge line (LHS) enabling direct transport of loads from the broad-gauge network to the GOP (Upper Silesian Industrial District) is particularly important. Due to the dynamic use of the terminal in Sławków, the LHS line serves the area of Upper Silesia as well as the Czech Republic. This is the only railway line of this type in Poland. PKP LHS—operator and infrastructure manager of this line, is going to electrification and modernization of the LHS line [23].

6 Summary

Belt Road Initiative (BRI) is an initiative to create an efficient connection between Western and Central Asia and Europe. Improvement of trade exchange is to serve the economic development of China, Europe and countries along the way. The above analysis indicates that in addition to the existing trade routes—the maritime and railway Silk Road running through Russia, Belarus and Poland, the south TRACECA corridor has a high potential.

The currently functioning corridor through Russia is insufficient. This is due to the fact that Russia is subject to trade sanctions, which means that cargo from East Asia is transported only one way. As a result, the return of empty containers and the need to send cargo in the Europe–China trade relationship exclusively by maritime transport causes an increase in the indebtedness of European countries towards China. Moreover, only energy resources and armaments constitute the potential of Russia.

Poland is currently the recipient of intermodal cargo arriving by sea to ports and a transit country for intermodal cargoes transported by rail.

The operating Silk Road through Russia and Belarus passes through the border crossing Brzesc–Terespol located on the external border of the European Union. Due to customs control, congestion is created at this transition. In this connection, Russia and Belarus are looking for other routes to transit Poland, which runs to the ports of the Baltic states. The challenge for Poland is to stop this process by building terminals and launching a second intermodal corridor running from Ukraine via Kraków, Katowice, Opole and Wrocław towards Germany.

Currently, Polish investment plans in railway infrastructure focus on improving access to Polish seaports and north–south corridors. This direction in intermodal transport will still be important. There is no doubt, however, that the share of rail transport in intermodal transport will grow. This is due to two factors:

1. China is betting on the economic recovery of the western part of the country from which rail transport is more competitive than combined transport based on maritime transport.
2. Rail transport is not much more expensive than maritime transport, but travel time is much more attractive.

Economically strong eastern China will continue to use mainly maritime transport, but the role of rail will increase, especially in the case of superior goods and foods that require shorter delivery times.

China is not a geopolitical threat for Europe, in contrast to Russia, for which, the BRI is a competitive and geopolitical threat. An important factor in shortening the travel time may have Ukraine's entry into the European Union, which will significantly reduce the time of customs clearance, which will take place during transhipment in the Ukrainian Black Sea ports instead of on the Polish–Ukrainian border.

The BRI also covers the extension of the seaports of Southern Europe, e.g. ports of Piraeus and Venice. This is due to the main target, which is to shorten the delivery time of goods [24].

The aim of the railway transport to the BRI should be to maximize its competitive advantage, which is the time and the flexibility of the rolling stock in the main cargo transport. However, there is no doubt, that in the case of transport last mile, complementarity with maritime transport is equally important, especially because of the high competitiveness of road transport and inland waterway transport, especially in Central and Western Europe.

References

1. https://www.bbc.co.uk/news/business-44706880. Access 25 July 2018
2. Deutsche Bahn 2016 Integrated Report, p. 77. https://www1.deutschebahn.com/file/ecm2-db-en/12210234/hWTFG8DCjKgtQrEuMLPsAqCyG4g/13620500/data/ib2016_dbgroup.pdf. Access 10 Mar 2018
3. http://logistyka.wnp.pl/pkp-sa-zbuduja-holding-problemem-moze-byc-pkp-cargo,313890_1_0_0.html. Access 8 Mar 2018
4. PKP Group 2016 Annual Report, p. 10. http://pkpsa.pl/grupa-pkp/raporty/01.Raport-Roczny-Grupy-PKP-2016_ENG.pdf. Access 10 Mar 2018
5. https://www.bankier.pl/wiadomosc/Zalamanie-kursu-PKP-Cargo-mimo-poprawy-wynikow-7577930.html. Access 27 Mar 2018
6. https://businessinsider.com.pl/wiadomosci/krzysztof-maminski-nie-jest-prezesem-pkp-cargo/nz9epwp. Access 27 Mar 2018
7. https://www.money.pl/gielda/wiadomosci/artykul/pkp-cargo-nowy-prezes-czeslaw-warsewicz,13,0,2401805.html. Access 27 Mar 2018

8. http://biznes.gazetaprawna.pl/artykuly/913238,koleje-pkp-cargo-przegralo-z-db-schenker-rail.html. Access 25 July 2018
9. National Railway Program up to 2023 (Krajowy Program Kolejowy do 2023 roku), p. 3. 2016. http://mib.bip.gov.pl/fobjects/download/194387. Access 15 Mar 2018
10. https://businessinsider.com.pl/finanse/handel/jedwabny-szlak-transport-kolejowy-z-chin-do-polski-przez-rosje/elve5d3. Access 25 July 2018)
11. International investors chase the red dragon. Financial Times (2017 Nov 11)
12. Refinance push China's ethical bond issuance takes it to top spot in world green paper league. Financial Times (2017 Nov 11)
13. Wobbling into the WTO. The Economist (2 Dec 2017)
14. Wojny handlowe. Chiny mają mocniejszą pozycję niż USA. Puls Biznesu No. 60 (5071). Mar 26, 2018, p. 5
15. Miecznikowski, S., Radzikowski, T.: Over capacity of container shipping as a challenge to rail silk road competitiveness. Res. J. Univ. Gdań. Transp. Econ. Logist. **70**, 121–132 (2017)
16. https://www.railwaypro.com/wp/five-countries-to-boost-south-west-corridor-project. Access 12 Mar 2018
17. Jakóbowski, J., Popławski, K., Kaczmarski, M.: Kolejowy Jedwabny Szlak. Połączenia kolejowe UE–Chiny: uwarunkowania, aktorzy, interesy, Centre for Eastern Studies (OSW). Prace OSW Nr 72 (2018)
18. http://www.traceca-org.org. Access 20 Mar 2018
19. http://www.traceca-org.org/fileadmin/fm-dam/Routes_Maps/MAP_TRACECA_ROUTES_10_09_2017_300DPI.png. Access 20 Mar 2018
20. https://www.lok-report.de/news/uebersee/item/2960-aserbaidschan-iran-grenzueberschreitende-verbindung-astara-astara-aufgenommen-finanzierung-zur-weiterfuehrung-gesichert.html. Access 20 Mar 2018
21. http://www.railways.by/76642-iran-i-azerbajjdzhan-sovmestno-profinansirujut.html. Access 20 Mar 2018
22. http://www.traceca-org.org/fileadmin/fm-dam/TAREP/58jh/EXPERT_GROUP_MODEL_GIS/UKRAINE_07_11_2011_300DPI.png. Access 20 Mar 2018
23. https://www.rynek-kolejowy.pl/wiadomosci/pkp-lhs-analizuje-mozliwosc-elektryfikacji-linii-szerokotorowej-84860.html. Access 1 Aug 2018
24. http://www.atimes.com/article/marco-polo-reverse-italy-fits-new-silk-roads. Access 22 Mar 2018

Bibliometric Analysis of Research on Green Shipping Practices

Natalia Wagner

Abstract The paper is to explore a bibliometric approach to quantitatively assessing current research trends on the idea of green shipping issues. The paper is especially looking for content, structure and connections between research on shipping and sustainability matters. The aim of the paper is to identify dominant subfields from green shipping practices research field and find out how they are connected with each other. The bibliometric tools are used—trend analysis as well as mapping and clustering of keywords. The findings show that the papers covering green shipping area can be divided into six clusters. The most important subfields focus on regulations imposed by International Maritime Organisation, new technologies (mainly connected with ship power) as well as productivity and outcome of shipping companies. Green shipping can be regarded as an umbrella which covers all processes from a ship lifetime—shipbuilding, environmentally friendly operations and sustainable ship recycling.

Keywords Green shipping · Shipping · Bibliometric analysis

1 Introduction

The topic of sustainable shipping is very actual and important. The idea of green shipping is more and more vital in EU policy and is visible in many various actions. Many different types of entities are engaged in such issues, for example, the EU institutions, maritime industry institutions, practitioners, academics and public opinion as well. There are shipowners whose involvement goes well beyond mere adherence to the relevant environmental requirements. Some shipping companies, especially the biggest and well-known container ship operators, have been adapting sustainability strategies which are more strict than environmental regulations imposed by

N. Wagner (✉)
Faculty of Economics and Transport Engineering, Maritime University of Szczecin, Szczecin, Poland
e-mail: n.wagner@am.szczecin.pl

M. Suchanek (ed.), *Challenges of Urban Mobility, Transport Companies and Systems*, Springer Proceedings in Business and Economics,
https://doi.org/10.1007/978-3-030-17743-0_27

International Maritime Organization (IMO). There are also entities and initiatives involved in encouraging shipping industry to perform green practices. Such institutions promote the sustainable development of ports and shipping industry by creating a number of tools useful for assessing shipowners or individual ships. They are used by several seaports in their eco-friendly strategies implemented towards ships calling at ports.

The aim of this paper is to identify dominant subfields from green shipping practices research field and find out how they are connected with each other. In order to achieve that goal, a bibliometric visualisation approach is applied. Study of literature using bibliometric tools can provide further insights into research topics not previously fully evaluated.

2 Literature Review

This paper refers to two main research fields. The first one relates to green shipping practice, the second is connected with the applied method—visualisation of bibliometric networks.

Green shipping practices are defined as environmental management practices undertaken by shipping companies with an emphasis on waste reduction and resource conservation in cargo handling and transporting [1]. Shipping firms undertake environmental business routines voluntarily with the belief that these initiatives will lead to improved environment performance and increased economic gains as well [2]. Research results show that expecting positive impact on company's profits thanks to adopting green shipping technologies and management may be right. Adoption of green shipping practices can improve the productivity performance of the shipping firms [3].

Green shipping practices consist of six dimensions, which are presented on Fig. 1.

These six dimensions were evaluated on the basis of survey performed among the biggest container shipowners (mainly Maersk and CMA CGM). These dimensions are focused on activities which are undertaken by shipowners in their everyday operations. This paper intends to verify if there are similarities between these practical aspects (Fig. 1) and the research subfields in the area of green shipping.

Despite the need for environmental management in shipping, there is no extant measurement scale that comprehensively captures green shipping practices (GSP) in shipping operations. It is not an easy task to proceed. There are some academic propositions; however, further studies are still needed [4].

Obviously, the essential element that is necessary when talking about shipowner's green shipping practices is to have a green ship at its disposal. Green ship can be defined in various ways. According to Lee, a green ship or eco-friendly ship means a ship that has reduced GHG emissions through the development of technologies related to fuel savings and alternative fuels [5]. The necessity of complying IMO regulations is stressed (especially Annex VI of MARPOL dedicated to prevention of air pollution from ships). Apart from IMO regulations, for many shipowners, it is

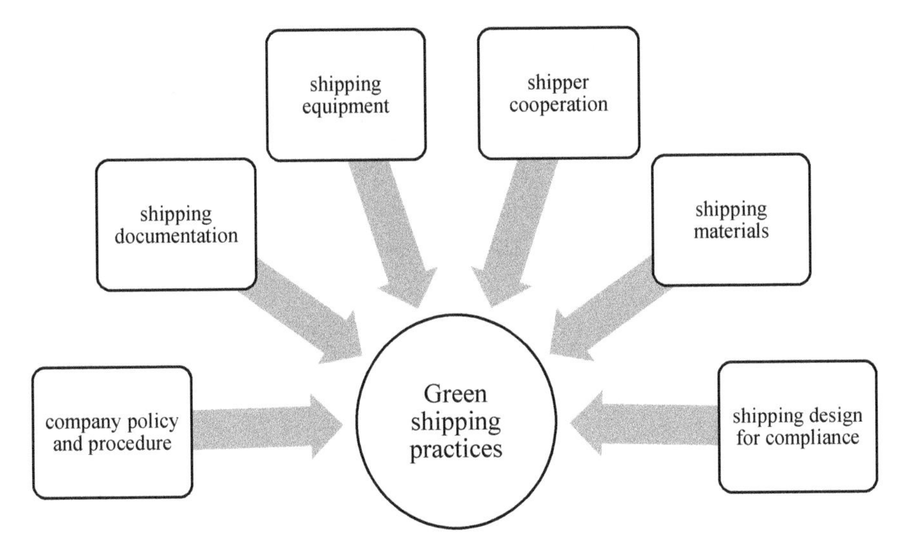

Fig. 1 Main green shipping practices dimensions. *Source* Own elaboration on the basis of [1]

important to operate green ships in order to meet the environmental requirements of some ports especially in USA and EU.

Trends in the field of green shipping are presented and analysed not only in research journals chosen by academics but also in maritime business and technology magazines intended for practitioners. A few benchmark technologies have already been developed to reach the goal of building a green ship, which would not only comply with the new environmental rules and regulations but would also leave least possible carbon footprints.

There are new technologies which if used together would result in the ultimate Green Ship of the Future. It is worth to name some of them as examples: LNG fuel for propulsion and auxiliary engine, sulphur scrubber system, advanced rudder and propeller system, hull paint, waste heat recovery system, no ballast system [6].

The second research field which is shaping this paper is the research methodology. Bibliometrics is a statistical analysis of publications to build their accurate formal representations for explanatory, evaluative and administrative purposes [7]. Such an analysis can be performed with the use of bibliometric mapping. Visualisation of bibliometric networks is applied in this paper.

Bibliometric methods have been used in few papers from the field of shipping research as well. They have been used to analyse the whole maritime industry [8] or its most dynamically developing segment—container shipping. Some papers do not use any visualisation methods [9], other, especially newer ones, consider them as a useful tool [10]. This paper shares that attitude and applies visualisation of bibliometric network. Research field like in [11] is constrained, however, in a different way. In this paper, the restriction is not imposed on a specific shipping market segment, but on a type of addressed issues—green practices. Combining a research field defined

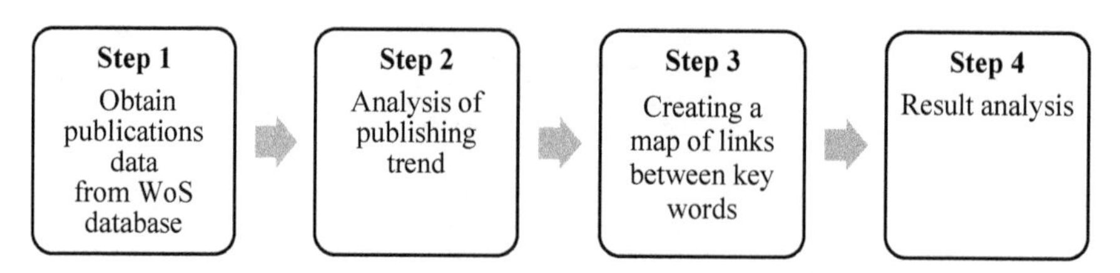

Fig. 2 Research procedure. *Source* Own elaboration

that way with a bibliometric visualisation tool, this paper fulfils a gap in existing research.

3 Research Methodology

In this paper, green shipping research area is analysed with the use of bibliometric approach. Applied research procedure is presented in Fig. 2. The first step was to obtain a relevant publications information from citation database. The data set used in this study was extracted and downloaded from Web of Science (WoS) Core Collection.

While searching for publications, it is very important to set the research criteria that suit the aim of the analysis. Not only was "green shipping" set as a key word, but other green-related words were also considered as well. Often words such as "green", "ecological" and "sustainable" are used as synonyms, that is why the range of keywords was extended to take into account them all when searching for publications. Eventually, a combination of keywords "green ship*" OR "sustainable ship*" OR "clean ship*" OR "ecologic* ship*" was applied to gather all the publications with those phrases in their titles, abstracts or keywords. The papers published during the period from 1980 to 2017 are included in the study. The analysis was performed with the use of the VOSviewer software [12] and the analytical tool available at the Web of Science.

The records extracted were analysed for bibliometric characteristics such as trend in publishing, belonging to the research category and mapping of frequent terms in publications pertaining to green shipping.

4 Results

Figure 3 plots the annual trends of publications related to green shipping. The first articles were published in 1995. Since then the papers were published in an irregular manner till 2008. There were years when this topic has not appeared in scientific journals even once. After 2008, systematically more and more scholars started doing

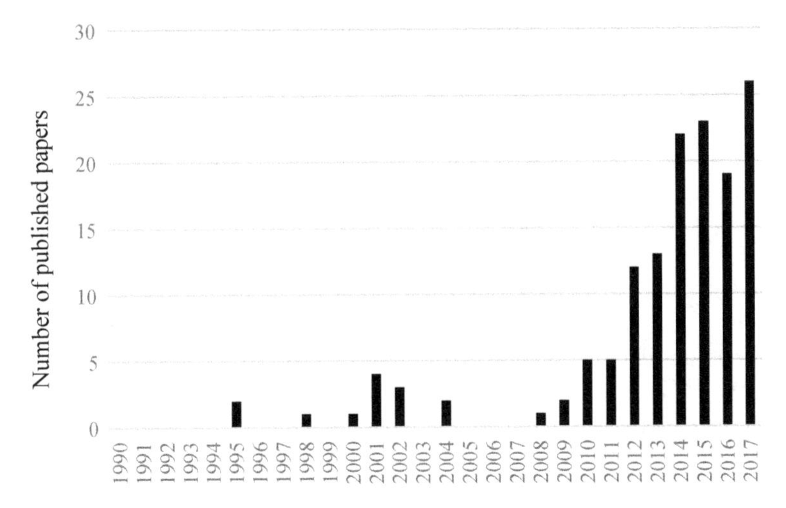

Fig. 3 Publishing trend in the area of green shipping. *Source* Own realisation based on Web of Science analytical tool

Table 1 The top ten research areas of green shipping-related publications

Research areas	Number of publications	Percent of 141 publications
Engineering	75	53.19
Transportation	36	25.53
Environmental sciences ecology	26	18.44
Business economics	14	9.93
Science technology other topics	12	8.51
Energy fuels	9	6.38
Oceanography	8	5.67
International relations	7	4.97
Materials science	7	4.97
Operations research management science	7	4.97

Source Own realisation based on Web of Science analytical tools

research in this field. This led to a jump in the number of publications in recent years. The number of research papers is growing with the increasing popularity of implementation green solutions among shipowners, as well as growing concern about these issues among institutions and public opinion.

Publications from green shipping area are mostly of technical nature and belong to "engineering" research area (Table 1). However, management-related aspects are present as well. Such papers can be found, for example, in "transportation" and "business economics" areas.

Table 2 The occurrence of key terms in green shipping-related publications

Term	Occurrence
Green ship	21
Fuel	13
Power	13
Material	13
Energy	12

Source Own realisation based on VOSviewer software

The top contributing authors in the field of green shipping are K. H. Lai, Y. H. V. Lun, T. C. E. Cheng. The organisations which are the most often engaged in this research topic are Hong Kong Polytechnic University, Wuhan University of Technology and Shanghai Jiao Tong University.

The top keywords with their number of occurrences are shown in Table 2. The keyword "green ship" has the highest frequency of 21. Other keywords with high occurrence include "fuel", "power" and "material".

The keyword co-occurrence network of green shipping was constructed with the use of the VOSviewer software. Binary counting method was applied, which means that if a noun occurs only once in the title and abstract of a publication it is treated in the same way as a noun that occurs several times. The size of the nodes and labels in Fig. 4 is determined by the weights of the item. The higher the weight of an item, the larger the label and the node. The distance between two nodes represents the strength of the link between them. A shorter distance generally reveals a stronger relation. The nodes with the same colour belong to a cluster. A cluster is a set of closely related nodes [13]. To make it more visible in black and white, the clusters are circled and numbered.

Results from the analysis done with VOSviewer divide the keywords into six clusters, which categorise the dimensions of green shipping research. For some items, the labels are not displayed, that is why it is worth to view the most interesting ones.

The first cluster is the strongest with the largest group of words (42 words). It concentrates on research field devoted to regulations which are imposed by International Maritime Organisation (IMO) in the area of fuel consumption and air emission restrictions. In this cluster, different kinds of fuel are mentioned such as Liquified Natural Gas and HFO. Air pollutions are also identified—GHG and NO_x. In two cases, separate nodes indicate words with the same meaning such as "International Maritime Organisation" and "IMO" as well as "Liquified Natural Gas" and "LNG". It means that in fact the weight of this noun phases is even higher than the size of node shows.

The first cluster is concentrated on regulations, whereas the second is focused on a shipping company. The second cluster is devoted to green shipping practice in the context of management and environmental company's policy. There are keywords connected with company's results (outcome and productivity) and environment (CO_2 emission, environmental impact). It is concentrated on implementation of green ship-

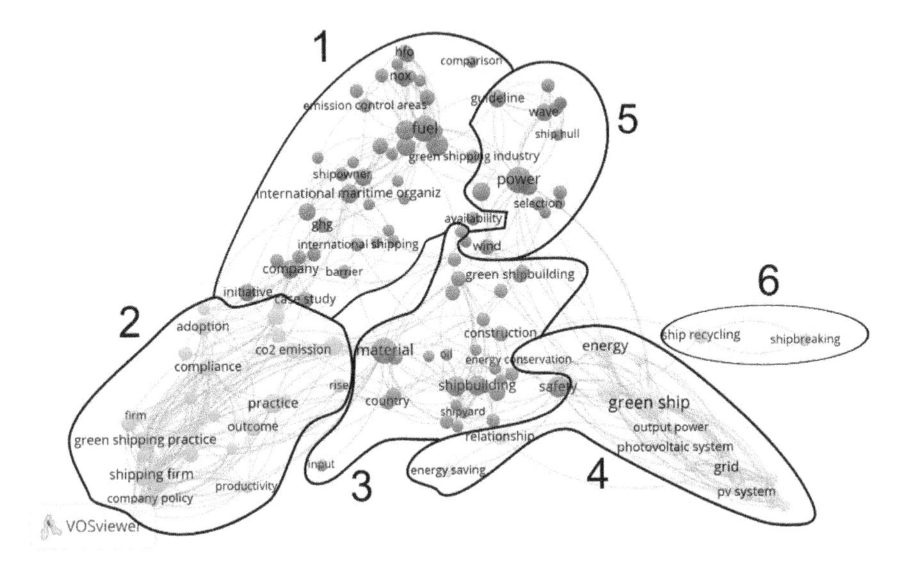

Fig. 4 Map of links between analysed keywords. *Source* Own realisation based on VOSviewer software

ping practices into shipping firm operations. The phrase “green shipping practice” is strongly connected with many other words in this cluster; however, it is rather isolated from keywords from other clusters.

Third cluster is devoted to green shipbuilding. The most popular keywords within this cluster are material, shipbuilding and safety.

The fourth cluster is focused on green ship in relation to energy saving and fuel consumption. The most frequent terms put in this cluster are green ship, energy and capacity. The most popular technology which appears in this cluster is photovoltaic system (also as a node “Pv system”).

The fifth cluster seems to contain terms from technical solutions area. The most popular keywords included in this cluster are “power”, “ship”, “wave”. Among keywords, there are “propeller” “LNG ships” “vibration” as well as “Energy Efficiency Design Index (EEDI)” which are guidelines imposed by IMO on newly built ships. There are also terms revealing the research methods such as “experimental data” and “model test”.

The last sixth cluster, the smallest one, is not of significant size, however, is very interesting. It consists of words “ship recycling” and “shipbreaking”. These words are connected only with terms “waste” and “safety” from third cluster. That cluster represents problems which are isolated from other clusters fields of research. Still, concern of safe shipbreaking belongs to the green shipping scientific area. Term Green shipping can be regarded as an umbrella which covers all processes from a ship lifetime—shipbuilding, environmentally friendly operations and sustainable ship recycling.

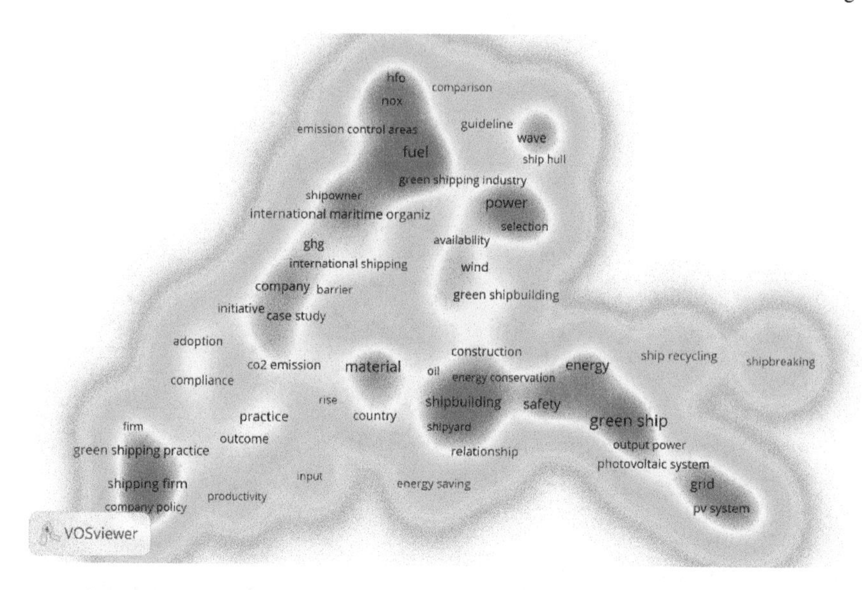

Fig. 5 Map of analysed keywords density. *Source* Own realisation based on VOSviewer software

Figure 5 presents visualisation of analysed keywords density. The largest number of items with the highest weight is located in the neighbourhood of terms: “shipping firm”, “fuel”, shipbuilding”, “green sheep” and “power”.

More words have technical than economical or managerial connotations. It is concise with the most popular research categories presented in Table 1. Green shipping can be defined mainly by two perspectives: operational and technological in accordance with the design of ships and ship’s equipment. However, the technological approach is much more extensive (Fig. 6).

Fig. 6 Main perspectives of green shipping practice research. *Source* Own elaboration

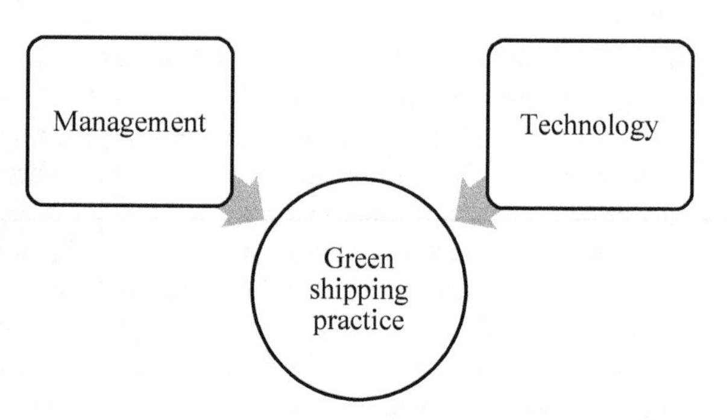

5 Conclusion

The analysis based on co-occurrence of terms in titles, abstracts and keywords allowed to identify dominant subthemes in the research field of green shipping. What is more, it also helped find new emerging trends. Most important subfields are similar to those, which are known from literature and research done by well-known scholars. Comparing known from literature green shipping practice dimensions (Fig. 1) with Fig. 4 obtained in this research reveals many common points. Among them, there are ship design, materials and several solutions, especially connected with propulsion of the ship. A lot of new papers are dedicated to alternative power and propulsion technology for ships as well as new IMO regulations concerning GHG emissions. These problems are important from technical and managerial point of view.

On the other hand, it is natural that both figures are not exactly the same because their perspective is different. Some shipping practices have operational character and they are not a popular research object. As an example can serve promoting reductions in resource utilising such as paper for documenting shipping activities, which is included in "shipping documentation" dimension [4]. However, in author's opinion, there is especially one subfield identified in this research that should be added as additional dimension of green shipping practice (Fig. 1). That is cluster 6 "ship recycling and shipbreaking". Application of green shipping practices throughout the whole ship life cycle—during shipbuilding process, fleet management and ship recycling—is extremely important area of sustainable shipping practice and one of the biggest challenges faced by the shipping industry nowadays.

Acknowledgements Research outcomes obtained as a result of a research study entitled "The concept of Corporate Social Responsibility in transport, forwarding and logistics market" no. 7/S/IZT/2017 financed with a subsidy from the Ministry of Science and Higher Education for financing statutory activity.

References

1. Lai, K.-H., Luna, V.Y.H., Wong, C.W.Y., Chenga, T.C.E.: Green shipping practices in the shipping industry: conceptualization, adoption and implications. Resour. Conserv. Recycl. **55**, 631–638 (2011)
2. Lun, Y.H.V., Lai, K.-H., Wong, C.W.Y., Cheng, T.C.E.: Green shipping practices and firm performance. Marit. Policy Manag. **41**(2), 134–148 (2014)
3. Chang, Y.T., Danao, D.: Green shipping practices of shipping firms. Sustainability **9** (2017)
4. Lai, K.-H., Lun, Y.H.V., Wong, C.W.Y., Cheng, T.C.E.: Measures for evaluating green shipping practices implementation. Int. J. Shipp. Transport Logist. **5**(2), 217–235 (2013)
5. Leea, T., Nam, H.: A study on green shipping in major countries: in the view of shipyards, shipping companies, ports, and policies. Asian J. Shipp. Logist. **33**(4), 253–262 (2017)
6. Marineinsight, https://www.marineinsight.com/green-shipping/13-technologies-to-make-the-ultimate-green-ship/ [13.06.2017]
7. De Bellis, N.: Bibliometrics and Citation Analysis: From the Science Citation Index to Cybermetrix. Scarecrow Press, Plymouth (2009)

8. Talley, W.K.: Maritime transportation research: topics and methodologies. Marit. Policy Manag. **40**(7), 709–725 (2013)
9. Shi, W., Li, K.X.: Themes and tools of maritime transport research during 2000–2014. Marit. Policy Manag. **44**(2), 151–169 (2017)
10. Lau, Y.-Y., Ng, A.K., Fu, X., Li, K.X.: Evolution and research trends of container shipping. Marit. Policy Manag. **40**(7), 654–674 (2013)
11. Lau, Y.-Y., Ducruet, C., Ng, A.K., Fu, X.: Across the waves: a bibliometric analysis of container shipping research since the 1960s. Marit. Policy Manag. **44**(6), 667–684 (2017)
12. van Eck, N.J., Waltman, L.: Software survey: VOSviewer, a computer program for bibliometric mapping. Scientometrics **84**(2), 523–538 (2010)
13. van Eck, N.J., Waltman, L.: Visualizing bibliometric networks. In: Ding, Y., Rousseau, R., Wolfram, D. (eds.), Measuring Scholarly Impact: Methods and Practice, pp. 285–320. Springer, Berlin (2014)

Rationalization of Energy Intensity of Road Transport of Member Countries of the International Energy Agency

Elżbieta Szaruga and Elżbieta Załoga

Abstract The objective of this article is to indicate the directions for the rationalization of energy intensity of road transport in the International Energy Agency's member countries. The directions for the rationalization were indicated on the basis of the maximized utility function due to the changing macroeconomic conditions. Secondary data were used from the Organization for Economic Co-operation and Development database. The study used taxonometric techniques (K-means method with Chebyshev distance) to determine similar economies of the IEA. General Linear Model was applied to obtain a model for rationalization of energy intensity of road transport.

Keywords Rationalization · Energy intensity · Road transport · IEA · Scenarios

Abbreviations

β	Beta; standardized form of coefficient b in regression
Con	Production in construction
Defl_gdp	Deflator of GDP
DF	Degree of freedom
EI	Energy intensity of road transport
Exp	Export
F	F-test
GDP	Gross domestic product

E. Szaruga (✉)
Department of Quantitative Methods, Faculty of Management and Economics of Services, University of Szczecin, Szczecin, Poland
e-mail: elzbieta.szaruga@usz.edu.pl

E. Załoga
Department of Systems and Transport Policy, Faculty of Management and Economics of Services, University of Szczecin, Szczecin, Poland
e-mail: elzbieta.zaloga@wzieu.pl

M. Suchanek (ed.), *Challenges of Urban Mobility, Transport Companies and Systems*, Springer Proceedings in Business and Economics,
https://doi.org/10.1007/978-3-030-17743-0_28

GHG	Greenhouse gas emissions
GLM	General Linear Model
IEA	International Energy Agency
Imp	Import
Ind	Industrial production
IR	Long-term interest rates
kg	Kilograms
Man	Manufacturing
MM	Million
MS	Mean squares
OECD	Organization for Economic Co-operation and Development
p	p-value; probability value; asymptotic significance
R	Correlation coefficient
R^2	Coefficient of determination
SS	Sum of squares
t	Tonnes
thous.	Thousand
TI	Transport intensity
tkm	Tonne-kilometer

1 Introduction

The recent decades have proven that the interest in energy and environmental policies to reduce energy consumption has been grown, especially focusing on rationalizing fossil fuels and reducing pollutant emissions, e.g. greenhouse gas emissions (GHG). Accordingly, the main areas of the interest are energy-intensive sectors, such as transport [1, p. 98]. It is worth noting that transport, notably road freight transport, is one of the largest oil's consumers. It also stimulates the economy and participates in all economic processes [2, p. 690]. Even though freight road transport needs energy to growth and stimulating economies, attention is paid to reducing demand for energy and decoupling from transport. Therefore, the energy intensity-oriented environmental policy may promote the harmonization of the economic growth and environmental protection [3, p. 875]. Despite the fact that transport is a key link in the development of the civilization, it negatively affects the environment through noise, vibrations, road accidents, land demand, congestion and energy intensity [4, p. 806, 5, 6]. That is significant from point of view of the balancing between economic and environmental objects for developing and developed economies. On the one hand, developing economies have problem with balancing economic growth and environmental issues [3, p. 875]. On the other hand, developed economies strive to ensure sustainable consumption [3, p. 875]. Multidirectional researches have been carried out on this background, inter alia: Talbi [7], Saidi et al. [8], Sierra [9], Załoga and Szaruga [10], Saisirirat and Chollacoop [11], Gerboni et al. [12], Szaruga et al. [13].

However, there is still a research gap in assessing the directions for rationalization of energy intensity of road transport due to macroeconomic criteria for the International Energy Agency's member countries. They have common goals in terms of the reducing energy consumption. The objective of this paper is to indicate the directions for rationalization of energy intensity of road transport of the International Energy Agency's member countries (with using the maximized utility function due to the changing macroeconomic conditions). The selection of the research sample is deliberate due to the implementation of the common provisions for this Agency focused on energy intensity. Problematic is the issue of the disparities members due to economic conditions. Thus, the directions of rationalization of energy intensity may differ substantially for countries with different economic and transport specifics. Taxonometric techniques were used to identify similarities and differences of the individual countries in view of the examined characteristics, classifying and grouping the objects to the appropriate concentration. The optimization and iterative method, i.e. K-means (with the V-fold cross-check), was used in this respect. Rationalization model was created for energy intensity of road transport with significant concentration. At the same time, parameterization with sigma restrictions was made. The utility function of the selecting directions for energy intensity of road transport was specified. The most advantageous rationalization scenarios were proposed on the basis of the obtained model. The conditions of distress were also identified for the economy from improperly selected directions of energy intensity of road transport. The paper consists of four sections: introduction, data and methodology, empirical results and discussion with conclusions.

2 Data and Methodology

The study used secondary data from statistics database of the Organization for Economic Co-operation and Development (OECD), i.e. 'OECD.Stats' [14] for 16 from 30 member countries of the International Energy Agency, viz. Czech Republic, Finland, France, Hungary, Ireland, Italy, Luxembourg, Netherland, Norway, Poland, Portugal, Slovak Republic, Spain, Sweden, Switzerland, United Kingdom from 2000 to 2016. The remaining 14 out of 30 of member countries of the IEA were not included in the study due to the poor quality or lack of data (insufficient number of observations to fill the gaps with one of the statistical and econometric methods).

For the purpose of the study, it was assumed that the dependent variable is energy intensity of road transport (abbrev. EI, in t/1 MM unit of current USD of GDP) and quantitative predictors are: transport intensity (abbrev. TI; in tkm/1 thous. units of current USD of GDP), long-term interest rates (abbrev. IR; in %), deflator of GDP (abbrev. Defl_gdp; in index 2010 = 100), import (abbrev. Imp; in % rate), export (abbrev. Exp; in % rate), production in construction (abbrev. Con; in index 2010 = 100), industrial production (abbrev. Ind; in index 2010 = 100), manufacturing (abbrev. Man, in index 2010 = 100), and qualitative factors are member countries of the IEA. In some cases, it was necessary to estimate the lack of the data by the method of K-nearest neighbors (including the number of neighbors = 5 and the number of patterns = 10). The number of the missing items was equal to: for TI—6 observations (2.2%), IR—6 observations (2.2%), Imp—2 observations (0.7%), Exp—2 observations (0.7%), Con—34 observations (12.5%) and EI—4 observations (1.5%).

In the pre-analysis, a cluster analysis was carried out using the K-means algorithm with the distance measure of Chebyshev. (Previously, data were standardized.) It was assumed that the initial cluster centers should be determined by the criterion of maximizing the distance of the clusters. In this study, a V-fold cross-check was helpful (number of sample sets = 10, nucleus of random number generator = 1, minimum number of clusters = 2, maximum number of clusters = 16, minimal decrease = 5%). Only Finland was identified as a separate cluster, different from the other 15 member countries of the IEA. And 15 other members of the IEA identified as one cluster (they were similar due to statistical properties). Therefore, Finland was not included in the study in subsequent stages. In the next stage, GLM was used to evaluate parameters and broaden the analysis. In the analysis process, we excluded the changes which were statistically insignificant (for which p-value $\geq$ 5%). Finally, the study includes variables: EI, TI, Defl_gdp, Imp, Ind, Man and 11 member countries of the IEA[1] (i.e. Czech Republic, Ireland, Italy, Luxembourg, Netherland, Poland, Portugal, Slovak Republic, Spain, Switzerland, United Kingdom (UK)—where UK was recognized as reference member of the IEA). The other predictors and factors, originally included in the pre-analysis, were not included in the proper analysis due to the lack of the desirable statistical properties. The empirical results are presented in the next section.

3 Empirical Results and Discussion

Tables 1 and 2 present the results of one-dimensional significance tests and the adequacy testing of the full model.

The variability of energy intensity of the IEA's road transport (see Table 1) was shaped to the greatest extent by: GDP deflator, constant, member countries of the IEA, transport intensity, industrial production, manufacturing and the worst extent by import. At the same time these variables could look for possibility of customize error. Quantitative predictors and qualitative factors varied in volatility, and their averages significantly differed.

In the case of the separating individual predictors and qualitative factors in the assessment of the strength of the effect on the dependent variable, it could be stated that the qualitative factors (the IEA's member countries) explain the variability of energy intensity of road transport in 86%. Explaining the variability of energy intensity of road transport of the IEA's member countries could be separately made

[1]The use of the abbreviation IEA in further parts of this paper will refer only to the 11 countries indicated above (finally included in the analysis).

Table 1 One-dimensional significance tests, the size of effects and powers for model of rationalization of energy intensity of road transport of the International Energy Agency (from 2000 to 2016)

Effect	SS	DF	MS	*F*	*p*	Partial eta-squared	Non-centrality	Observed power (alpha = 0.05)
Intercept	2628.58	1	2628.58	156.19	0.00	0.48	156.19	1.00
TI	545.14	1	545.14	32.39	0.00	0.16	32.39	1.00
Defl_gdp	3405.58	1	3405.56	202.36	0.00	0.54	202.36	1.00
Imp	116.37	1	116.37	6.91	0.01	0.04	6.91	0.74
Ind	184.65	1	184.65	10.97	0.00	0.06	10.97	0.91
Man	158.23	1	158.23	9.40	0.00	0.05	9.40	0.86
Member States of IEA	17,841.41	10	1784.14	106.02	0.00	0.86	1060.17	1.00
Error	2877.74	171	16.83					

Parameterization with sigma restrictions. Decomposition of effective hypotheses; standard error rating: 4.10
Source: Own computations based on: OECD.Stats [14]

Table 2 *R* of the full rationalization model of energy intensity of road transport of the International Energy Agency (from 2000 to 2016)

Dependent variable	Multiple *R*	Multiple R^2	Adjusted R^2	SS of model	DF of model	MS of model	*F*	*p*
EI	0.95	0.91	0.90	29,686.97	15	1979.13	117.60	0.00

Source: Own computations based on: OECD.Stats [14]

through: GDP deflator (explaining in 54%), constant (explaining in 48%)[2], transport intensity (explaining in 16%), industrial production (explaining in 6%), manufacturing (explaining in 5%) and imports only (explaining in 4%). Thus, member countries (their specificity) most differentiate the level of energy intensity of road transport of the IEA, by imposing a permanent effect on it. One could assume that the national and economic conditions have the greatest impact. The GDP deflator quite differentiates the level of energy intensity of road transport—it refers to the level of prices for the entire economies of the IEA's member countries. In contrast, the transport intensity, industrial production, manufacturing or import without additional aggregates would explain the differentiation of energy intensity of road transport to a negligible extent (in the full model, their inclusion improves the quality of the assessment; however, they are not sufficient quantitative predictors as only/separable predictors). This could be confirmed by the results, as presented in Table 2.

[2]It is high value for a constant. However, it is known that energy intensity of road transport maintains its level from previous years. These were not drastic changes—compared to short-term analysis.

Based on the results in Table 2, it could be concluded that the quantitative predictors and qualitative factors were statistically significantly correlated with the energy intensity of the IEA's road transport (correlation at the level of 95%). All quantitative predictors together with qualitative factors explained energy intensity of road transport in 91% (90% after corrections).

Detailed assessment of the parameters of energy intensity of the IEA's road transport is presented in Table 3. It presents how changing the quantitative predictors or qualitative factors will affect the energy intensity of the IEA's road transport.

As can be seen from Table 3, transport intensity, import, industrial production, inclusion of certain countries, such as Czech Republic, Ireland, Italy Luxembourg, Poland, Portugal and Spain, have had positive impact on the increase in energy intensity of the IEA's road transport. The GDP deflator and presence of countries such as Netherland, Slovak Republic and Switzerland in the model have had a negative impact on increase in energy intensity of the IEA's road transport. It can be assumed that the obtained estimation errors are acceptable. Estimate errors equal to or greater than 50% should be seen as warning signals. In this case, they can be treated as rationalization margins. The equation for energy intensity forecast is presented below (see Eq. 1).

Table 3 Assessment of the model parameters of rationalization of energy intensity of the IEA's road transport (from 2000 to 2016)

Level of effect	EI parameter	EI standard error	EI p	EI β	EI standard error of β
Intercept	80.73	6.46	0.00		
TI	0.03	0.01	0.00	0.34	0.06
Defl_gdp	−0.61	0.04	0.00	−0.42	0.03
Imp	0.15	0.06	0.01	0.20	0.07
Ind	−0.62	0.19	0.00	−0.79	0.24
Man	0.51	0.17	0.00	0.75	0.25
Czech Republic	8.78	1.91	0.00	0.28	0.06
Ireland	4.07	1.19	0.00	0.13	0.04
Italy	4.14	1.55	0.01	0.13	0.05
Luxembourg	10.53	1.19	0.00	0.34	0.04
Netherlands	−8.83	1.46	0.00	−0.29	0.05
Poland	5.48	1.55	0.00	0.18	0.05
Portugal	7.92	1.22	0.00	0.26	0.04
Slovak Republic	−28.79	2.31	0.00	−0.93	0.07
Spain	3.36	1.36	0.01	0.11	0.04
Switzerland	−5.74	1.42	0.00	−0.19	0.05

Source: Own computations based on: OECD.Stats [14]

Equation for the forecast:

$$\begin{aligned} \mathrm{EI} = {} & 80.7319 + 0.0287 * \mathrm{TI} - 0.6095 * \mathrm{Defl}_{\mathrm{gdp}} + 0.1503 * \mathrm{Imp} \\ & - 0.6189 * \mathrm{Ind} + 0.5110 * \mathrm{Man} + 8.7803 * \text{Czech Republic} \\ & + 4.0656 * \text{Ireland} + 4.1395 * \text{Italy} + 10.5266 * \text{Luxembourg} \\ & - 8.8338 * \text{Netherlands} + 5.4804 * \text{Poland} + 7.9194 * \text{Portugal} \\ & - 28.7874 * \text{Slovak Republic} + 3.3629 * \text{Spain} - 5.7440 * \text{Switzerland} \\ & + 0.0000 * \text{United Kingdom} \end{aligned} \tag{1}$$

Interpreting the prediction equation for energy intensity of road transport and the assuming that the UK is the reference country, it should be stated that the level of energy intensity of the IEA's road transport is constant (in total 11 analyzed countries) at the level of 80.73 t/MM USD. In addition, several important conclusions can be derived from this equation:

- increase in transport intensity by 1 tkm/1 thous. USD will increase the energy intensity of the IEA's road transport by 28.7 kg/MM USD (*ceteris paribus*);
- increase in GDP deflator by 1 unit (1 point of index) will decrease the energy intensity of the IEA's road transport by 609.5 kg/MM USD (*ceteris paribus*);
- increase in import by 1 percentage point will increase the energy intensity of the IEA's road transport by 150.3 kg/MM USD (*ceteris paribus*);
- increase in industrial production by 1 unit (1 point of index) will decrease the energy intensity of the IEA's road transport by 618.9 kg/MM USD (*ceteris paribus*);
- increase in manufacturing by 1 unit (1 point of index) will increase the energy intensity of the IEA's road transport by 511 kg/MM USD (*ceteris paribus*);
- the increase in the energy intensity of the IEA's road transport by 8.78 t/MM USD is caused by the Czech Republic (*ceteris paribus*);
- the increase in the energy intensity of the IEA's road transport by 4.07 t/MM USD is caused by Ireland (*ceteris paribus*);
- the increase in the energy intensity of the IEA's road transport by 4.14 t/MM USD is caused by Italy (*ceteris paribus*);
- the increase in the energy intensity of the IEA's road transport by 10.53 t/MM USD is caused by Luxembourg (*ceteris paribus*);
- the decrease in the energy intensity of the IEA's road transport by 8.83 t/MM USD is caused by Netherlands (*ceteris paribus*);
- the increase in the energy intensity of the IEA's road transport by 5.48 t/MM USD is caused by Poland (*ceteris paribus*);
- the increase in the energy intensity of the IEA's road transport by 7.92 t/MM USD is caused by Portugal (*ceteris paribus*);
- the decrease in the energy intensity of the IEA's road transport by 28.79 t/MM USD is caused by the Slovak Republic (*ceteris paribus*);
- the increase in the energy intensity of the IEA's road transport by 3.36 t/MM USD is caused by Spain (*ceteris paribus*);

– the decrease in the energy intensity of the IEA's road transport by 5.74 t/MM USD is caused by Switzerland (*ceteris paribus*).

The importance of the predictors and qualitative factors (similar assessment to correlation assessment) can be assessed after making econometric transformations, i.e. excluding the constant, taking into account the influence of the individual predictors and qualitative factors, then standardization of the *b* regression coefficients (see β coefficients). Therefore, it is stated that energy intensity of the IEA's road transport is very much dependent on the economic specificity of the Slovak Republic, industrial production and manufacturing. However, it is moderately dependent on the GDP deflator, transport intensity and economic specifics of Luxembourg, Netherlands, the Czech Republic and Portugal. Import, economic specifics of Switzerland, Poland, Ireland, Italy and Spain, the United Kingdom had small impact on energy intensity.

However, the most interesting in this analysis is the scenario for rationalization of energy intensity of the IEA's road transport (see Fig. 1). These scenarios were created on the basis of the approximation of the utility function for energy intensity of the IEA's road transport. It was defined by adopting a utility equal to 1.00 for low energy levels of road transport (0.00 t/MM USD is considered as low level); utility equal to 0.50 for intermediate levels of energy intensity of road transport (23.5134 t/MM USD) and utility equal to 0.00 for high levels of energy intensity of road transport (49.4769 t/MM USD). The optimal values were assigned to the factors shaping the rationalization ranges and the rationalization criterion consisted in the maximization of the utility functions. Fields shaded by darker colors (black or dark gray) indicate high utility from energy efficiency of the IEA's road transport when the range is adopted for selected reference scales. Light colors (light shades of gray or white) mean low utility or even negative, i.e. an unfavorable solution from the point of view of the economic reasons. Negative utility means the necessity to bear negative consequences from the choice of a given scenario. Scenarios include an infinite number of the directions of rationalization, and selection of a particular may depend on the target level for the individual aggregates.

4 Conclusions

The directions of rationalization of energy intensity of the IEA's road transport have been identified from the point of view of the macroeconomic conditions. Not only the desired rationalization ranges were presented in the scenarios, but attention was also paid to the negative side of rationalization (negative utility). From the point of views of the common goals or challenges for the IEA's members (included in the study) and creating a common policy of the sustainable development, one should keep in mind that the most desirable conditions for rationalization of energy intensity of road transport are:

- GDP deflator should fluctuate around 100–120 (index; additional favorable conditions are: lower transport intensity, lower import growth and higher industry),

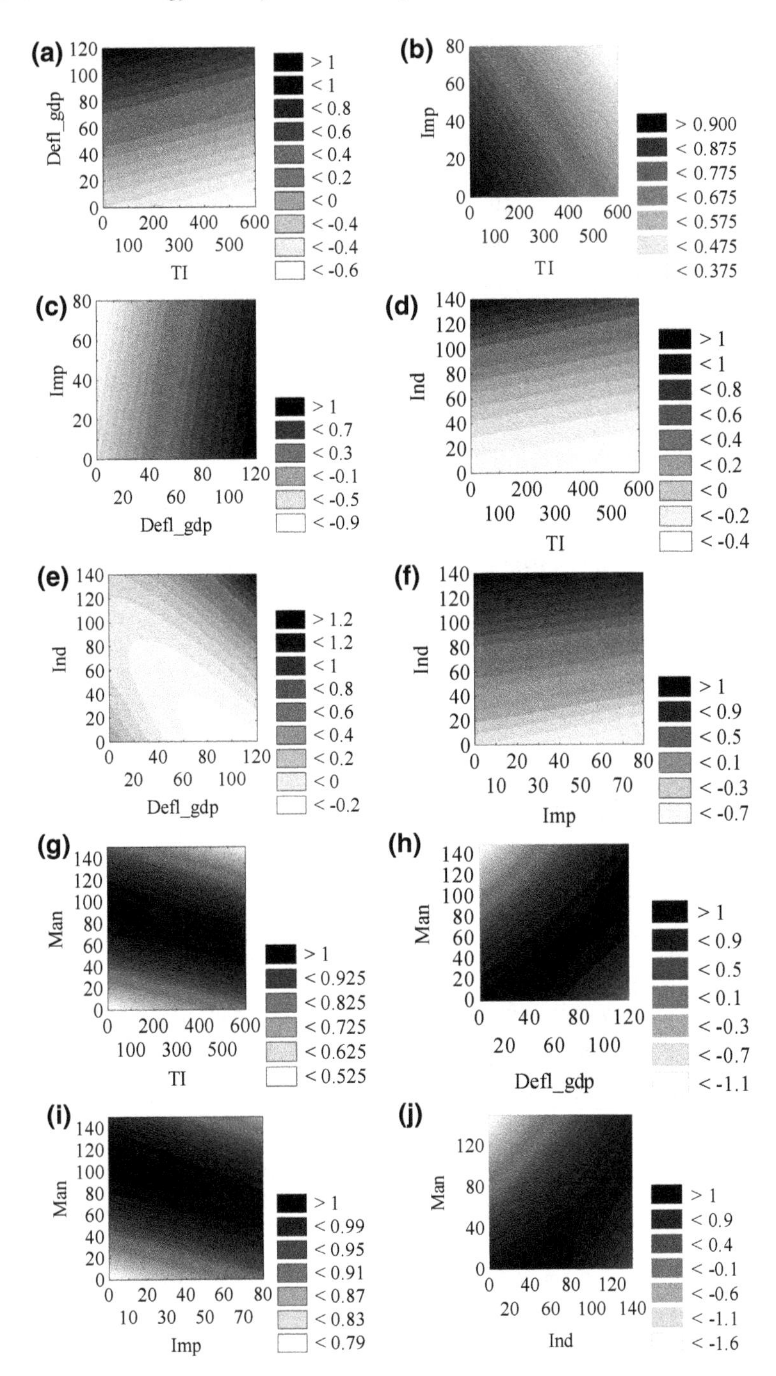

Fig. 1 Scenarios of the rationalization of energy intensity of the IEA's road transport from 2000 to 2016. *Source:* Own computations based on OECD.Stats [14]

- the upper limit of the increase in the level of imports is 40 p.p. and transport intensity is 200 tkm/1 thous. units of current USD of GDP,
- production in industry should be at least 120 (index), and in manufacturing 80–120 (index),
- in the case of other ranges of the values for rationalization conditions, the desired directions (dark fields) can also be determined for a given country, but then the priorities for specific conditions should be set.

As a result, in the rationalization programming one can avoid the negative consequences of the macro or even global decision.

References

1. Llorca, M., Jamasb, T.: Energy efficiency and rebound effect in European road freight transport. Transp. Res. Part A Policy Pract. **101**, 98–110 (2017)
2. Mulholland, E., Teter, J., Cazzola, P., McDonald, Z., Gallachóir, B.P.: The long haul towards decarbonising road freight—a global assessment to 2050. Appl. Energy **216**, 678–693 (2018)
3. Wu, J., Wu, Y., Cheong, T.S., Yu, Y.: Distribution dynamics of energy intensity in Chinese cities. Appl. Energy **211**, 875–889 (2018)
4. Skrucany, T., Kendra, M., Skorupa, M., Grencik, J., Figlus, T.: Comparison of chosen environmental aspects in individual road transport and railway passenger transport. Procedia Eng. **192**, 806–811 (2017)
5. Dolinayova, A.: Factors and determinants of modal split in passenger transport. Horiz. Railw. Transp. Sci. Papers **2**(1), 33–39 (2011)
6. Nedeliakova, E., Nedeliak, I.: Comparison of transport modes and their influence to environment. In: TRANSCOM 2009: 8th European Conference of Young Research and Scientific Workers, Žilina, 22–24 June 2009, pp. 43–46. Žilina, Slovak Republic (2009)
7. Talbi, B.: CO_2 emissions reduction in road transport sector in Tunisia. Renew. Sustain. Energy Rev. **69**, 232–238 (2017)
8. Saidi, S., Shahbaz, M., Akhtar, P.: The long-run relationships between transport energy consumption, transport infrastructure, and economic growth in MENA countries. Transp. Res. Part A Policy Pract. **111**, 78–95 (2018)
9. Sierra, J.C.: Estimating road transport fuel consumption in Ecuador. Energy Policy **92**, 359–368 (2016)
10. Załoga, E., Szaruga, E.: Analiza zależności pomiędzy intensywnością emisji dwutlenku węgla a energochłonnością, jednostkowym zużyciem energii i produktywnością ciężarowego transportu samochodowego. Logistyka **4**, 6973–6982 (2015)
11. Saisirirat, P., Chollacoop, N.: A scenario analysis of road transport sector: the impacts of recent energy efficiency policies. Energy Procedia **138**, 1004–1010 (2017)
12. Gerboni, R., Grosso, D., Carpignano, A., Chiara, B.D.: Linking energy and transport models to support policy making. Energy Policy **111**, 336–345 (2017)
13. Szaruga, E., Skąpska, E., Załoga, E., Matwiejczuk, W.: Trust and distress prediction in modal shift potential of long-distance road freight in containers: modeling approach in transport services for sustainability. Sustainability **10**(7), 2370 (2018). https://doi.org/10.3390/su10072370
14. OECD.Stats. http://stats.oecd.org/. Accessed 17.02.2018

CPSIA information can be obtained
at www.ICGtesting.com
Printed in the USA
LVHW020358040619
619991LV00002B/15/P

9 783030 177423